MOLECULAR CHARACTERIZATION OF COMPOSITE INTERFACES

MOLECULAR CHARACTERIZATION OF COMPOSITE INTERFACES

Edited by

Hatsuo Ishida

Case Western Reserve University
Cleveland, Ohio

and

Ganesh Kumar

Vistakon Inc. (a Johnson & Johnson Company)
Jacksonville, Florida

Springer Science+Business Media, LLC

Library of Congress Cataloging in Publication Data

Symposium on Polymer Composites.
   Molecular characterization of composite interfaces.

   (Polymer science and technology; v. 27)
   "Proceedings of the Symposium on Polymer Composites: Interfaces, held at the
American Chemical Society meeting in March 1983, in Seattle, Washington"—T.p.
verso.
   Bibliography: p.
   Includes index.
   1. Polymeric composites—Congresses. 2. Surface chemistry—Congresses. I.
Ishida, Hatsuo. II. Kumar, Ganesh. III. American Chemical Society. Meeting (185th:
1983: Seattle, Wash.) IV. Title. V. Series.
   TA418.9.C6S93 1983   620.1′18   84-20648

ISBN 978-3-662-27597-9   ISBN 978-3-662-29084-2 (eBook)
DOI 10.1007/978-3-662-29084-2

Proceedings of the Symposium on Polymer Composites: Interfaces, held at the
American Chemical Society meeting in March 1983, in Seattle, Washington

©1985 Springer Science+Business Media New York
Originally published by Plenum Press, New York in 1985.
Softcover reprint of the hardcover 2nd edition 1985

PREFACE

This book is an extended version of the proceedings of the Symposium on Polymer Composites, Interfaces, which was held under the auspices of the Division of Polymer Chemistry, American Chemical Society (ACS) during the annual ACS meeting in Seattle, March, 1983.

The importance of the interface in composite materials has been recognized since the inception of modern composite technology. Specifically, silane coupling agents were developed for glass fiber reinforced composites at a very early date. Ever since then the diversity of composite materials and the development of various surface treatment methods have led to the establishment of an "interface art." A trial-and-error approach has dominated the interfacial aspects of composite technology until very recently. With the advent of modern analytical techniques for surface characterization, it became possible to study detailed surface and interface structures.

It was hoped that this symposium would catalyze such a fundamental and scientific approach in composite studies. For this reason, the symposium was structured to verify the influence of interfacial structures on the mechanical and physical performance of composites and to improve our knowledge of the microstructure of composite interfaces. As the word "composite" indicates, interdisciplinary interaction is indispensable for proper understanding of multiphase systems.

The symposium consisted of four sessions each of which is represented by the titles in Part II through Part V. However, in order to provide a rather in depth introduction to the field, some papers are rearranged into Part I, general overviews. All papers are reviewed by leading scientists in this field with standards similar to those of well-respected journals. The rest of this volume is divided into: Part II. Influence on Physical Properties; Part III. Structure of Coupling Agents and Interfaces; Part IV. Influence on Matrix Structure; and Part V.

Surfaces of Reinforcements. These divisions are instrumental in
identifying the dominant structural factors. True understanding
of the role of the interface, however, must come from an intricate
combination of the findings made for each region.

This book is an important addition to the field since little
has been written on the subject. It should be useful for those
who want to manufacture more reproducible and reliable composites.
For beginners, this may provide a milestone as the book represents
the forefront of the field. The book discusses composites made of
glass fibers, carbon fibers, organic fibers and particulate
fillers. Characterization techniques include FT-IR, NMR, ESCA,
ESR, SIMS, emission, microbalance and others. It is tempting to
imagine that the knowledge accumlated with all these techniques
will soon be used to control and tailor the interfacial structures
required for specific composite properties.

No book is complete without making proper acknowledgement to
those deserving it. We are grateful to the Polymer Chemistry
Division and ACS for encouraging us to organize the symposium.
Anonymous reviewers are the judge of high standards. Their hidden
effort and help must be highly praised. Special thanks are due to
Mr. R.T.Graf, S.R.Culler and J.D.Miller for careful proof reading
of many of the manuscripts. We are very thankful for Ms.
D.Waldron for typing the manuscripts in spite of the
heart-breaking tragedy in her family during the entire period of
the preparation of this book. Many thanks are also due to the
interest in our project and the patience of Mr. P.J.Alvarez of
Plenum Publishing Co. Whole-hearted support of our family members
is always special to us. We wish that someday our children will
advance beyond this milestone. As long as this book remains, the
authors' commitment in their scientific pursuit will be
remembered. Their cooperation and encouragement are gratefully
acknowledged.

H.Ishida   G.Kumar

Department of Macromolecular Science Vistacon Inc.
Case Western Reserve University   P.O.Box 10157
Cleveland, Ohio 44106   Jacksonville, Florida 32247

CONTENTS

THE ROLE OF THE INTERFACE IN POLYMER COMPOSITES - SOME MYTHS,
MECHANISMS, AND MODIFICATIONS

J. L. Kardos

Materials Research Laboratory and
Department of Chemical Engineering
Washington University
St. Louis, MO 63130

ABSTRACT

A considerable effort has been made over the past 20 years to understand the reinforcement-matrix interface, to control it, and even to specifically modify it. It is at the interface where stress concentrations develop because of differences between thermal expansion coefficients of the reinforcement and matrix phases, because of loads applied to the structure, and because of cure shrinkage (in thermosetting matrices) and crystallization (in some thermoplastic matrices). The interface can also serve as a nucleation site, a preferential adsorption site, and a locus of chemical reaction. This paper attempts to clarify some common misconceptions regarding the effects of interfacial adhesion on mechanical properties including stiffness and toughness, the presence of covalent bonding of silanes to glass fibers, and the characterization of thermal stability of a polymeric composite. Two specific approaches to interface modification are also discussed, one for thermoplastic matrices and one for thermosetting matrices.

INTRODUCTION

The interface between reinforcement and matrix has always been considered as a crucial aspect, if not the Achilles' heel, of polymer composites. It is at the interface where stress concentrations develop because of differences between the reinforcement and matrix phase thermal expansion coefficients, because of loads applied to the structure, and because of cure shrinkage (in thermosetting matrices) and crystallization (in some thermoplastic matrices). The interface can also serve as a nucleation site, a preferential adsorption site, and a locus of chemical reaction. Accordingly, a considerable effort has been made over the past 20 years to understand the interface, to control it, and even to specifically modify it.

THE INTERFACE AND MECHANICAL PERFORMANCE – SOME MYTHS AND MECHANISMS

Structural parts made from composites are often designed to stiffness criteria or to strength criteria, but rarely to both. One common misconception is that excellent adhesion at the interface is necessary to provide the maximum realizable stiffness from the composite. However, the elastic stiffness is defined as the strain approaches zero (for example, Young's modulus in a simple unidirectional tensile test); thus the degree of adhesion has no bearing on the stiffness of the system. Obviously, there cannot be voids at the interface; but, all that is required of the composite to realize its maximum stiffness potential is that there be contact between the two phases so that the load may be transferred between them. An excellent example of this principle may be found in the work of Kenyon and Duffy [1] and Kenyon [2], who demonstrated for a glass bead-filled epoxy that the tensile stiffness was unchanged regardless of whether a coupling agent or a debonding agent was used. Figure 1 is taken from their work and shows the tensile stress-strain curves for a glass-bead filled epoxy. At the same volume fraction beads, all different treatments yield exactly the same Young's modulus. The top curve, marked A-1100, is for an aminosilane treatment which yields excellent adhesion, while the lowermost curve is for a silicone oil debonding agent. Note that the debonding agent curve yields the largest ultimate strain although the ultimate strength is by far the lowest. Clearly, the strength and ultimate strain are radically affected by the degree of adhesion, and therefore by the chemical and physical nature of the interface.

Fracture toughness of polymeric composites is probably one of the least understood of all the mechanical responses. For most polymeric composite systems, a sometimes espoused rule of thumb is that as the strength increases the toughness decreases. Thus, it might be implied that as the degree of adhesion increases, the toughness should decrease. While this is true generally for continuous fiber reinforced brittle matrices, it is not the case for bead-filled systems, nor for short fiber-reinforced ductile matrices.

Figure 2 qualitatively summarizes some of the results obtained by DiBenedetto and coworkers [3-5]. Improving the adhesion in a short glass fiber/polyphenylene oxide system actually increases the fracture toughness as measured in a double-

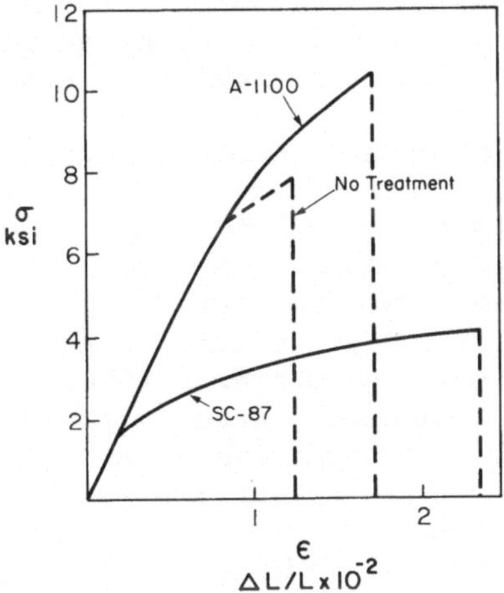

Fig. 1.   Effect of degree of adhesion on the stress-strain curve of glass bead/epoxy composites. Note that there is no effect on the initial slope (Young's Modulus).

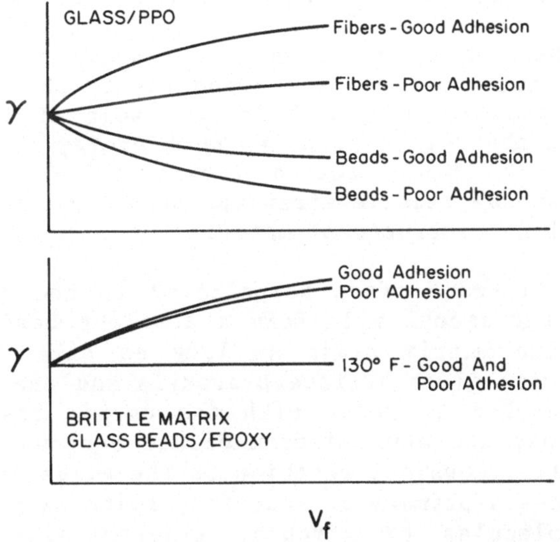

Fig. 2.   Qualitative effects of reinforcement geometry, matrix ductility, degree of adhesion, and volume loading on the fracture toughness of glass reinforced plastics.

-edge-notched tensile test. The same trend is clear in the glass bead/PPO system. Thus, the reinforcement geometry and the matrix ductility are important fracture toughness considerations.

Atkins [6] has attempted to utilize the best of both worlds by alternately coating glass fibers with a debonding agent along the fiber length in "barber-pole" fashion. Figure 3 displays some of Atkins' experimental results. The x-axis represents the degree of intermittent bonding, that is, the fraction of the boron fiber interfacial surface which is coated with either polyurethane varnish (PUV) or silicone vacuum grease (SVG). Basically, Figure 3 shows that as the debonded area increases, the toughness increases. The tensile strength can be maintained to some degree as the debonded area increases, but analysis of the results shows that as the x-axis of Figure 3 approaches unity, a drop in tensile strength of about 20% occurs. Since tensile strength was not measured as a function of the degree of bonding, it is impossible to ascertain where this dropoff begins. Furthermore, the tensile strengths reported were for continuous fiber specimens in the fiber direction (longitudinal). Any off-axis or transverse tensile test would have likely shown a much more serious degradation of strength than the 20% seen for the longitudinal test. Nonetheless, there may indeed be a value for the degree of bonding where the off-axis strength is relatively unaffected while the toughness is increased.

In one of the most unusual demonstrations of increased fracture toughness in composites, Jones, Suh, and Sung [7] demonstrated that power-law shearing of a viscous oil layer at the interface during fiber pull-out can provide a mechanism to absorb energy and thus increase toughness. Figure 4 depicts some of their results which clearly show that up to twice as much energy can be absorbed during fracture for thinly coated fibers. Unfortunately, the off-axis and transverse tensile strengths were very low, making the system impractical as a structural material.

One other half-truth often perpetuated in the literature is that silane coupling agents will form a covalent bond bridge from glass fibers to the matrix resin as long as silanol groups are present to react with glass surface hydroxyls and the other end of the coupler is matched to react with the matrix resin functionality. When silanes are applied from dilute aqueous solutions as happens industrially, covalent reaction to the glass does not immediately occur unless a primary or secondary amine is present either on the silane molecules (γ-aminopropyltriethoxysilane) or as a catalyst. Kaas and Kardos [8-9] conclusively demonstrated this using infrared spectroscopy in a series of studies on silica, glass beads, and quartz fibers in various matrices. Table 1 demonstrates the amine catalysis effect on ultimate tensile strength for two

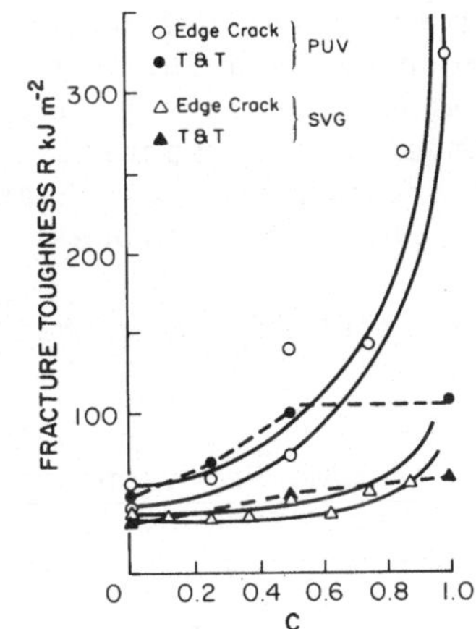

Fig. 3. The effect of intermittant bonding on the fracture tough-
         ness of boron/epoxy composites. C is the fraction of
         fiber area coated with either polyurethane varnish (PUV)
         or silicone vacuum grease (SVG). Toughness was measured
         with both edge-crack (EC) and Tattersall and Tappin (TT)
         methods.

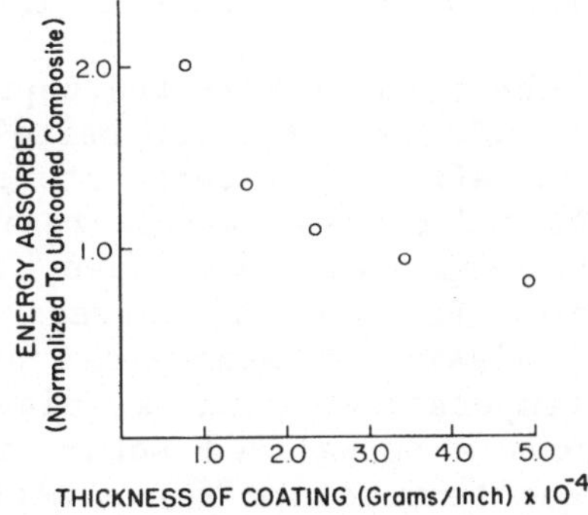

Fig. 4. Effect of oil layer thickness on the fracture toughness
         of composites containing oil-coated fibers.

different systems. The standard deviations on the tensile strength
data are about $\pm$ 5%; thus, improvements exhibited in Table 1 are
indeed real and significant. The quartz cloth samples were tensile
tested with the weave directions at $\pm$ 45° to the test direction, so
that the shear stress at the interface was maximized. Gent and Hsu
[10] reached the same conclusion using near infrared spectroscopy.
It may be possible to achieve covalent bonding between glass and
silane coupling agent under rather special conditions using exotic
solvent systems or high-temperature gas-phase reactions, but these
conditions do not exist under normal processing conditions.

Table 1. Surface Treatment Effect on Reinforced Polyester
          Strength.

Surface Treatment	Volume Fraction and Reinforcement Type	Tensile Strength MPa (psi)
Pure Resin	----------------	54.6 (7920)
Untreated	0.21-glass beads	40.6 (5890)
Vinyl Silane	0.22-glass beads	40.5 (5880)
Amine plus Vinyl Silane	0.22-glass beads	50.9 (7380)
Methacryloxy Silane	0.44-quartz cloth	174 (25,300)
Amine plus Methacryloxy Silane	0.45-quartz cloth	192 (27,900)

Others who have contributed significantly to the understanding of
precisely how and why coupling agents work are Plueddeman [11], who
first described the interaction of water and coupling agents at the
interface in terms of a dynamic equilibrium, and Koenig and
coworkers [12-14] who have pioneered the use of FTIR and laser
Raman spectroscopy in explaining specific coupling agent mechanisms
of interaction with both the resin and the glass surface.

Near the top of the myth list is the supposition that one can
determine the thermal stability of a composite's mechanical properties by observing the effect of temperature on the mechanical
properties of unreinforced or neat matrix resin, assuming that the
matrix is the only component to have temperature-dependent
properties. Aside from the fact that stress concentrations arise
when two phases with dissimilar mechanical properties are cooled
from the processing temperature, chemical dissimilarities can also
result. Figure 5 shows a normalized schematic plot of composite
tensile strength versus temperature for a neat epoxy resin and an
inorganic/epoxy composite made from the resin and cured under
precisely the same conditions. Clearly, the composite system
behaves much worse than the neat resin. In this particular case

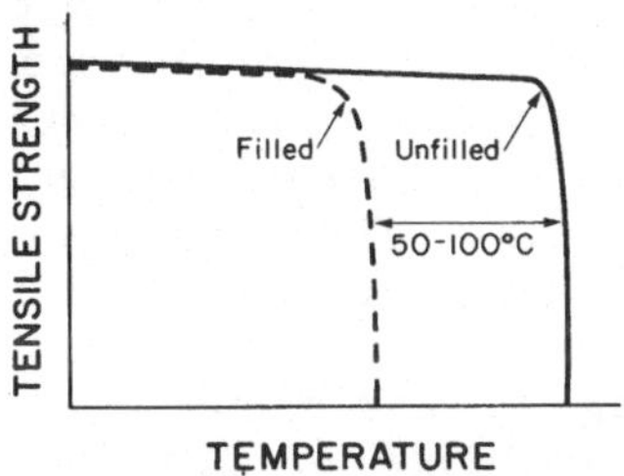

Fig. 5.   Schematic plot of the effect of temperature on the tensile
          strength of a neat epoxy resin and an inorganic/epoxy com-
          posite (normalized to the neat resin strength).

curing agent has preferentially migrated to the filler surface
yielding a very lightly crosslinked (low glass transition) region
exactly where it should not be--at the interface where the high
shear stress concentrations develop. Heterogeneity near the interface for thermosetting matrix composites has also been demonstrated
by Kardos for an E-glass/Epon 828-Z system [15]. Kenyon and
Nielsen [16] and Cuthrell [17] have also demonstrated that two
phases can exist in various epoxy systems depending on the
resin-curing agent stoichiometry.

INTERFACE MODIFICATIONS

     Since the interface is the most highly stressed region of a
composite material, it makes sense to lower these stress concentrations either by placing a material of intermediate modulus between
the reinforcement and matrix or by placing a ductile material
there. In the first case the concept involves lowering the modulus
ratio of any two neighboring components and thus lowering the
stress concentrations which can arise because of modulus mismatch.
This approach is sometimes called a graded modulus interphase. In
the second case, local deformation capability is built into the
interfacial region so that the modulus mismatch stress concentrations are damped out at least partially.

     Cheng, Kardos, and Tolbert [18] applied the intermediate
modulus concept by causing polycarbonate to preferentially crystallize around short graphite fibers. This was achieved in a two-step
process wherein the composite system was compression-molded at
275°C, considerably above the normal 190°C molding temperature for
polycarbonate, and then annealed at 245°C for three hours. The
275°C temperature is slightly above the melting point of polycarbonate and, even though there is no crystallinity initially present
in the matrix, micro-order in this amorphous polymer is apparently
destroyed and better wetting by the melt occurs. This phenomenon
alone caused an increase in composite strength. The subsequent

annealing temperature of 245°C is the temperature which most favors crystallization on the graphite fibers which themselves are excellent nucleating agents for crystalline polymers. This approach was applied to the thermoforming of right-angle and semi-circular channel specimens. These shapes were formed at 190°C and the temperature subsequently raised to 235°C for three hours to develop the crystallinity [19] and raise the strength and stiffness. The latter step is analogous to heat treating in metals. This approach takes advantage of the increased processability of the unannealed material and then provides an increased modulus and higher strength in the final part (see Table 2).

Although it was impossible to find similar effects in a polysulfone/graphite system, two new thermoplastic resins (PPS and PEEK), now being considered as matrices in graphite composites, should be examined for such advantageous effects.

Table 2. Flexural Loading of Hot-Formed Polycarbonate Composites

	Flexural Load to Break (Pounds Force*)	
Processing Cycle and Material	Half Cylinder	90° Channel
Molded at 190°C		
Unfilled	30-40	20-30
Filled (10 v%)	15-30	20-35
Molded at 190°C, annealed at 235°C		
Unfilled	†	†
Filled	46-62	54-64
*Range of breaking loads observed.		
†Excessive creep prevented annealing of unfilled specimens.		

The concept of reducing stress concentrations at the interface between a brittle matrix and glass fibers by applying a "rubber bumper" innerlayer to the glass was first described by Lavengood and Michno in 1975 [20] and later amplified in detail by Tryson and Kardos [21]. The innerlayer was a flexibilized epoxy (Epon 815-Versamide 140) and was applied from solution in thicknesses ranging from one to four percent of the fiber diameter. The innerlayered fibers were then filament-wound using a brittle epoxy (Epon 828-Z). Measurements of mechanical properties of the fibers alone (both uncoated and innerlayered) as well as the composites indicated that the innerlayer improves mechanical properties through three distinct mechanisms. First, the layer acts as a spacer and prevents fiber-fiber contacts during filament winding, which are sources of very high stress concentrations in the final composite. Second, the layer is much less stiff than the brittle matrix and

Table 3. Fiber Tensile Strength (K.S.I.)

Material	Unhandled	Handled
No innerlayer	72.0 ± 8.3	54.5 ± 3.1
Innerlayer #2	77.0 ± 8.9	75.0 ± 6.7
Innerlayer #3	77.6 ± 9.5	75.5 ± 8.5

provides a local deformation mechanism to reduce interfacial stress concentrations. Finally, the layer when applied from solution surprisingly heals flaws in the glass fiber and increases the intrinsic fiber strength. The latter effect is depicted in Table 3 which summarizes some preliminary results obtained for E-glass monofilament.

The designation "handled" in Table 3 refers to monofilament run through the process dry (without innerlayer application) to purposely abraid it. The as-received filament is unaffected by the innerlayer within the standard deviations of the data. It is clear that the innerlayer acts to protect the filaments and, indeed, seems to be healing flaws and increasing the filament tensile strength in the case where the filament was purposely abraided.

Unidirectional continuous filament composite specimens were fabricated by filament winding on flat-sided mandrels. The flat plates were B-staged at 40°C for 18 h, then cured at 100°C for 30 min followed by a post cure of 1-1/2 h at 180°C.

A summary of preliminary composite mechanical property data is displayed in Table 4. Clearly, the results provide significant incentive to proceed with development and optimization of the process.

The apparent flaw healing which occurs upon application of the innerlayer to an abraided filament is an advantage not initially predicted. The exact mechanism is not known at this writing, but must involve surface energetics changes at the crack tip such that crack growth is resisted.

The major innerlayer mechanical property parameter is the modulus. Ideally, this stiffness should be the same as the stiffness of a bulk resin containing the same proportion of resin and curing agent. However, innerlayers are applied from concentrated solutions and some solvent could be trapped within the coating as it cures. Also, during composite fabrication, Curing Agent Z can diffuse into the innerlayer. These factors can cause the actual

Table 4.   Composite Performance Improvement with Innerlayers on
           Monofilament.

Property	Change
Transverse Tensile Strength	67% Increase
Transverse Strength Loss After Water Exposure (2-hr. boil)	6% with innerlayer 40% without innerlayer
Torsional Fatigue Life	1000-fold Increase
Interlaminar Shear Strength	40% Increase

innerlayer modulus to be much different than the bulk material
modulus. Hence, great care must be taken to correlate the modulus
with the innerlayer glass transition so that the _in-situ_ innerlayer
modulus may be correctly determined.

A systematic study is now under way to establish the relative
importance of innerlayer stiffness and thickness as well as how
these parameters are affected by changes in process variables.

ACKNOWLEDGEMENTS

Some of the work reviewed above was supported by the National
Science Foundation, Division of Materials Research, Polymers
Program, under grant No. DMR-75-06795, as well as by the Industrial
Consortium for Composite Materials Research at Washington
University, and by the Advanced Research Projects Agency, Depart-
ment of Defense under Contract No. N00014-67-C-0218, for all of
which I am grateful. The work of former graduate students R. L.
Kaas, F. S. Cheng, and L. D. Tryson, now busily solving problems in
U.S. industry, forms an important part of the literature in this
particular area and should be a source of personal pride for them;
I gratefully acknowledge their efforts.

REFERENCES

1.   A. S. Kenyon and H. J. Duffy, Polym. Eng. Sci., 7, 1 (1970).
2.   A. S. Kenyon, J. Colloid Interfac. Sci., 27, 761 (1968).
3.   A. Wambach, K. Trachte, and A. T. DiBenedetto, J. Comp. Mater.,
     2, 266 (1968).
4.   A. T. DiBenedetto and A. D. Wambach, Intern. J. Polymeric
     Mater., 1, 159 (1972).

5.   K. L. Trachte and A. T. DiBenedetto, Intern. J. Polymeric
     Mater., $\underline{1}$, 75 (1971).
6.   A. G. Atkins, J. Mater. Sci., $\underline{10}$, 819 (1975).
7.   T. Jones, N. P. Suh, and N.-H. Sung, Preprints, 34th Annual
     Tech. Conf., Soc. Plast. Engrs., Atlantic City, April 26,
     1976, p. 458.
8.   R. L. Kaas and J. L. Kardos, Polymer Eng. Sci., $\underline{11}$, 11 (1971).
9.   R. L. Kaas and J. L. Kardos, Preprints, 34th Annual Tech.
     Conf., Soc. Plast. Engrs., Atlantic City, April 26, 1976,
     p. 22.
10. A. N. Gent and E. C. Hsu, Macromolecules, $\underline{7(6)}$, 933 (1974).
11. E. P. Plueddemann, Proc. 25th Ann. Tech. Conf., Reinf. Plas-
     tics Div., SPI, Section 13-D (1970).
12. J. L. Koenig and P. T. K. Shih, J. Colloid Interface Sci., $\underline{36}$,
     247 (1971).
13. H. Ishida and J. L. Koenig, J. Colloid Interface Sci., $\underline{64}$,
     555 (1978).
14. C. H. Chiang and J. L. Koenig, Pol. Comp., $\underline{2}$, 192 (1981).
15. J. L. Kardos, Trans. N. Y. Acad. Sci., $\underline{35}$, 136 (1973).
16. A. S. Kenyon and L. E. Nielsen, J. Macromol. Sci.-Chem., $\underline{A3}$,
     275 (1969).
17. R. E. Cuthrell, J. Appl. Pol. Sci., $\underline{12}$, 1263 (1968).
18. F. S. Cheng, J. L. Kardos, and T. L. Tolbert, SPEJ, $\underline{26}$, 62
     (1970).
19. J. L. Kardos, J. Adhesion, $\underline{4}$, 1 (1972).
20. R. E. Lavengood and M. J. Michno, Jr., Proc. Div. Techn.
     Conf., Engrg. Props. and Structure Div., Society of Plas-
     tics Engineers, 1975, p. 127.
21. L. D. Tryson and J. L. Kardos, Preprints 36th Ann. Conf.,
     Reinforced Plast./Comp. Inst., SPI, Section 2-E, 1 (1981).

BONDING THROUGH COUPLING AGENTS

Edwin P. Plueddemann

Dow Corning Corporation
Midland, Michigan 48640

GENERAL

Fiberglass-reinforced plastic composites have been developed to high performance materials since their introduction in 1948. Along with modest improvements in resins and glass fibers, the greatest advance has been in development of "coupling agents" to improve the bonding of resin to glass [1] (Table 1). Since unsaturated polyester resins were the most common organic matrix material, various unsaturated compounds of silicon and other elements were first tested as coupling agents. Only unsaturated silanes and methacrylatochrome complexes (DuPont Volan®) have reached commercial importance for these resins (Fig. 1).

As fiberglass reinforcements were used with other resins, new coupling agents were introduced with specific reactivity for these resins. In general, the best silane coupling agents are those where the organofunctional group on silicon has maximum reactivity with the particular thermosetting resin during cure.

The same silane coupling agents that were used on glass were also effective on particulate minerals in filled polymers (Fig. 2), and as primers on metal or ceramic surfaces for adhesion of sealants, paints, and adhesives. In some cases, silane coupling agents even promote adhesion of an organic polymer to another organic polymer.

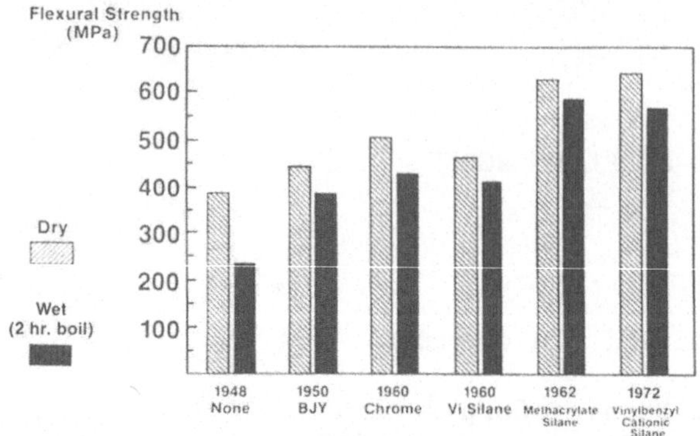

Fig. 1.   Progress in Fiberglass-Polyester Composites.

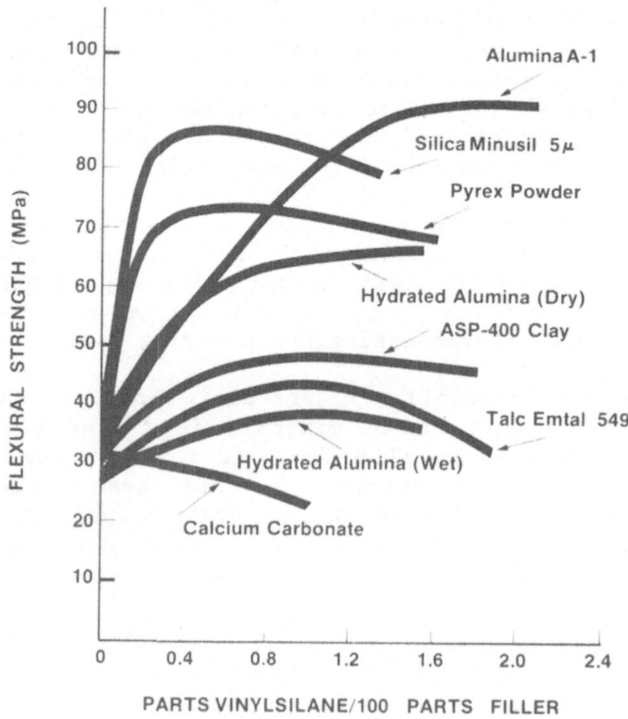

Fig. 2.   Flexural strength of particulate-filled composites as a
function of vinylsilane loading.

Table 1. Representative Commercial Coupling Agents

Organofunctional Group	Chemical Structure
A. Vinyl	$CH_2=CHSi(OCH_3)_3$
B. Chloropropyl	$ClCH_2CH_2CH_2Si(OCH_3)_3$
C. Epoxy	$\overset{O}{\overset{/\backslash}{CH_2CH}}CH_2OCH_2CH_2CH_2Si(OCH_3)_3$
D. Methacrylate	$CH_2=\overset{CH_3}{\overset{\vert}{C}}-COOCH_2CH_2CH_2Si(OCH_3)_3$
E. Primary Amine	$H_2NCH_2CH_2CH_2Si(OC_2H_5)_3$
F. Diamine	$H_2NCH_2CH_2NHCH_2CH_2CH_2Si(OCH_3)_3$
G. Mercapto	$HSCH_2CH_2CH_2Si(OCH_3)_3$
H. Cationic styryl	$CH_2=CHC_6H_4CH_2NHCH_2CH_2NH(CH_2)_3Si(OCH_3)_3 \cdot HCl$
I. Chrome complex	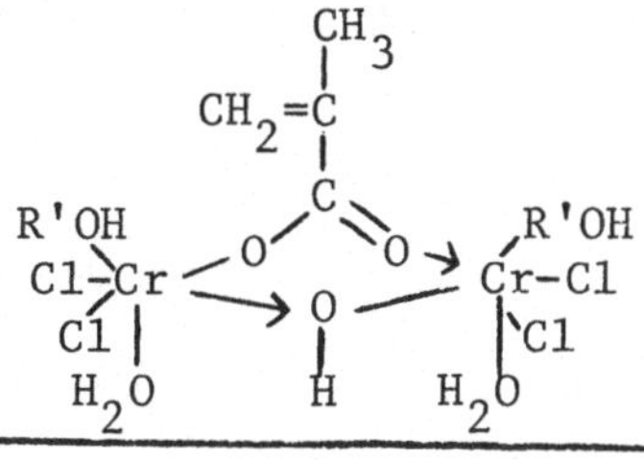 (see structure below)

Chrome complex structure:

$$
\begin{array}{c}
CH_2=\overset{CH_3}{\overset{\vert}{C}} \\
\vert \\
C
\end{array}
$$

R'OH, Cl–Cr, Cl, H_2O — O — C=O — O → R'OH, Cr–Cl, Cl — O — H — H_2O

THE NATURE OF ADHESION THROUGH SILANE COUPLING AGENTS

Coupling agents are hybrid organic-inorganic compounds that bridge the interface between resin and reinforcement. Organofunctional silanes are selected with R-groups that can form covalent bonds with the resin while hydroxyl (or alkoxy) groups on silicon are available to form oxane bonds to the material [2] (Fig. 3). The bonds to polymer and to mineral may be considered separately.

A. Adhesion to Mineral Surfaces

Organic polymers with polar functional groups can form oxane bonds with mineral hydroxide surfaces. These covalent bonds can provide adhesive strength greater than the cohesive strength of the materials involved, but they tend to have poor resistance to water. Such hydrolysis has some degree of reversibility. Oxane bonds through silicon also can be hydrolyzed, but they appear to have a better degree of reversibility, [3] or a more favorable equilibrium constant for bond formation.

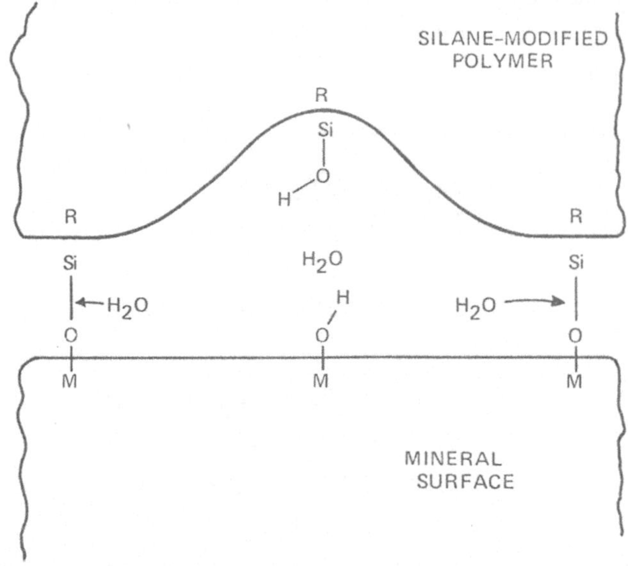

Fig. 3. Nature of bonding through a silane coupling agent in the presence of water.

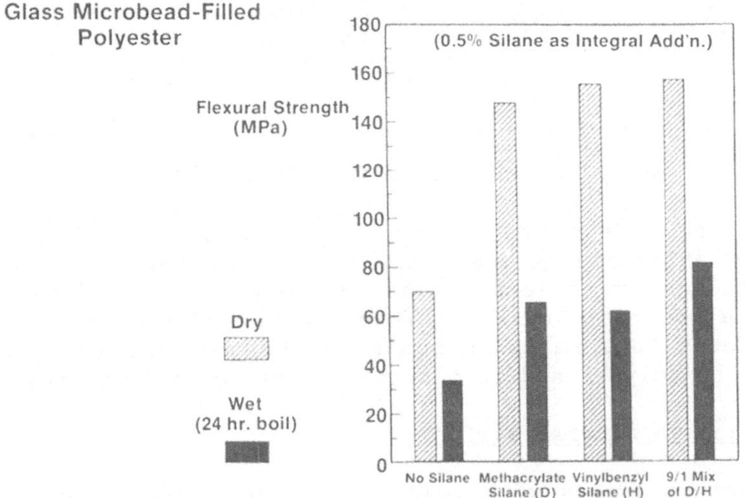

Fig. 4. Wet and dry flexural strength of glass microbead-filled polyester with different silane treatments.

$$M-O-Si-R + H_2O \; \rightleftharpoons \; M-O-H + H-O-Si-R$$

Recognition of the equilibrium character of oxane bonds to mineral surfaces suggests some obvious means for preparing more water-resistant composites:

1. Drive the oxane-forming reaction to completion.

2. Provide a hydrophobic interphase region with minimum water absorption.

3. Develop optimum polymer morphology in the interphase region to hold silanol groups in contact with the surface.

1. Complete Oxane Formation

Condensation of neutral alkoxysilanes with mineral surfaces is catalyzed by tertiary amines, alkyltitanates or tin compound [4]. Amine-functional silanes (H of Table 1) are self-catalytic for bonding with the surface. A small proportion of silane H, or a tertiary amine, should be mixed with neutral silanes used as integral additives or applied from organic solvents (Fig. 4).

When a silane is applied from aqueous solution, excess of volatile acid (CO_2 or acetic acid) is added to adjust the pH to 4-5 for bath stability. As the treated mineral is dried, it loses the acid, leaving an amine (e.g., benzyldimethylamine) to catalyze condensation of silanols. Finally, the amine is also lost, leaving a neutral modified surface.

2. Low Water Absorption/Hydrophobic

Hydrophobic resins (hydrocarbons) and hydrophobic coupling agents on water-resistant minerals like quartz produce composites that show essentially no deterioration after seven days in boiling water [5]. The coupling agent itself should not contribute a hydrophilic interphase region as illustrated by a series of methacrylate-functional silanes on glass in polyester composites [6] (Table 2). All of the methacrylate-functional silanes of Table 2 presumably had comparable reactivities with the resin and with glass, but performance of composites dropped off as the coupling molecules became more hydrophilic. Mixtures of hydrophobic coupling agents (A and B of Table 1) with more reactive silanes may also improve the water resistance of appropriate polymer composites.

Table 2. Glass Cloth–Polyester Laminates

Silane Coupling Agent on Glass $(MeO)_3Si-$	Flexural Strength (MPa) of Laminates		
	Dry	2 hr. Water Boil	% Retention
$-CH_2CH_2CH_2$ methacrylate (MA)	647	628	97
$-(CH_2)_3OCH_2CH_2OCH_2CH_2MA$	626	610	97
$(CH_2)_3(OCH_2CH_2)_{12}MA$	243	174	72
$-(CH_2)_3OCH_2\underset{\mid}{C}HCH_2OCH_2CH_2MA$ OH	581	531	91
$-(CH_2)_3\underset{\mid}{N}CH_2\underset{\mid}{C}HCH_2MA$ Me OH	513	342	67

3. Optimum Morphology

Equilibrium conditions at the interface are lost if silanol groups resulting from hydrolysis of oxane bonds are physically removed from the interface. Practical experience has correlated performance of composites with resin morphology at the interface. Silanols that are part of a rigid, or of a viscoelastic polymer provide optimum water resistance in bonding to minerals. Flexible rubbery interphase regions do not provide water-resistant bonds even with optimum silanol modification [7]. In all of these cases, oxane bonds with the mineral will hydrolyze. The resulting silanol groups at a viscoelastic interface will flow back to the mineral and allow reformation of oxane bonds. Silanols generated at a rigid interface will be held close to mineral hydroxyl groups so they can also reform as oxane bonds. Adjacent silanols formed at a flexible rubbery interface can retract from the surface and condense with each other to siloxanes, thus being effectively removed from equilibrium bonding across the interface (Fig. 5).

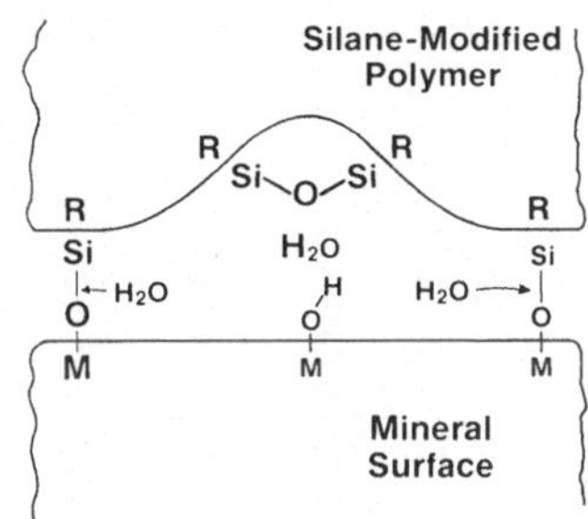

Fig. 5. A model for hydrothermal degradation of a rubber bonded to a mineral surface by a silane coupling agent.

B. Silane Bonds with Polymers

Although simplified representations of coupling through organofunctional silanes often show a well-aligned monolayer of silane forming a covalent bridge between polymer and filler, the actual picture is much more complex. Coverage by hydrolyzed silane is more likely to be equivalent to several monolayers. The hydrolyzed silane condenses to oligomeric siloxanols that initially are soluble and fusible, but ultimately can condense to rigid cross-linked structures. Contact of a treated surface with polymer matrix is made while the siloxanols still have some degree of solubility. Bonding with the matrix resin, then, can take several forms:

*The oligomeric siloxanol layer may be compatible in the liquid matrix resin and then form a true copolymer during resin cure.

*The oligomeric siloxanol may have partial solution compatibility with the matrix resin and form an interpenetrating polymer network as the siloxanols and matrix resin cure separately with a limited amount of copolymerization.

*A siloxanol layer may also diffuse into a non-reactive thermoplastic layer and then crosslink at fabrication temperature to form pseudo-interpenetrating networks.

*Polymeric siloxane segments may interdiffuse with a polymer with no crosslinking in either phase.

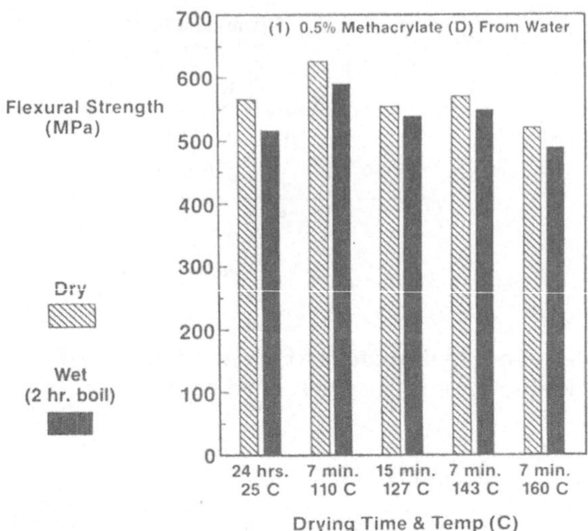

Fig. 6.   Performance vs. Drying of Silane[1]-Treated Glass

Performance of coupling agents in reinforced composites may depend
as much on physical properties resulting from the method of application as on chemistry of the organofunctional silane. Physical
solubility or compatibility of a siloxanol primer layer will be
determined by the nature and degree of siloxane condensation on a
mineral surface. Undercured coupling agent layers may diffuse so
far into the matrix that they are lost for potential crosslinking
at the interface. Silane coupling agent D applied to E-glass
fibers from water gave its best performance when dried 7 min at
110° C (Fig. 6). More vigorous drying may have caused loss of some
methacrylate groups through oxidation or polymerization [8].

A prehydrolyzed amino-functional silane (F) is a very good
primer for adhesion of PVC plastisols to glass or metals. More
water-resistant bonds to glass were obtained if the primer was pre-
cured for 15 min at 150 °C, even though more crosslinking no doubt
occurred during fusion of the plastisol for 20 min at 150° C (Table
3). Overcuring the primer at 175-200 °C provided a smooth, extremely
hard film that apparently was not penetrated by the plastisol.

C. Organic-to-Organic Adhesion

Silane primers that are effective in bonding two separate
polymers to a mineral substrate are also effective in bonding the
two polymers to each other. The lower melting polymer is generally
fused against the primed, higher melting polymer. Alternatively, a
third low melting polymer may be used as a hot-melt adhesive
between the two primed thermoplastics. In this way, sheets of

Table 3. How Much Should Primer Be Pre-Cured? (PVC Plastisol On Glass With Aminosilane Primer) Plastisol Fused 20 min at 150°C.

Dry Primer 15 min at Temperature (°C)	Peel Adhesion of PVC Film (N/cm)	
	Dry	1 day in 50°C water
100°C	(c)	1.08
125	(c)	1.27
150	(c)	15.8
150 (30 min)	(c)	13.1
175	4.35	---
200	2.50	---
No primer	0.3	---

(c) = cohesive failure in film at about 25 N/cm

Table 4. Bonding Polyethylene to Mylar® Through "Hot Melts" at 150°C.

"Hot Melt" Polymer	Primer on PE	Peel Str. (N/cm) PE	Primer on Mylar®	Peel Str. (N/cm) to Mylar®
CXA 1025[1]	None	1.8	None	0.8
CXA 1025	A	*	B	*
Vitel 5571[2]	None	3.5	None	11.5
Vitel 5571	A	*	B	*

*Cohesive failure in adhesive at greater than 30N/cm.

1. EVA terpolymer, product of DuPont
2. Elastomeric polyester, product of Goodyear

Primer A = 1% dicumyl peroxide in silane H, (10% in methanol)
Primer B = 10% silane C in Cymel® 303, (10% in isopropanol)

polyethylene and polyester (Mylar®) were primed with the silane
primers and bonded with hot-melt adhesives (Table 4).

METHODS OF APPLYING SILANES TO FILLERS

Coupling agents can give good or poor performance in filled
composites depending on uniformity of treatment. Silanes may be
dry-blended with fillers at room temperature or at elevated temperature. Neutral trimethoxysilanes (A, B, C, D and G of Table 1)
disperse most readily over the filler and may benefit from addition
of a trace of tertiary amine or alkyltitanate to catalyze the reaction with filler surface. Amino-functional silanes (E, F and H)
are self-catalytic in reacting with filler surfaces, but may benefit from dilution with an alcohol solvent to aid dispersion of a
trace of silane over the large surface of a typical filler. The
treated filler should then be dried before it is used in polymer
mixes.

The effectiveness of silane treatment may be measured by a
Daniel flow point test [9] and by measuring performance of a
finished composite. Ground quartz (Minusil 5 μ) was treated with
0.5% of two silanes by several variations in technique and evaluated by Daniel flow point test, and by properties of polyester
composites. In the Daniel flow test, 6g of filler were titrated
with a 25% solution of polyester (Paraplex P-43) in styrene while
stirring with a spatula until the mixture became fluid enough to
flow from the spatula. A lower titration is indication of better
dispersion resulting from more uniform coverage of the filler.
Flexural strengths of polyester castings containing 50% of these
fillers correlated well with Daniel flow points (Table 5).

Table 5. Evaluation of Silane-Treated Silica (Minusil® 5μ)

Solvent Silane Treatment	Daniel Flow point (ml)	Flex. Str. of Composite (MPa) Dry	2-hr. Boil
No silane (control)	7.0	103	65
Silane D			
from methanol	5.9	140	108
from n-butanol	5.0	169	144
No solvent	3.9	172	172
Silane H			
from i-propanol	5.0	179	124
from Dowanol EM	3.7	166	146
No solvent	5.3	165	90

REFERENCES

1. E. P. Plueddemann, "Silane Coupling Agents," Plenum Press, New
    York (1982).
2. E. P. Plueddemann, J. Adhesion $\underline{2}$, (July) 184 (1970).
3. P. Walker, J. of Coatings Technol., $\underline{52}$, (670) 49 (1980).
4. R. L. Kaas and J. L. Kardos, Proc. SPE 32nd ANTEC paper 22,
    (1976).
5. B. M. Vanderbilt, SPI, 17th Ann. Tech. Conf. Reinf. Plast. 10-D
    (1962).
6. E. P. Plueddemann, H. A. Clark, L. E. Nelson, K. R. Hoffman,
    Mod. Plast. $\underline{39}$ (8), 139 (1962).
7. E. P. Plueddemann, SPI, 29th Ann. Tech. Conf. Reinf. Plast.
    24-A (1974).
8. H. Ishida and J. L. Koenig, J. Colloid and Interface Sci., $\underline{64}$,
    (3), 565 (1978).
9. F. K. Daniel, Nat. Paint, Varnish & Lacquer Assn., Scientific
    Section Cir. 744 & 745, Oct. (1950).

STRUCTURAL GRADIENT IN THE SILANE COUPLING AGENT LAYERS AND ITS

INFLUENCE ON THE MECHANICAL AND PHYSICAL PROPERTIES OF COMPOSITES

Hatsuo Ishida

Department of Macromolecular Science
Case Western Reserve University
Cleveland, Ohio 44106

ABSTRACT

Studies of the structural gradient within the silane coupling
agent interphase are reviewed. Origins of the structural variation
in the silane layers are discussed in terms of the silane treating
solution and variations in the substrates. The solution pH,
concentration, solvent, hydrolysis time and aging time are some of
the important parameters identified. Consideration is given to the
following categories: the first monolayer on the substrate, the
tightly chemisorbed layers near the substrate, the loosely
chemisorbed layers in the middle and finally the physisorbed
silanes in the outermost layers. The role of these regions in
terms of the reinforcement mechanism of composites and the rheology
of polymer melts is discussed. Complex formation of silanes with
metallic substrates is briefly mentioned.

INTRODUCTION

Silane coupling agents have been used to improve interfacial
adhesion at the glass fiber or particulate filler matrix interface,
reduce viscosity during processing, protect glass fibers from
damage during handling, alter the catalytic effect of surfaces and
improve the dispersion of particulate fillers. An excellent
monograph has been published on the chemistry, structure and
application technology of silane coupling agents [1].
Historically, the application of coupling agents has been done by a
trial-and-error basis with little fundamental understanding of the
interfacial region or the structure of the applied coupling agent.
The use of modern surface spectroscopic techniques has led to a

remarkable improvement in the molecular understanding of the interfacial structures. It is clear that the glass matrix interface
region is not a simple layer but consists of complex structures.
The importance of the interface in composite performance has been
well documented [1,2].

     The structural variations that exist at the glass matrix
interfacial region have been studied in recent years both qualitatively and quantitatively. Several factors influence the
interfacial structures of the silane coupling agent including the
pH, concentration, organofunctionality, solvent, application
method, and temperature of the silane treating solution. Other
factors related to the reinforcement material that influence the
silane coupling agent structure are the average pH, local acid-base
character, surface coverage, surface functionality and topology.
Drying conditions such as thermal treatment accelerate the rate at
which the predetermined structure affected by the abovementioned
factors is achieved. A summary of interfacial structural
parameters and their origin is listed in Table I.

Table I. Interfacial Silane Structures that Influence the
          Mechanical Performance of Composites

FACTORS	ORIGIN
1. Degree of curing	Time, Temperature, Concentration
2. Orientation	Isoelectric point of surface, pH of the silane solution
3. Organization	Isolated silanetriol in the treating solution
4. Physisorbed silane	Concentration, pH of the silane solution, Surface functionality, Isoelectric point of surface
5. Chemisorbed silane	Surface functionality, pH of the silane solution
6. Thickness Concentration, Topology	

It is possible in many cases to control the interfacial structure by properly adjusting the treating conditions. The mechanical and physical properties of composites are strongly influenced by the interfacial structures. If the interfacial structure can be quantitatively linked with the mechanical and physical behavior of composites, then their performance may be predicted and controlled. The few studies that have been reported in this area indicate the usefulness of this approach.

It is possible to simulate some of the silane structures present on various surfaces without using the reinforcing materials. Unique aspects of the surface effects arise from the ability of the surface of solids to catalyze some reactions, restrict the molecular mobility and orientation, and influence the packing of silane molecules. Combined studies of the bulk simulation and surface structures are helpful to extract the specific influence of the surface of reinforcement material.

It is the purpose of this review article to discuss the structural gradient in the silane coupling agent layers on the surface of reinforcement materials. Special emphasis is placed on the role that specific silane structures have on the reinforcement mechanisms and rheological properties of the composite. Little attention has been given to the existence and the role of physisorbed silanes in the past. Those who are interested in background information on the molecular structure of interfaces should refer to the aforementioned monographs [1,2] and recent review articles [3,4].

THE STRUCTURE OF THE PHYSISORBED SILANES

Physically adsorbed silane molecules exist in the outermost layers of the silane interphase. These silanes can be removed from the surface by washing with an organic solvent which does not cleave the siloxane linkages. Small oligomers predominate physisorbed silanes and the molecular weight is believed to be a function of the thickness across the interface. Thus, the amount of the physisorbed silane determined by an organic solvent is a function of the length of time washed. However, the majority of physisorbed silane desorbs relatively quickly.

A radioisotope-labeled aminosilane was used by Schrader et al. [5] to study the desorption of the coupling agent by hot and cold water from a glass plate. Their work was the earliest attempt to demonstrate the existence of a structural gradient in the silane interphase. The use of water as a solvent prevented the observation of the well defined physisorbed silane because the aminosilane self-catalyzes the hydrolysis of the Si-O-Si linkages. A combination of cold and hot water extractions probed the structural profile of the silane interphase. Based on the resistance to desorption, the silane layers were conveniently divided into three regions. The few layers closest to the glass surface was reported to be the most difficult to remove. Johannson et al. [6] using radioisotope-labeled methacryl-functional silane showed that a portion of the silane can be washed away by toluene and ethyl acetate confirming the existence of physisorbed silane.

The hydrolytic stability study of various silanes with and without matrix resin using Fourier transform infrared spectroscopy by Ishida and Koenig [7] showed that some silane molecules desorbed during the polymerization with the matrix resin. These results suggested the formation of a matrix interphase consisting of physisorbed silane and matrix copolymer near the glass surface. The amount of the physisorbed silane was measured quantitatively as a function of the concentration of the silane treating solution. While the chemisorbed hydrolyzate of γ-methacryloxypropyltrimethoxysilane (γ-MPS) showed surface induced polymerization of the C=C group, the physisorbed silane showed no signs of such effects.

The following authors [8-11] used the so-called dry-blending technique to treat particulate fillers. The dry-blending method uses a silane organic solution which is sprayed onto the filler powder avoiding complete wetting and subsequent caking problem of the filler.

The first extensive attempt to study the structure of the physisorbed γ-MPS on particulate fillers was reported by Nakatsuka et al. [8] using IR and gel permeation chromatography (GPC). They reported a number of important observations including the molecular weight variation on different fillers, evaporation of the silane during drying when an alcoholic solvent was used, and desorption of the physisorbed silane by a styrene wash. They showed that the molecular weight of the γ-MPS oligomer was lower on calcium carbonate surface than the clay sample used. Also, the molecular weight of the oligomer increased upon phospholic acid treatment of the calcium carbonate surface. Their mechanical study clearly demonstrated that calcium carbonate, a filler historically considered to be nonreinforcing in terms of strength, can reinforce a rubbery matrix when proper surface treatments are applied.

Subsequently, Ishida and Miller [9] studied the structure of physisorbed and chemisorbed silane on many particulate fillers using GPC and diffuse reflectance FT-IR. They determined that the siloxane network structure is influenced by the long-range acid-base effects of the filler surface. Fillers that had a similar slurry pH, yielded oligomeric siloxanes that had the same network structures and molecular weights as shown in Figures 1 and 2. As long as the extent of surface coverage is similar, the

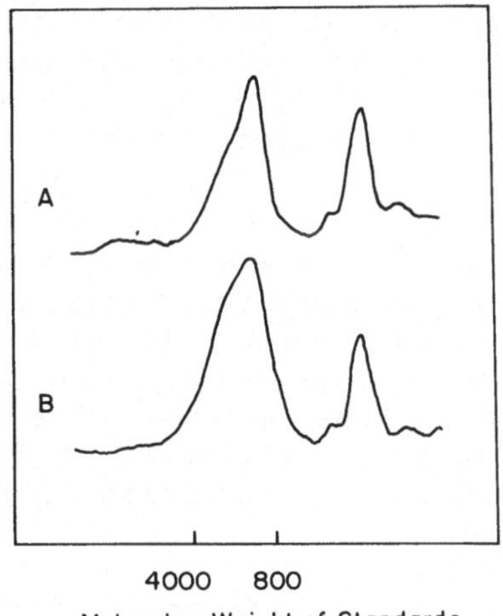

Fig. 1.   Gel permeation chromatograms of the physisorbed γ–MPS
          collected from the surface of (A) tungsten oxide (slurry
          pH 4.8) and (B) clay (slurry pH 4.1).   Polystyrene was
          used as molecular weight standards.

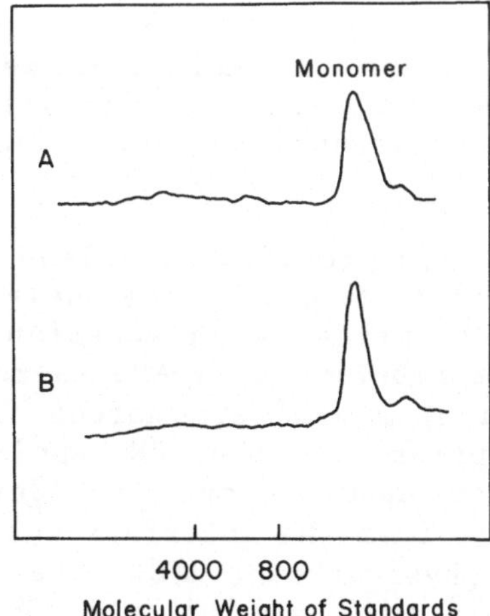

Fig. 2.   Gel permeation chromatograms of the physisorbed γ–MPS
          collected from the surface of (A) Kaolin (slurry pH 7.1)
          and (B) Zinc oxide (slurry pH 7.6).

surface coverage determines in part the concentration and availability of the silanol for homocondensation reaction influencing the kinetics and resultant structure of siloxane networks. The fillers were classified based on a unified concept, the slurry pH.

Miller et al. [10] simulated the siloxane network structure by adjusting the pH of the silane treating solution. They were able to duplicate the siloxane structures that appeared on all fillers of interest without using the fillers. These results implied that the pH of the treating solution predetermines to some extent the structure of the silane. The combination of the solution pH and the acid-base character of the filler determines the final siloxane structure in addition to the surface coverage effects already mentioned.

An attempt to control the structure of the coupling agent on filler surfaces by mixing different silanes with γ-MPS was reported [11]. When tetraethoxysilane, vinyltrimethoxysilane and γ-aminopropyltriethoxysilane (γ-APS) were mixed with γ-MPS in a 1:10 mole ratio, all combinations showed a remarkable increase in molecular weight of the bulk hydrolyzates upon room temperature drying. For a particular combination of γ-APS/γ-MPS, the alcoholic aqueous solution gelled relatively quickly due to the catalytic effect of the amine in γ-APS. Quite surprisingly, negligible effects were seen when the same solution was used to treat fillers and subsequently examined on the physisorbed silane molecules by GPC. This observation indicates that there is a strong driving force for cyclization of the siloxane chain on the filler surface possibly by influencing the mobility and availability of the silanol groups. These results are illustrated in Figures 3 and 4, where the molecular weights of the mixed silane with and without a filler powder were measured by GPC. Thus, the filler surface induces the formation of the physisorbed silane.

Migration of an amine-functional silane was demonstrated by Sung et al. [12] using x-ray microprobe analysis. When γ-APS was used to treat an Al_2O_3 plate as an adhesion promotor for polyethylene (PE), the distribution of γ-APS as measured by the x-ray signal of the silicon atom was very narrow if the silane layers were heat treated prior to the PE application. When non heat-treated γ-APS was used, a much broader distribution with respect to the distance from the interface was obtained indicating the migration of the physisorbed γ-APS. The spacial distribution of the silane molecule with and without the heat treatment is shown in Figures 5 and 6. These results indicated that the physisorbed silane migrates out of the substrate surface to a great distance compared to the initial thickness of the silane layers.

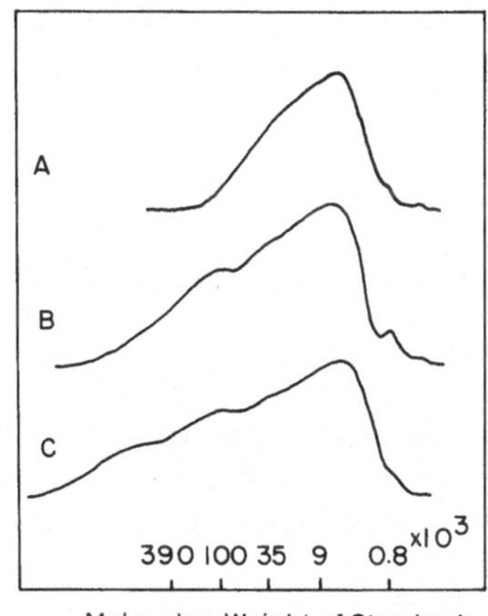

Fig. 3.   Gel permeation chromatograms of the bulk hydrolyzates of
γ-MPS containing a second component and heated at 80°C
for 85 h.   (A) no second component, (B) 10 mole % tetra-
methoxysilane and (C) 10 mole % vinyltrimethoxysilane.
The silanes were hydrolyzed in n-butanol containing small
amount of water.

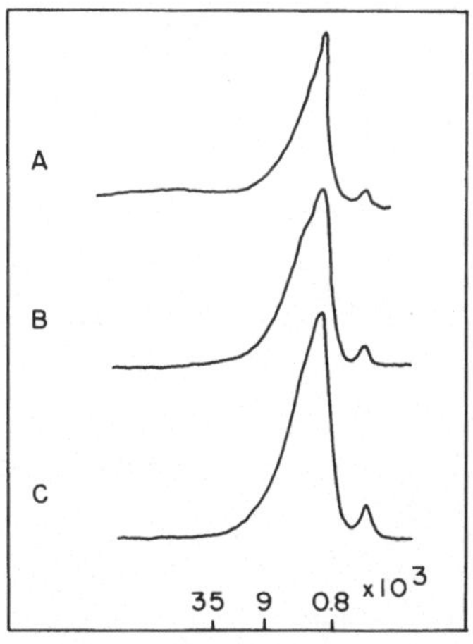

Fig. 4.   Gel permeation chromatograms of γ-MPS hydrolyzate on
clay.   (A) no second component, (B) 10 mole % tetrame-
thoxysilane and (C) 10 mole % vinyltrimethoxysilane.
The same silane solutions, hydrolysis conditions and
heating conditions as in Figure 3 were used.

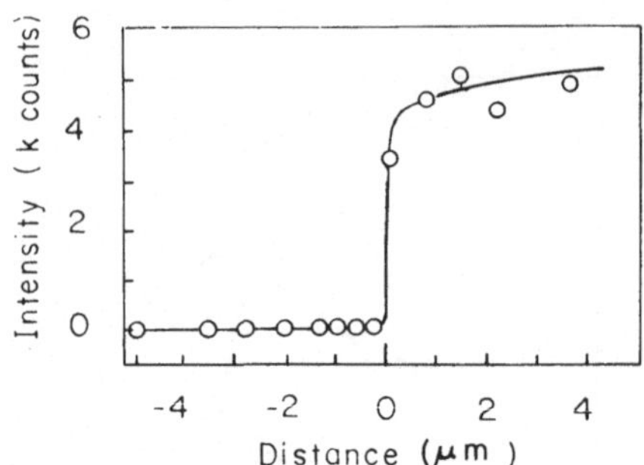

Fig. 5. X-ray microprobe analysis of γ-APS hydrolyzate on Al_2O_3 as a function of the distance from the substrate surface. The negative values indicate within the substrate.

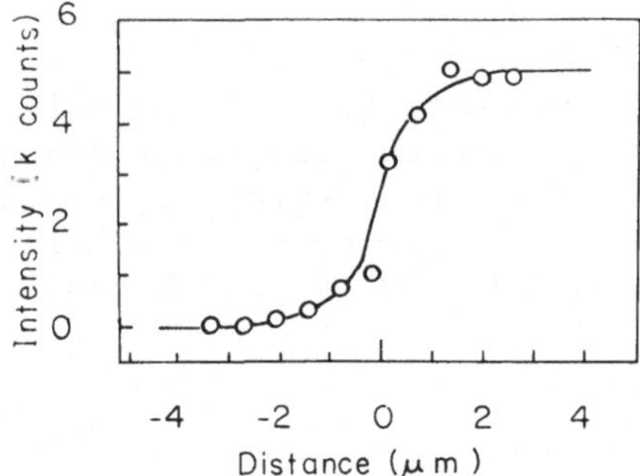

Fig. 6. X-ray microprobe analysis of γ-APS hydrolyzate on polyethylene as a function of the distance from the substrate surface.

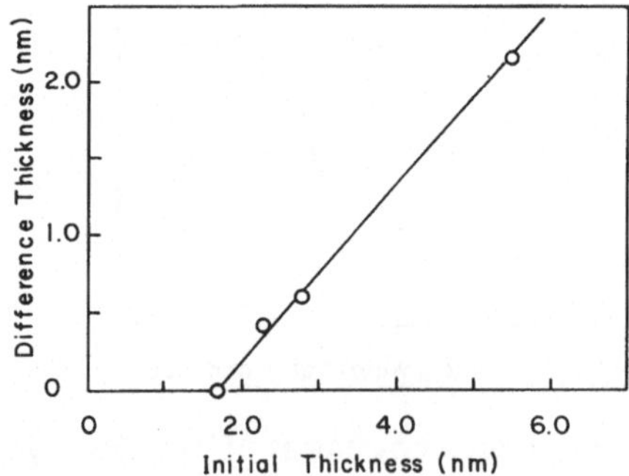

Fig. 7. Thickness of γ-APS films on a silicon wafer measured by ellipsometry. The difference thickness between the initial and solvent washed film is plotted against the thickness of the initial γ-APS film.

It has been proposed that the structure of the silane in the treating solution strongly influences the network structure within the silane layers [13]. This is supported by Belton et al's [14] work using ellipsometry of γ-APS as an adhesion promotor for a polyimide film on a silicon wafer. They observed that the thickness of the γ-APS layer was reduced the most for the thickest silane layers while, for a very thin layer (2.3 nm), no reduction in thickness was observed. The reduced thickness as a function of the original silane thickness is replotted in Figure 7. Since the thickness variation was produced by changing the silane structure in the solution, the desorbed silane is predominately due to the physisorbed silane. As described in the chemisorbed silane section regarding Culler et al's [15] results, this portion of silane obtained from very dilute solutions shows a great sensitivity to thermal treatments being converted to chemisorbed silane.

Based on the above discussion, the qualitative and quantitative nature of the physisorbed silane is better understood. The question as to why the relative amount of the physisorbed and chemisorbed silane varies from one substrate to another remains. Silanes with large and flexible organofunctional groups tend to form more cyclic structures than do silanes with smaller and more rigid substituents [16]. Many useful silanes for composite applications belong to the former category. Effects of pH on the structure of silane coupling agent has been studied by GPC using γ-MPS [9]. When silanes condense at pH ranges below 7 or above 9, the molecular weights tend to be small. If there are no surface functional groups, such as the metal hydroxide groups normally present on metal oxides, these small oligomers easily desorb. The low surface coverage tends to restrict the availability of the silanol groups for the condensation reaction causing more cage-like structures to form. It is possible that this cage-like structure incorporates the surface as a major part of the cage. Within the pH ranges described above, the amount of physisorbed silane is influenced by the surface functionality, surface coverage of the silane and the topology of the filler.

When the pH range is 7 to 9, the situation is quite different. The molecular weight of the silane is much higher and these large molecules have an open structure, a double chain ladder-like structure, that tend to form hydrogen bonding between themselves resulting in the formation of a physically crosslinked gel structure. It requires a long time for this gel to be separated and dissolve into a solvent. Thus, these gels appear to be chemisorbed, even if there are no primary bonds connecting the substrate and the silane molecules.

THE ROLE OF THE PHYSISORBED SILANE

Although the existence of the physically adsorbed silane has

been known for a long time and now more detailed characteristics
have been revealed, few attempts have been made to elucidate its
role in the reinforcement mechanisms of composites. Fragmental information reported to date, however, indicate the unique role of
the physisorbed silane at the glass matrix interface. It is
intuitively obvious that the physisorbed silane, a low molecular
weight silicone oil, would have a quite different role from the
chemisorbed silane which is a part of the reinforcement materials.
In this section, an attempt will be made to elucidate the effects
of the physisorbed silane on the mechanical and rheological
properties of composites.

The influence of these physisorbed silanes on the mechanical
properties of composite is not known. Nevertheless, there have
been a few occasions where removal of the physisorbed silane
improved the flexural strength of the composite. Kokubo et al.
[17] reported that extraction of the silane by methanol from a mica
surface increased the flexural strength of the composite.
Similarly, Graf et al. [18] observed an improved flexural strength
after removal of the physisorbed silane by tetrahydrofuran (THF).

It is commonly observed [6,12,18,19,20] that there is a
certain silane concentration at which optimum flexural strength is
achieved. Concentrations higher than this resulted in inferior
strengths. It was found that the higher the content of physisorbed
silane, the lower the flexural strength for the E-glass
fiber γ-MPS polyester composite [18]. Many more systems have to be
examined before generalizations can be made.

It is thought that the copolymer between the physisorbed
silane and the matrix resin influences the mechanical behavior of
the composite. Judging from the distance to which the physisorbed
silane is capable of migrating, the thickness of the matrix interphase can be substantial. More quantitative information has been
obtained by a simulation study of the interfacial materials by Graf
et al. [18]. They studied the copolymer of the γ-MPS hydrolyzate
and polyester resin at various silane contents. The flexural
strength of the copolymers linearly decreased as the silane content
increased during the first 40% by weight silane content. This
simulation study suggests that the matrix near the glass fiber, the
matrix interphase, is one of the weakest portions of the
composite. Naturally, wash with an organic solvent prior to the
mixing with the matrix resin minimizes this weak matrix interphase
without altering the integrity of the chemisorbed silane layers,
resulting in an overall improvement in the flexural strength.
Illustrated in Figure 8 is the improved flexural strength of
fiber-glass reinforced polyester upon removal of the physisorbed
silane by THF. The importance of this data lies in its mechanistic

understanding of the reinforcing action rather than the magnitude of the improvement.

Physisorbed silane also has an important role in composite processing. A major reason why surface treatment by silane coupling agents reduces, in many cases, the viscosity of particulate-filled polymer melts is that the physisorbed silane acts as a lubricant during processing. An additional reason includes the inhibition of the preferred interaction between the surface functional groups and the specific sites on the polymer chain.

Han et al's [21] results are consistent with the structural scheme described in the previous section. They studied the effects of surface treatments on the rheological properties of $CaCO_3$/polypropylene (PP) and glass beads/PP systems with amine- and octyl-functional silanes as coupling agents. The viscosity of the particulate-filled PP melt was reduced markedly for both silanes when $CaCO_3$ was used as depicted in Figure 9. The trends are explainable by Miller et al's work [9,10] demonstrating that the molecular weight of the silane would be low and the content of the physisorbed silane very high, since there are no surface functional groups. Therefore, the silane oligomers act as a lubricant and reduce the viscosity of the PP melt.

Little reduction in the viscosity was observed over the untreated filler when the octyl-functional silane was used on glass beads. Furthermore, the viscosity even increased when an aminosilane was used. The results are shown in Figure 10. Possible reasons are that the amount of the physisorbed silane would be much lower on glass beads than on $CaCO_3$, because the glass beads have surface functional groups, and that the molecular weight of the silane may be higher than on $CaCO_3$ since the glass beads are more neutral than $CaCO_3$. These factors may contribute to the small reduction of the viscosity. As for the aminosilanes, there may be a small number of chemical bonds or molecular entanglement with the PP, which increases the viscosity. This statement is supported by the fact that the increment of the viscosity was higher at a higher experimental temperature.

STRUCTURE OF THE CHEMISORBED SILANE

In addition to the physisorbed silane, it is apparent that chemisorbed silane also exists. Within the chemisorbed silane layers there exist structural variations depending on the treatment conditions. This portion of silane may be mainly responsible for the reinforcement mechanisms. The chemisorbed silane may not be as easily amenable to ordinary analytical techniques as the physisorbed portion. The molecular weight is, by definition,

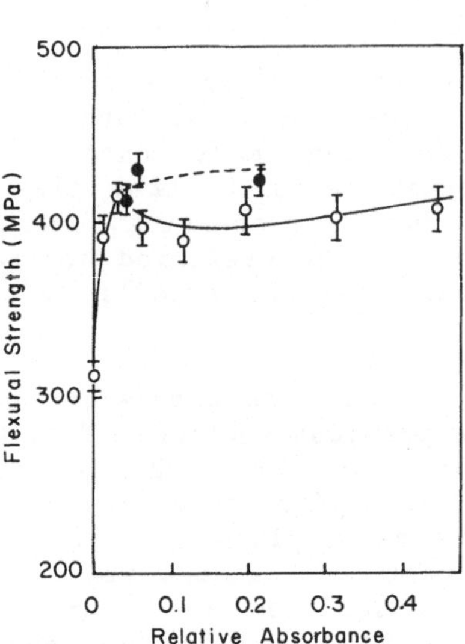

Fig. 8.   Flexural strength of fiber-glass reinforced polyester as
a function of the amount of γ-MPS as measured by infrared
band intensity. Open circle represents the composite
prepared from as-treated E-glass cloth while the closed
circle utilized E-glass cloth washed with tetrahydrofuran
prior to the composite manufacturing.

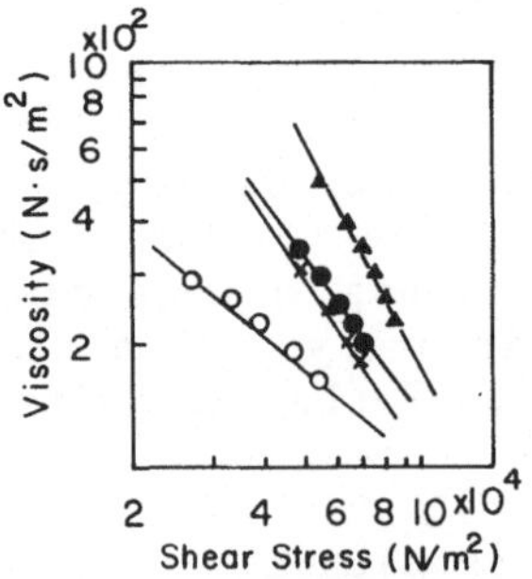

Fig. 9.   Viscosity of polypropylene melt with $CaCO_3$ at 200°C. The
surface of $CaCO_3$ is treated with various silanes. The
polypropylene contains: (O) no filler ($\blacktriangle$) untreated $CaCO_3$
($\bullet$) $CaCO_3$ treated with octyltrymethoxysilane and (X) $CaCO_3$
treated with γ-APS. The filler content is 30% by weight.

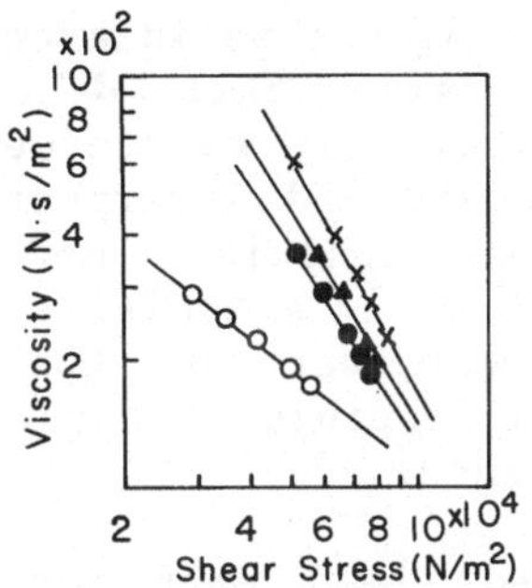

Fig. 10. Viscosity of polypropylene melt with glass beads at 200°C.
The surface of glass beads is treated with the same silane
as in Fig. 9. (O) no filler (▲) untreated glass beads
(●) glass beads treated with octyltrimethoxysilane and (X)
glass beads treated with γ-APS. The filler content is 30%
by weight.

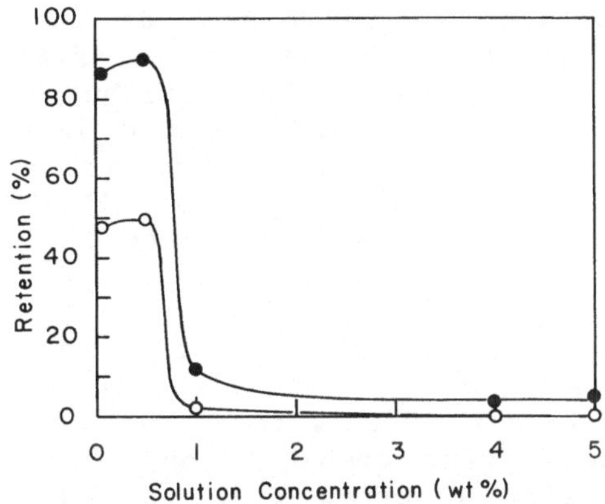

Fig. 11. Amount of chemisorbed γ-APS on E-glass fibers as a function of the concentration of silane treating solutions.
Water is used as a washing solvent. (O) dried at room
temperature and (●) dried at 120°C for 2 hr.

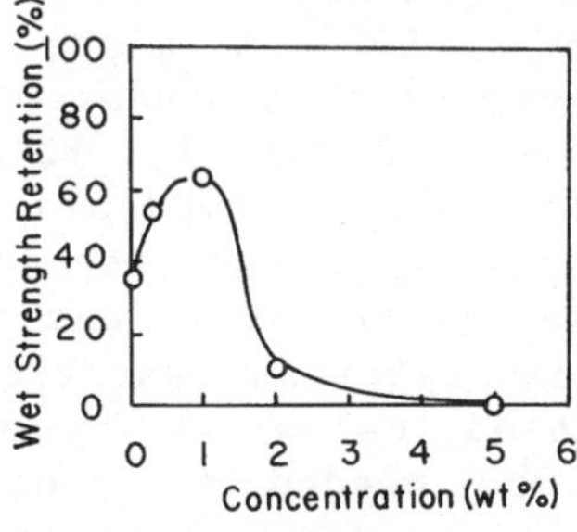

Fig. 12. Wet peel strength retention of polyethylene coating on
Al_2O_3 whose surface is treated with γ-APS at various
concentrations of aqueous solution.

infinite since it is a part of the solid even though there seems to exist distinct segmental size that determine the mechanical consequence of the composite. The existence of the substrate may complicate the data collection and interpretation of the results. In spite of all these difficulties, modern surface sensitive spectroscopies have made these interfacial studies much easier. As we shall see in the following section, the wealth of information reported now enables us to distinguish structural differences even within the chemisorbed silanes.

It should be realized that, when the substrate surface is molecularly smooth as is the case for glass fibers, the chemisorbed silane may not be a monolayer thick but may form thick multilayers. The first molecular layer is unique in its structure and interaction. The organofunctional group may be attracted to form hydrogen bonds or repelled depending on the chemical nature of the organic group and the surface. The silanol groups of the silane molecules usually condense with the surface hydroxyl groups and the silane molecule becomes a part of the substrate. The mobility of the silane molecule within the first monolayer is, therefore, uniquely restricted by the distribution and the nature of the surface active sites. From the second layer and above, the bulk nature increases as the distance from the surface increases although the surface effects may still be felt. In this section, the existence of loosely chemically bound layers are also proposed.

Schrader [5] reported a very thin, tightly bound γ-APS layer near the substrate surface. It was difficult to desorb even by boiling water. The origin of this tenacious nature is not known, however it is believed that chemical bonding to the surface is responsible. They also reported that as much as 98% of the adsorbed silane desorbed by cold water wash. The film thickness was approximately a few hundred monolayer equivalents which corresponds to a relatively high concentration of silane treating solution. Observation of such a high content of physisorbed silane is in good agreement with the study by Culler et al. [15] on the hydrothermal stability of γ-APS on glass fibers, where they reported a strong concentration dependence on the amount of physisorbed silane as shown in Figure 11. Below 1% by weight, the amount of the chemisorbed silane suddenly increases and this portion of silane is susceptible to the thermal treatment while the silane from relatively concentrated solutions show little effect to heat treatment. It has been reported that the silanetriol content of γ-APS suddenly increased below 1% by weight [22]. These silanetriol monomers may be needed to produce the chemisorbed silane. Consequently, it can be concluded that the self-catalyzation of the amine group of γ-APS leads to oligomer formation in the silane solution prior to adsorption. These oligomers do not produce a high yield of chemisorbed silane resulting in the very high content of the physisorbed silane as observed by Schrader et al. [5] and Culler et al. [15]. This type of concentration dependence is unique to amine-functional silanes where the catalytic effects produce a high content of oligomers. However, the role of oligomers may be similar regardless of the type of organofunctionality. The low oligomer content of other neutral silanes do not produce such a dramatic effect as the aminosilanes.

A detrimental effect of the physisorbed oligomer on the wet strength retention of Al_2O_3/γ-APS/PE joint system is seen in Figure 12. Since little chemical reaction is expected in this system, the oligomeric physisorbed silane cannot provide good hydrothermal stability while the chemically adsorbed silane can. A remarkable similarity in the shape of the curves in Figures 11 and 12 indicates the reasonableness of this statement.

In addition to the quantity of silane molecules extant on the substrate, structural information can also be obtained. Interaction of organic molecules with high-surface-area metal oxides has been the subject of long and extensive investigation and many excellent monographs have been published [23-25]. We shall mainly discuss silanes on relatively low surface area substrates, though one can observe many similarities among these substrates in terms of silane structures.

It is well known that more than a monolayer of silane is needed to yield the optimum strength of a composite material. This is likely due to the necessity of the interfacial chemical bonds, interpenetrating network formation in the chemisorbed silane layers, and proper orientation of the organofunctional groups. Many silanes used in composite materials especially for the purpose of improving the strength have reactive, polar groups as exemplified in amine-, methacryloxy- and epoxy-functional silanes. These silanes tend to form a bridge-like structure in the first monolayer where the silanol groups covalently bond to the surface and the organofunctional groups hydrogen bond with the surface active sites reducing the availability of the reactive groups for copolymer formation with the matrix resin.

Hydrogen bonding of carbonyl-containing molecules with mineral surfaces is well-known. As expected, the methacryloxy-functional silane also shows evidence of hydrogen bonding with glass fibers [26] and various particulate fillers [27]. This silane is a very useful molecular probe for studying the interfacial structure because of the sensitivity of the infrared carbonyl frequency to different environments. Definition of a monolayer equivalent of a silane on a substrate is not as straightforward as a monolayer film deposited on a flat surface by the Langmuir-Blodgett technique. Nonetheless, the amount of silane at which the surface property of the substrate changes drastically from the initial inorganic nature to the organic nature is of great technical and fundamental importance. Monolayer equivalence may be defined as the unimolecular layer which consists of only uniquely interacting molecules with the substrate surface. The molecular probe described above provides a convenient means of determining the monolayer coverage. An example is shown in Figure 13 where the relative intensity variation of the hydrogen bonded and free carbonyl groups as a function of γ-MPS loadings is illustrated. Determination of surface energetics may be useful to study the orientation of the molecule once the surface coverage is determined by other means. The same silane molecule may yield different values depending on the orientation of the organofunctional groups. A typical example is that an aminofunctional silane yields contact angle data similar to the hydrocarbons rather than the amine [28] due to the bridge-like structure mentioned above [15,26,27].

In relation to the definition of monolayer equivalence, the area occupied by a single molecule becomes an important quantity. To date, only a few experimental determinations of the molecular size of the silane on the surface of substrates have appeared in the literature. Ishida et al.[26] have reported that γ-MPS occupies approximately 0.48 nm^2/molecule on an E-glass fiber surface. This is in accord with the expected molecular size when the molecule is flatly adsorbed. Favis et al. [29,30] reported that the areas of silanes and a titanate on mica were 0.24 nm^2/molecule for γ-MPS, 0.33 nm^2/molecule for vinylbenzylamino-functional silane and 1.43 nm^2/molecule for isopropyl-tri(dioctyl-pyrophosphato)titanate. Their values are consistent with the expected molecular size for the vertical orientation. Miller and Ishida [31] also measured the occupied area of a γ-MPS hydrolyzate on the surface of clay and lead oxide, and obtained the value to be 0.60 and 0.59 nm^2/molecule, respectively. Again this value is in agreement with the flat adsorption, which is consistent with the observation of the hydrogen bonded carbonyl groups to the surface. The difference in orientation observed above is believed to be due to the treatment conditions such as the concentration of the treating solution.

The discussion of the chemisorbed silane is extended to the silanes near the surface but beyond the first monolayer. There seems to be tightly bound layers and loosely bound layers within the chemisorbed multilayers which can be distinguished by the hydrothermal desorption studies [7,14]. Extensive siloxane networks are statistically and chemically unfavorable for the desorption. Thus, the shape of the desorption curve yields quali-

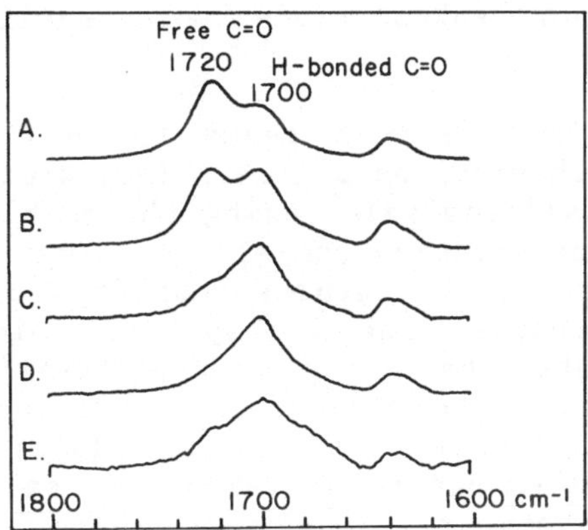

Fig. 13.   Fourier transform infrared spectra of γ–MPS on clay with
silane loadings of (A) 4 wt. % (B) 2 wt. % (C) 1 wt. %
(D) 0.5 wt. % and (E) 0.2 wt. % showing the free carbonyl
(1720 cm^{-1}) and the hydrogen bonded carbonyl (1700 cm^{-1}).

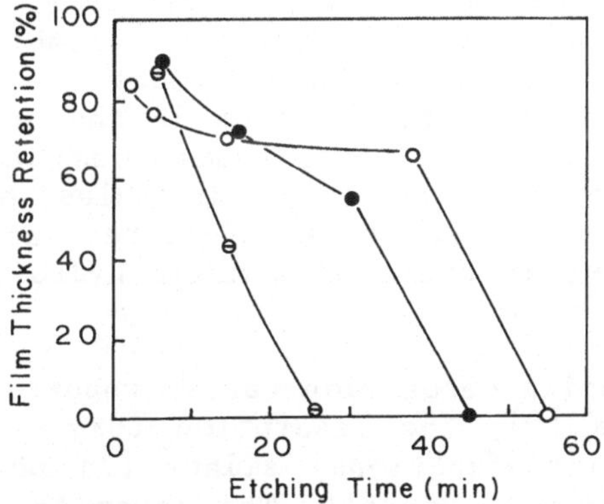

Fig. 14.   Etching time of γ–APS on a silicon wafer as a function of
the initial thickness. An alkaline aqueous solution is
used as an ethant. The thickness of the film is determined by ellipsometry. The γ–APS concentration is
(O) 0.02 vol %, (●) 0.04 vol % and (⊖) 0.10 vol %.

tative information on the extent of open structure. The cyclic or caged structures tend to isolate the segmental units and between these units little covalent bonding exists, thus in this case hydrolytic desorption takes place more easily than the open structure.

Belton et al.'s [32] work seems to be consistent with the above model of the chemisorbed silane. They studied an amine-functional silane on a silicon wafer using the combination of chemical etching with alkaline solution and ellipsometry. Their results are replotted in Figure 14. The results indicate that the thicker the silane film, the quicker the desorption. Since the thickness variation was created by using silane treating solution with varying concentration, the higher concentration solution may be said to yield more loosely bound silane layers. The amount of physisorbed silane also has to be taken into account. The results of water wash were shown previously in Figure 7 where the thicker layers had more physisorbed silanes. Even though the physisorbed silanes are taken into account, the etching rate for the thicker silane film cannot be accounted for unless the etchability of the thicker film is easier. The γ-APS film remained the longest on the silicon wafer when treated from the most dilute solution. As it has been reported in the literature [13], this is due to the effects on the siloxane network structure by the silane structure in the treating solution. It should be emphasized that the hydrolytic stability of the individual siloxane linkage may not be affected by the specific network structure. In fact, very different silanes show similar hydrolytic stability in terms of bond cleavage by water [7]. What affects the desorption characteristics in such a dramatic fashion is the organization of the siloxane bonds or extent of three dimensional network formation. In other words, cyclic cage-like molecules have fewer bonds connecting each cage unit and this structure leads to poor overall hydrolytic stability, in spite of similar hydrolytic stability of individual siloxane bond.

Unlike exponential desorption curves reported for thick γ-APS films on substrates [5], the desorption curves of thinner films resemble that of vinyl-functional silane [7] which tend to form more open, non-cyclic structures. The lower the concentration of the silane treating solution, the longer the threshhold period for quick desorption. The thicker films show almost no threshhold periods. This indicates a more open (or extensive) siloxane network structure for the silane from lower concentrations and a more closed (or cyclic) siloxane structure for the silane from higher concentrations.

Molecular order in the silane interphase was first proposed by Ishida and Koenig [13] using vinyltrimethoxysilane as an

example. They also observed a higher rate and extent of condensation of the silanol groups when adsorbed from a more dilute
solution. The amount of the residual silanol increased
dramatically above the concentration termed onset of association.
General tendency of the head-to-head orientation was also proposed
[13] and experimental support was obtained from the study of x-ray
crystallographic data of a single crystal of cyclohexylsilanetriol
[33]. It should be noted that the degree of order and head-to-head
adsorption vary considerably depending on the chemical structure of
the organofunctional group. In general, a more flexible group
tends to deviate from the trend described above. These trends may
influence the structural hierarchy within the silane interphase.

     Molecular order in the chemisorbed silane layers has also
been reported by Favis et al. [29,30] where they studied a few
silanes on phlogopite mica utilyzing carbon-hydrogen-nitrogen
analyzer. They have reported that silane molecules adsorbed on a
mica surface in a stepwise fashion thus there was a time lag before
the next molecular layer of silane started adsorbing. Figure 15
shows some of their results where γ-MPS was hydrolyzed for 25 min
and then adsorbed onto the mica surface. A stepwise adsorption
scheme is clearly seen. Two factors may be especially significant
for interpreting the observed results. First, the mica surface is
molecularly smooth and it has been observed that a smooth surface
allows better molecular stacking [13]. Second, the concentration
of the silane solution used is approximately 0.025% by weight which
is extremely low compared with the concentrations studied by many
other researchers. At this concentration level, all silane
molecules are isolated and adsorbed onto the surface individually.

     It was also noted that when the hydrolysis time was increased
to 90 min, no γ-MPS was adsorbed onto mica, whereas the cationic
styryl-benxyl amino-functional silane did not show this effect over
the experimental uncertainty. Since the concentration is so low
that no silane molecules can associate with each other, condensation of silanol groups in such a short period of time is
negligible. Thus, oligomer formation is an unlikely cause. An
alternative explanation is that the methacryloxy moiety is
hydrolyzed in 90 min to form methyl methacrylate acid and no longer
the silanetriol of γ-MPS. Alkali catalyzed hydrolysis product has
been reported [10] and indication of acid-catalyzed product has
also been reported [34]. It is not surprising since methyl
methacrylate can be hydrolyzed both in acidic and alkaline media.

     Substrate effects on the structure and adsorption character-
istics of silane is an interesting subject and essential to the
understanding of the structure of the silane film. Silanes on
metals have been studied extensively by Boerio et al. [34-37] and
others [38,39]. Very little depth profile information has been

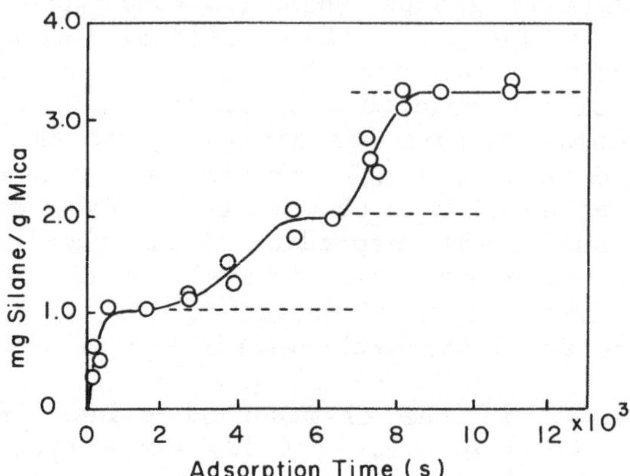

Fig. 15. Amount of γ-MPS adsorbed on mica surface as a function
of absorption time. Each plateau corresponds approximately to monolayer equivalent when vertical adsorption
is assumed.

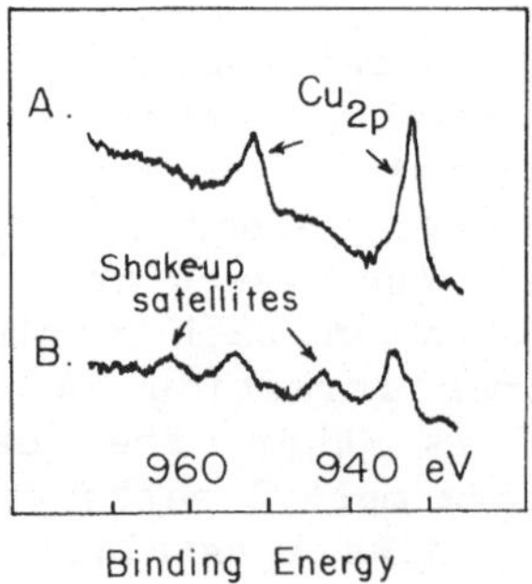

Fig. 16. X-ray photoelectron spectra of γ-APS on a copper substrate. The samples are dried at room temperature and
stored (A) in a decicator until spectral examination,
(B) in laboratory atmosphere for 4 days.

reported. Nonetheless, it is believed that the structure of silane on metallic substrates follow general trends described above for the nonmetallic substrates except for the complex formation of the organofunctional group and the metal ions. Boerio et al. [37] reported that copper (II) ion exists in the silane interphase using x-ray photoelectron spectroscopy as shown in Figure 16. Copper (II) ions show the same fundamental core electron energies in ESCA spectrum with copper (I) which normally exists on the polished copper surface. However, the shake-up satellites observed at the lower energy side are characteristic to copper (II). Thus, they have concluded that the amine group of γ-APS forms a complex with the copper ion after dissolution of the surface oxide film. The etching effect of the surface oxide films by γ-APS has also been observed on aluminum substrates.

Allen and Stevens [38] reported the infrared study of γ-glycidoxypropyltrimethoxysilane (γ-GPS) on aluminum substrate as a function of the amount adsorbed. A very thin silane film lowered the OH stretching frequency indicating stronger hydrogen bonding. They reasoned that the frequency shift was due to the complex formation with the aluminum substrate, though only inconclusive evidence was shown.

Silane coupling agents are sometimes reported to inhibit corrosion of metals. An amine containing silane has synonymous function as anticorrosion agents such as benzotriazole and imidazole derivatives for copper, as evidenced by the observation of complex formation of γ-APS with copper. These anticorrosion agents are also known to form complexes which are a necessity for their anticorrosion action. Hence it is reasonable to expect the anticorrosion action of γ-APS on copper and possibly on other metals by a similar mechanism of these azoles. Once a complex is formed, the interaction between the amine group and the metallic ion is strong and inhibits the availability of the organofunctional groups for copolymer formation with the matrix resin. If the complex is known to catalyze the cure of the matrix resin, another effect may be expected. Thus, the metal/silane coating interface can be quite different than those corresponding interfaces with non-metallic substrate. The glass/silane interface is relatively well defined as compared to the probable diffuse boundary of the metal/silane interface. The silane/matrix interface is diffusely defined for both glass and metal substrates.

THE ROLE OF THE CHEMISORBED SILANE

The role of the chemisorbed silane in the reinforcement mechanisms of composites can be discussed based on the studies by Schrader et al. [40], Emadipour and Koenig [41], and Belton et al. [32,42]. Schrader et al. [40] investigated the effect of silane desorption on the joint life during hydrothermal aging. Adhesive joints were prepared after the hydrothermal desorption of the silane. They observed a relatively insensitive decrease of the joint life in spite of the major desorption of the physisorbed silane as shown in Figure 17. On the other hand, the hot water extraction of the silane leads to a significant reduction of the joint life indicating the importance of the chemisorbed silane. It should be pointed out that the hot water extraction not only etches the outer layer of the silane but also degrades the integrity of the siloxane networks which still remain on the substrate surface. Thus, the loss of network integrity has to be taken into account. If the silane layers are subjected to postcuring after the hydrothermal desorption, this effect will be minimized. This statement is supported by Emadipour and Koenig [41] as shown in Figure 18 where they measured the pull-out strength of a glass rod from an epoxy matrix. The silane layers were etched by hot water and given subsequent heat treatment prior to the preparation of the fiber pull-out samples. The interfacial shear strength after the hydrothermal treatment improved. Furthermore, the improved strengths were nearly constant regardless of the concentration of the silane treating solution in the relatively concentrated range (1-10% by weight), which is compared with the strong concentration dependency of the strength for the as prepared samples.

More direct observation was made by Belton et al. [4,32,42] using etching experiments of silicon wafer/γ-APS/polyimide system. A thin polyimide film cured on a silane treated silicon wafer was etched by an alkaline solution and the time required to etch the film was studied as a function of the thickness of the initial silane interphase. Although they did not distinguish the contribution of the chemisorbed and physisorbed silanes, the thickness range and the concentration of the silane treating solutions produce predominantly chemisorbed silanes upon heat treatment. Thus, the physisorbed component at room temperature drying has different configurations than the truly physisorbed silane which remains physisorbed even after the heat treatment. This type of silane will provide a favorable situation for the intermixing between the silane and the polyimide precursor. Upon curing of the polymeric film, the silane is also cured to form more extensive networks than the truly physisorbed silane. Also the availability of the organofunctional groups of this potentially chemisorbed silane may be quite different than the truly physisorbed silane. Their results are replotted in Figure 19 where the etching time is plotted as a function of the silane film thickness. As the silane thickness increased, the time required for complete removal of the polyimide film increased to a certain thickness, again showing that a monolayer silane does not yield an optimum property.

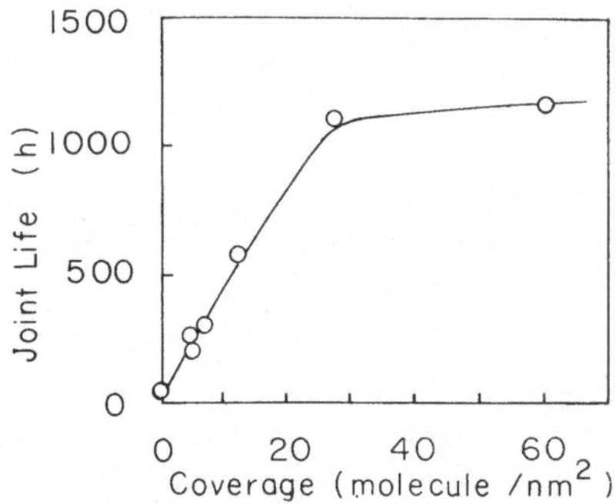

Fig. 17.   Joint life of the glass/epoxy adhesive/glass system where
the glass surface is treated by a radioisotope-labeled
γ-APS.   The variation of surface coverage is produced by
extracting the silane with cold water (in the 60-30 mole-
cule/nm^2 range) and with hot water (in the 30-0 molecule
/nm^2 range).   The joint is immersed in hot water with a
constant load.

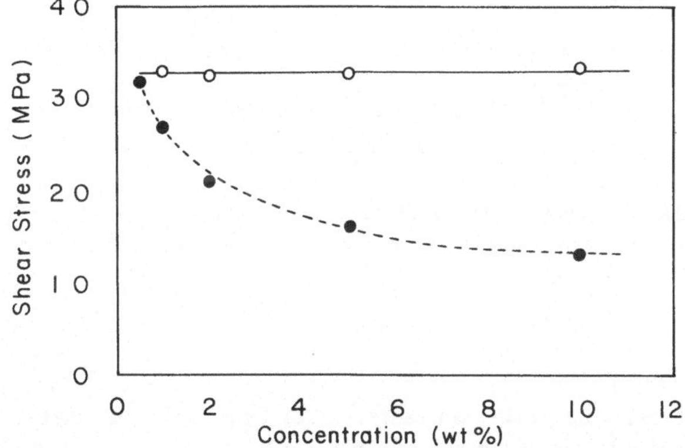

Fig. 18.   Shear stress of a single glass rod determined by the
fiber pull-out test from an epoxy matrix.   The glass surface is treated with γ-APS at various concentrations.
Closed circle represents the glass rod as treated and
the open circle shows the silane-treated glass rod which
is boiled in water for 4 hr and dried.

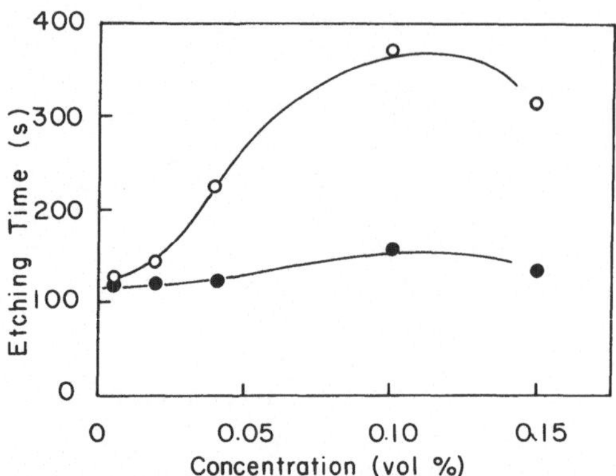

Fig. 19.   Etching time for complete film removal of polyimide
           against γ-APS solution concentration for different silane
           application techniques.   The polyimide is cured on γ-APS-
           treated silicon wafer at 150°C for 45 min.   The alkaline
           aqueous solution is used as an ethant.   Silane is spin
           coated at 5000 rpm by (O) immersing a silicon wafer in the
           silane solution for 15 min followed by spin coating and
           (●) placing the silane solution on a silicon wafer fol-
           lowed by spin coating.

     With all these observations, one can postulate the existence
of loosely chemisorbed silane layers whose structure is favorable
for interpenetration of the matrix resin.   Only indirect observations of the existence and function of this portion of the silane
have been made to date and further rigorous studies are needed to
elucidate the structure.

CONCLUSION

     The gradient in the silane interphase on various substrates
has been discussed in terms of the chemical structure as well as
the role of various portions of the silane in the reinforcement
mechanisms of composite and rheological behavior of composite
melts.   It is proposed that silane interphase be subdivided into
two clearly distinguishable regions using an organic solvent wash
that does not induce the hydrolytic scission of the siloxane
bonds.   These are termed the physisorbed and chemisorbed silanes.
Within the chemisorbed silane layers, it is also proposed that this
portion of silane be again subdivided into at least two regions and
possibly three regions depending on the interaction with the
substrate.   The first monolayer is uniquely interacting with the
substrate surface and the nature of the interaction is totally

dependent on the combination of the chemical structure of the organofunctional group and the substrate surface. Above this first layer but very near the surface of the substrate, there are tightly chemically adsorbed layers. Further out layers near the physisorbed silanes, another structure exists where loosely bound though chemisorbed silane dominates.

The role of the physisorbed silane on the mechanical properties of the composite is dependent upon the mechanical properties of the particular copolymer between the physisorbed silane and the matrix resin. In some limited examples, these copolymers are reported to be unfavorable with respect to the improved mechanical performance. This silane acts as a processing aid by reducing the viscosity of the reinforcement polymer systems. The chemisorbed silane is thought to be mainly responsible for the reinforcement mechanisms through chemical bond formation along with interpenetrating network formation.

ACKNOWLEDGEMENT

This research was in part supported by the Office of Naval Research.

REFERENCES

1. E. P. Plueddemann, "Silane Coupling Agents," Plenum, New York (1982).
2. E. P. Plueddemann, ed., "Interfaces in Polymer Matrix Composites," Academic Press, New York (1974).
3. H. Ishida and J. L. Koenig, Polym. Eng. Sci., 18, 128 (1978).
4. H. Ishida, Polymer Composite, in press.
5. M. E. Schrader, I. Lerner, and F. J. D'Oria, Mod. Plast., 45, 195 (1967).
6. O. K. Johanson, F. O. Stark, and R. Baney, AFML-TRI-65-303, Part 1 (1965).
7. H. Ishida and J. L. Koenig, J. Polym. Sci.-Phys., 18, 1931 (1980).
8. T. Nakatsuka, H. Kawasaki, K. Itadani, and S. Yamashita, J. Appl. Polym. Sci., 24, 1985 (1979).
9. H. Ishida and J. D. Miller, Macromolecules, in press.
10. J. D. Miller, K. Ho and H. Ishida, Polym. Eng. Sci., in press.
11. H. Ishida and J. D. Miller, J. Polym. Sci.-Phys., submitted.
12. N. H. Sung, A. Kaul, I. Chin, and C. S. P. Sung, Polym. Eng. Sci., 22, 637 (1982).
13. H. Ishida and J. L. Koenig, J. Polym. Sci.-Phys., 17, 1807 (1979).
14. D. J. Belton and A. Joshi, Polymer Preprints, 24, 206 (1983).
15. S. R. Culler, H. Ishida and J. L. Koenig, Proc. 15th Nat. Tech. Conf., SAMPE, Oct. (1983).

16. K. A. Andrianov and B. A. Izmaylov, J. Organomet. Chem., $\underline{8}$, 435 (1967).
17. M. Kokubo, H. Inagawa, M. Kawahara, D. Terunuma, and H. Nohira, Kobunshi Ronbunshu, $\underline{38}$, 201 (1981).
18. R. T. Graf, J. L. Koenig and H. Ishida, J. Adhesion, in press.
19. R. Wong, J. Adhesion, $\underline{4}$, 171 (1972).
20. Y. Eckstein and E. J. Berger, Proc. Org. Coat. Appl. Polym. Sci., $\underline{48}$, 23 (1983).
21. C. D. Han, C. Sanford, and H. J. Yoo, Polym. Eng. Sci., $\underline{18}$, 849 (1978).
22. H. Ishida, S. Naviroj, S. K. Tripathy, J. J. Fitzgerald and J. L. Koenig, J. Polym. Sci.-Phy., $\underline{20}$, 701 (1982).
23. L. H. Little, "Infrared Spectra of Adsorbed Species," Academic Press, London (1966).
24. M. L. Hair, "Infrared Spectroscopy in Surface Chemistry," Dekker, New York, (1967).
25. A. V. Kiselev and V. Il Lygin, "Infrared Spectra of Surface Compounds," Keter Publishing House, Jerusalem (1975).
26. H. Ishida, S. Naviroj and J. L. Koenig, in "Physicochemical Aspects of Polymer Surfaces," K. L. Mittal, ed., Plenum, New York 91 (1983).
27. H. Ishida and J. D. Miller, Proc. 38th Ann. Tech. Conf., Reinforced Plastics/Composites Inst., SPI, Section 4-E (1983).
28. B. Arkles, Chemtech, $\underline{7}$, 766 (1977).
29. B. D. Favis, L. P. Blanchard, J. Leonard, and R. E. Prud'homme, Polym. Eng. Sci., in press.
30. B. D. Favis, L. P. Blanchard, J. Leonard and R. E. Prud'homme, Polym. Eng. Sci., in press.
31. J. D. Miller, and H. Ishida, J. Phys. Chem., submitted.
32. D. J. Belton, P. van Pelt, A. E. Morgan, Silicon Processing, ASTM STP 804, D. C. Gupta, ed., American Society for Testing and Materials, 273 (1983).
33. H. Ishida, J. L. Koenig and K. H. Gardner, J. Chem. Phys., $\underline{77}$, 5748 (1982).
34. F. J. Boerio and S. Y. Chen, J. Colloid Interface Sci., $\underline{68}$, 252 (1979).
35. F. J. Boerio and J. W. Williams, Surf. Sci., $\underline{7}$, 19 (1981).
36. F. J. Boerio, C. A. Gosselin, R. G. Dillingham and H. W. Lin, J. Adhesion, $\underline{13}$, 159 (1981).
37. F. J. Boerio, J. W. Williams and J. M. Burkstrand, J. Colloid Interface Sci., $\underline{91}$, 485 (1983).
38. K. W. Allen and M. G. Stevens, J. Adhesion, $\underline{14}$, 137 (1982).
39. A. K. Hays and D. M. Haaland, Proc. Org. Coat. Appl. Polym. Sci., ACS, $\underline{47}$, 383 (1982).
40. M. E. Schrader, and A. Block, J. Polym. Sci. Part C, $\underline{34}$, 281 (1971).
41. H. Emadipour, P. Chiang and J. L. Koenig, Res. Mechanical, $\underline{5}$, 165 (1982), and H. Eamdipour, Masters Thesis, Case Western Reserve University, Cleveland, Ohio (1982).
42. D. J. Belton and A. Joshi, this monograph.

SURFACE MODIFICATION OF CALCIUM CARBONATE FOR POLYMER COMPOSITES

Takuo Nakatsuka

Industrial Technology Center of Okayama Prefecture
3-18, Ifuku-cho 4 chome
Okayama 700, JAPAN

INTRODUCTION

Ground calcium carbonate is one of the oldest manufactured powders. It is reported that British whiting was produced at Brandon in Suffolk, England in the pre-Roman era [1]. Thereafter, it has been prepared by dry- or wet-grinding methods. Developments in grinding and classifying techniques have enabled us to obtain commercially dry-ground ultrafine grade calcium carbonate having average particle diameter as small as 0.5 ~ 0.7 μm [2].

Modern chemistry brought about a different kind of calcium carbonate powder; precipitated calcium carbonate [3]. In 1898, synthetic calcium carbonate was prepared by the reaction of calcium chloride and sodium carbonate (calcium chloride process). Use of calcium hydroxide in place of calcium chloride also gives precipitated calcium carbonate. However, this process has been unavailable because of the difficulty in excluding the by-product of sodium hydroxide (soda-lime process). Recarbonation of calcium hydroxide slurry with carbon dioxide gas has been an alternatively useful process for precipitated calcium carbonate production recarbonation process [4,5].

Calcium chloride process:

$$CaCl_2 + Na_2CO_3 \rightleftharpoons 2NaCl + CaCO_3 \qquad (1)$$

Soda-lime process:

$$Ca(OH)_2 + Na_2CO_3 \rightleftharpoons H_2O + CaCO_3 \qquad (2)$$

Recarbonation process:

$$Ca(OH)_2 + CO_2 \; \rightleftharpoons \; H_2O + CaCO_3 \qquad\qquad (3)$$

Early studies on the preparation and physical properties of calcium carbonate have been reviewed in the literature [6].

In polymer composites, calcium carbonate is one of the important particulate fillers; however, until several years ago the surface modification of calcium carbonate had not attracted attention outside the production industry. Therefore, much of the available research can be found in recent patent literature.

PHYSICAL AND CHEMICAL PROPERTIES OF CALCIUM CARBONATE FILLER

Calcite is a stable phase under ambient conditions. Natural and synthetic calcium carbonate generally have a calcite structure, and may contain some aragonite of metastable phase [7]. Grinding of the calcite crystal is known to result in the aragonite structure [8]. The phase transformation has accounted for the phenomenon that a prolonged continuous grinding of calcium carbonate does not lead to continuous increase in the surface area. Preparation of vaterite always requires specific reaction conditions. Yamaguchi et al. [9] prepared vaterite calcium carbonate by the reaction of calcium methoxide and aqueous sodium carbonate. It may be said that the crystalline structure of calcium carbonate does not significantly affect the reinforcing ability of the particulate in polymer composite materials.

Average particle size and particle size distribution are important factors when calcium carbonate is used as a filler in rubbers or plastics [10,11]. These parameters are closely correlated with the manufacturing process. Ikegami [12] illustrated the relation in Figure 1. Recently, ultrafine particles of 0.005 ~ 0.02 μm (5 ~ 20 nm), [13] or 0.015 ~ 0.07 μm (15 ~ 70 nm) [14] average diameter have been prepared by the recarbonation process. It was also reported that an ultrasonic wave [15,16] or a magnetic field [17] applied during calcium carbonate precipitation affects the physicochemical properties.

Shape of particulate fillers is a significant factor in determining the mechanical properties of the particulate-filled polymer composite. Ground calcium carbonate has the shape of sponge-like aggregates. Precipitated calcium carbonate having calcite structure forms cube, spindle, or rod shapes and the aragonite structure exhibits a rod-like shape. Aspect ratio is

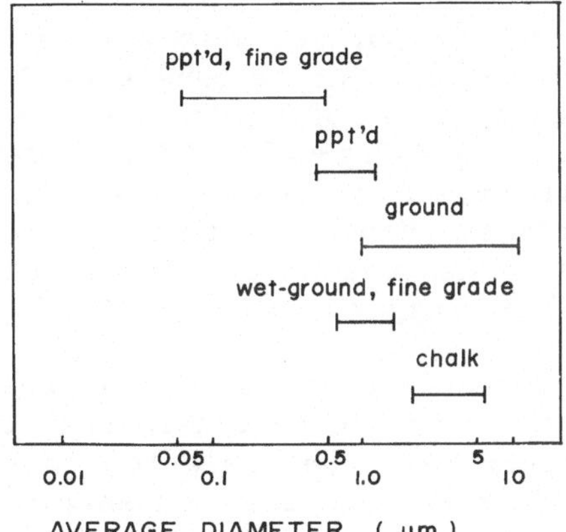

Fig. 1.   Schematic correlation between calcium carbonate particle
size and producing methods.

often used to denote the shape factor in composite technology.
Efforts to increase the aspect ratio of the precipitated calcium
carbonate by forming chain structure have been made [18-20]. Quite
differently, spherical particulates were obtained by interfacial
reaction of potassium carbonate aqueous solution emulsified in
benzene with calcium chloride aqueous solution [21].

The physical and chemical properties of the calcium carbonate
surface have been characterized by both traditional and recently
developed methods [22]. Measurement of the water adsorption isotherm is an effective approach to obtain information concerning the
surface hydroxyl groups [23,24]. Goujon and Mutaftscheiev [25]
combined this measurement with mass spectroscopy. They also determined the heats of immersion of the calcite crystal and a ground
crystal to be 480 and 540 erg/cm^2, respectively.

Thermogravimetric approaches provide information mainly about
the thermal decomposition of calcium carbonate [26]. Diffuse
reflectance infrared spectroscopy [27-29], and Raman spectroscopy
[30] have been applied to the studies on calcium carbonate and its
surface modification. Electron spectroscopy for chemical analysis
(ESCA) [31] shall be an effective device to identify surface
species on filler particles. However, ESCA requires high vacuum
conditions and is unavailable for surface characterization under
ambient conditions.

Acid-base properties of calcium carbonate surface should be
taken into account when the surface is to be modified. Yamanaka
and Tanabe [32] determined acid-base properties of 28 inorganic
powders in benzene. These values are significant under anhydrous
conditions. Inorganic mineral particulates generally have adsorbed

multilayers of water on the surface when exposed to the surrounding
atmosphere. For example, calcium carbonate used in our laboratory,
having average surface area of 7.09 m^2/g, was found to have
adsorbed water of 8.18 $mg-H_2O/g-CaCO_3$ (38.5 molecules/nm^2) by Kirl
Fischer method. This value corresponds to four molecular layer
formation on the surface. Hence, it should be more significant as
acid-base parameters to measure pH of aqueous suspension of filler
[29,36] as ASTM (D 1208) suggests.

The effects of carbon dioxide in air on the surface modification of calcium carbonate should be considered, too. Johnstone and
Williamson [33] determined the relation between the partial
pressure of carbon dioxide and the concentrations of hydroxide,
carbonate, bicarbonate, and calcium ions in the system of
$CaO-H_2O-CO_2$. If a partial pressure of carbon dioxide in the air is
3×10^{-4}, stable solid phase and predominant anion are calcium carbonate and bicarbonate ion, respectively.

Somasundaran and Agar [34] measured streaming potential,
solution equilibrium, and flotation response of calcite particles
as a function of pH. The pH values of the calcite-water-air system
comes to 8.2 after equilibration. The major ions at this pH, are
Ca^{2+} and HCO_3^- based on the thermodynamic data as shown in Figure
2. In the higher pH region, CO_3^{2-} becomes more important. Therefore, it is reasonably assumed that calcium and bicarbonate ions
are the principal components of the calcium carbonate filler
surface, after being exposed to air for a long period of time.

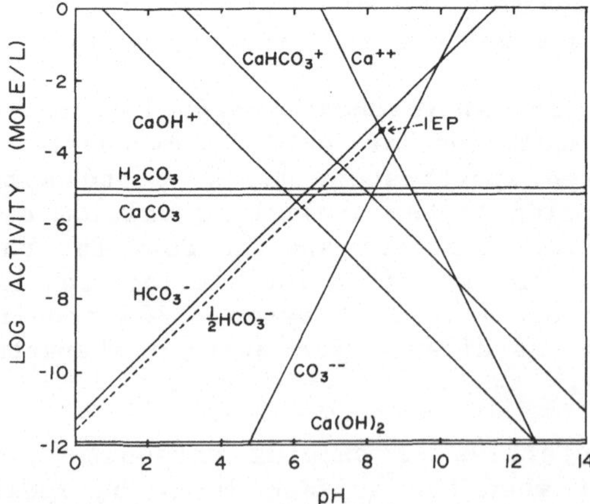

Fig. 2. Determination of isoelectric point (IEP) and potential-
determining ions for calcite-aqueous solution-air system,
using thermodynamic data.

SURFACE MODIFICATION OF CALCIUM CARBONATE

When particulate fillers are used in polymer composites, it is of critical importance to decrease the free energy of the filler surface to reduce agglomeration. Particulate filler surfaces are often modified with coupling agents. One might be overwhelmed with the voluminous patents and literature on carboxylate modifications of calcium carbonate filler. Recently, studies have investigated the modification of polymer matrices by introducing polar monomeric units through copolymerization or graft polymerization. These modifications, as well, improve the filler/matrix interaction and consequently the particulate dispersion in the matrix (vide infra).

Modification strategies for calcium carbonate filler may be classified into two groups. One is to condense modifying molecules with surface hydroxy groups, such as the well-known silane or titanate coupling schemes. The other is to use acid-base interaction followed by an acid salt deposition on the surface, such as fatty acid or acid phosphate modification.

Silane Modification

Surface modification with silane coupling agents has given satisfactory results not only for glass fibers but also inorganic mineral particulate such as silicas or silicates. Such fillers exhibit excellent reinforcing ability in rubbery and plastic composites [35]. Calcium carbonate filler, however, does not respond well to silane coupling agents, although it may have some surface hydroxy groups. Molecular understandings of this low effectiveness for calcium carbonate are far from enough.

Studies on the desorbable silane from the filler surface were initiated. Nakatsuka et al. [36] pointed out the relation between the aqueous suspension pH value and the molecular weight of desorbable silane coupling agent. Figure 3 and Table I show the gel permeation chromatograms and aqueous suspension pH values of calcium carbonate, phosphoric acid-pretreated calcium carbonate ($P-CaCO_3$), and clay. They also suggested the desorbable silane affected the physical properties of the particulate-filled vulcanizates (Table II).

Ishida and Miller [29] established acid-base properties of the particulate slurry as a long range acid-base effect, which exerts on the molcular structure of the desorbable silane. They also determined the amount of chemisorbed silane and suggested the chemical bond formation between the silane coupling agent and surface hydroxy group. Nevertheless, the real molecular structure is still not clear.

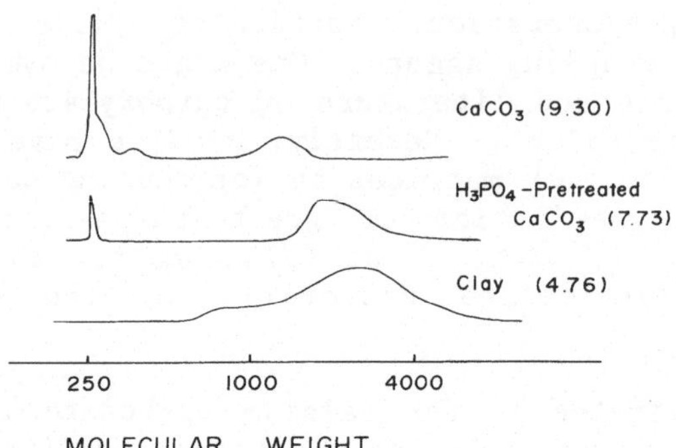

Fig. 3. Gel permeation chromatograms of tetrahydrofuran extracts
         from γ-methacryloyoxypropyltrimethoxysilane-modified
         calcium carbonate, H_3PO_4-pretreated calcium carbonate, and
         clay. The values in parentheses are pH of the aqueous
         suspension at 25.0°C.

Table I. Acid-Base Properties of Fillers (1)

Filler	pH of aqueous suspension at 25.0°C
$CaCO_3$	9.30
$P-CaCO_3$	7.73
Clay	4.76

(1) Five gram of filler was suspended and boiled in 100 ml water
     for 5 min.

Table II. Effects of Silane (1) Modification on Vulacanizate Properties of $CaCO_3$-Loaded SBR (2)

Filler	$CaCO_3$	$CaCO_3$/ silane	P-$CaCO_3$/ silane	P-$CaCO_3$/(EtO)$_4$Si/ silane (3)
Tensile strength, MPa	4.6	3.8	4.6	5.5
Elongation, %	520.0	480.0	440.0	440.0
300% Modulus, MPa	1.9	2.3	3.2	3.9
Hardness (4)	60	61	64	64
Tear (JIS–B), kN/m	13	15	15	19
Set, (4) %	8.7	7.4	5.2	4.3
Swelling, (5) %	321.0	279.0	257.0	252.0

(1)   γ–Mercaptopropyltrimethoxysilane (A–189, Union Carbide Corporation).
(2)   Recipe: SBR (Nipol 1502) 100, filler 100, zinc oxide 5.0, stearic acid 1.0, sulfur 2.0, 2–benzothiazyl disulfide 1.5, diphenyl guanidine 0.5. Cure: 150°C, 10 min.
(3)   One ml of (EtO)$_4$Si was treated as a methanolic solution (8.0 ml MeOH).
(4)   Tested by the method of Japanese Industrial Standards (JIS K 6301).
(5)   Weight increase after immersion in benzene for 48 h at room temperature.

Rosen and Goddard [37] characterized the desorption of silane coupling agents from the silane-modified particulates by measuring the surface tension change of water, on which the modified particulate was placed. From calcium carbonate surface, silane was desorbed away within a few minutes.

As a practical approach, surface pre-modifications of calcium carbonate with siliceous compounds were attempted. For example, calcium silicate [38], silica hydrosol [39], or polysiloxane [40] was formed before the treatment of silane coupling agents. These pre-treatments might offer more stable silanol groups and make bonds with the silane agents.

New types of silane coupling agents having polymeric organo groups have appeared in patents such as polyoxyethylenylpropyltrimethoxysilane [41], vinyltrimethoxysilane-diethyl fumalate copolymer [42], and vinyltrimethoxysilane-maleic acid or anhydride copolymer [43]. Introduction of polar units into silane coupling agents was reported to improve their coupling effects toward calcium carbonate. Trimethoxysilane-terminated polybutadiene [44] is an effective coupler for calcium carbonate as well as chopped glass strands. Table III demonstrated the improvement of calcium carbonate-filled polyethylene Izod impact strength.

Carboxylate Modification

It has been generally believed that carboxylic acids or salts react with calcium carbonate to form their calcium salts on the

Table III.   Notched Izod Impact Strength of $CaCO_3$-Filled
             Polyethylene (1)

Coupling Agent	Izod Impact Strength, kg-cm/cm
control	1.8
polybutadiene	2.2
$(MeO)_3Si$-terminated polybutadiene	3.9

(1) High density polyethylene 100, modified or non-modified
     $CaCO_3$ 50.

solid surface. In aqueous medium, this reaction scheme is true.
Szczypa et al. [45] measured the adsorption isotherm of sodium
laurate from water onto calcium carbonate and concluded that the
sodium laurate diffused and chemically bonded to calcium carbonate
as calcium laurate precipitates, according to a basic reaction of:

$$CaCO_3 + 2RCOO^- \rightleftharpoons Ca(RCOO)_2 + CO_3^{2-} \qquad (5)$$

The precipitates were thought to be closely packed and strongly
bound to the calcium carbonate surface. Suito [46] observed a peak
in the infrared spectrum of stearate-modified calcium carbonate at
$1580 \ cm^{-1}$ and assigned it to calcium stearate formed on calcium
carbonate surface.

Ivanishchenko and Gladkikh [47] studied aliphatic acid
adsorption from isopropyl alcohol onto calcium carbonate to conclude the formation of multimolecular layers. In nonpolar
solvents, the reaction of equation 5 seems to scarcely take place.
Scarifkhodzhaeva et al. [48] suggested that naphthenic acid
reacted with $Ca(OH)_2 \cdot CaCO_3$ or $CaSiO_3 \cdot H_2O$ to give calcium naphthenate but not with calcium carbonate.

In the reaction of calcium carbonate with fatty acids or
salts, an aqueous medium performs an essential role. Moreover, the
effects of carbon dioxide in air should be taken into consideration
for practical modification of calcium carbonate. A relationship
between the partial pressure of carbon dioxide and calcium
carbonate solubility is of great interests in the regions of geolo-
gical science and soil science. Cruz-Romero and Coleman [49] were
interested in the reaction among calcium carbonate, carbon dioxide,
and sodium adsorbate, and determined the following equilibrium:

$$Na_2X + 2CaCO_3 + 2H_2O + 2CO_2 \rightleftharpoons CaX + 2NaHCO_3 + Ca(HCO_3)_2$$
$$(6)$$

where X is the ion exchange resin carrying a sulfate functional
group.

Kasai [50] has reviewed the surface modifiers of calcium carbonate employed in the recarbonation process before 1968. Thereafter, a wide variety of modifiers have been examined and used. In
recent years, modifiers having relatively large molecular weights
have been attracting increasing attention. For example,
carboxylated polybutadiene [51], carboxylated polyethylene [52],
polyoxyethylenylstyrene-styrene-sodium maleate copolymer [53],
acrylic acid-modified polyvinylalcohol [54], hydroxy-terminted
polybutadiene [55], metal acrylate-butyl acrylate copolymer [56],
maleic acid-i-octene copolymer [57], sodium polyacrylate or metha-

crylate [60], and polyethyleneimine [61] have appeared in patents
and literature. To elucidate these polymer effects on the coupling
mechanism at a molecular level, much more basic studies are
required.

Phosphate Modification

In order to enhance acid-resistancy and flame-retardency of
polymer composites, calcium carbonate fillers have been treated
with phosphorous compounds. Matsushima and Miyata [62] prepared a
calcium carbonate surface covered with calcium hydrogen phosphate
hydrated by adding ortho phosphoric acid to a disodium hydrogen
phosphate-containing calcium carbonate aqueous slurry. They confirmed the salt deposition on the calcium carbonate surface by
scanning electron microscopy combined with selected-area diffraction.

Avnivelech [63] dealt more theoretically with the surface
complex formation derived from the reaction of calcium carbonate
and dipotassium hydrogen phosphate. The complex is claimed to have
a formula of $Ca_3(HCO_3)_3PO_4$. In patents, phosphoric acid or meta-
phosphoric acid have been reported as useful for calcium carbonate
modification.

Some acidic phosphorous compounds having organic groups have
been proven to be effective surface modifiers of calcium carbonate
filler. Nakatsuka et al. [64] studied the reaction of calcium
carbonate with alkyl dihydrogen phosphates. They provided a reaction scheme to explain the reaction products (Table IV) and medium
pH values.   (7)

$$CaCO_3 + ROPO_3H_2 \longrightarrow [ROPO_3H \cdot Ca \cdot HCO_3] \longrightarrow ROPO_3Ca + CO_2$$

$$ROPO_3H_2 \longrightarrow \downarrow$$

$$Ca(ROPO_3H)_2 \qquad ROPO_3Ca \cdot 2H_2O$$

$$+ \qquad\qquad (solid)$$

$$CO_2$$

When R is a relatively large alkyl group such as octyl, sparingly
water soluble calcium salts precipitate. Scanning electron micrographs in Figure 4 demonstrate the morphology of octyl dihydrogen
phosphate-modified calcium carbonate surface.

Table IV. Reaction of Phosphate Coupling Agents and Calcium Carbonate (1)

Alkyl Group in $ROPO_3H_2$	Treated $ROPO_3H_2$ (mmol)	CO_2 and HCO_3^- (mmol)	Ca^{2+} (mmol)	Recovered $ROPO_3H_2$ (mmol)
Ethyl	0.306	0.211 (69) [2]	0.141 (46) [2]	0.309 (101) [2]
Butyl	0.276	0.197 (71)	0.068 (25)	0.117 (42)
Hexyl	0.153	0.146 (95)	0.042 (27)	0.052 (34)
Octyl	0.224	0.221 (99)	0.043 (19)	0.043 (19)

(1) One gram of calcium carbonate (10.0 mmol) was used.
(2) The values in parentheses stand for the percentages of the products towards the phosphate coupling agents treated.

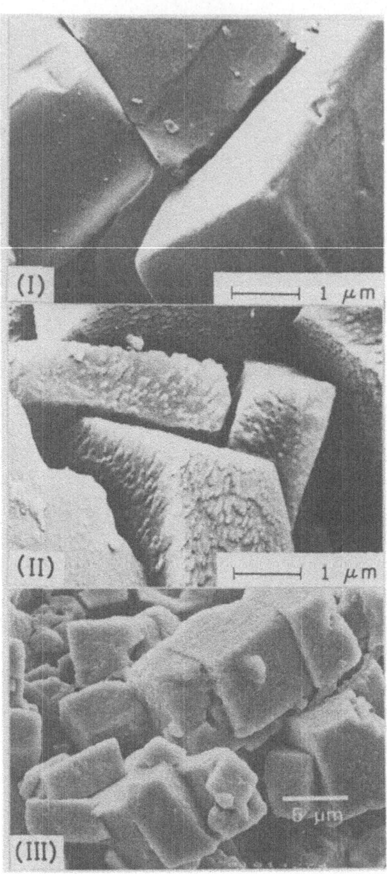

Fig. 4. Scanning electron micrographs of (I) original, (II) octyl
         dihydrogen phosphate-modified, and (III) polystyrene-
         grafted calcium carbonate.

Nakatsuka et al. [65] also prepared several dihydrogen phosphate of functional alcohols (Table V), and proved their significant coupling effects on the physical properties of calcium carbonate-filled vulcanizates (Tables VI and VII).

It can be found in the patent that organic phosphorus compounds have been effective in the surface modification of calcium
carbonate; phenylalkyl diester of phosphoric acid [66], mono or
diester of phosphoric acid containing acryloyoxy or methacryloyoxy
group [67], and phenyl phosphonic acid [68]. Dihydrogen phosphate
of hydroxy-terminated polybutadiene [69] was found to be an
excellent modifier of calcium carbonate. Its coupling effects on
the mechanical properties of calcium carbonate-loaded styrene-buta-
diene rubber are collected in Table VIII.

Table V. Phosphate Coupling Agents

Organic Group in Phosphate Coupling Agent	Abbrev.	R_f (1)
Ethyl	EP	0.24 (2)
Butyl	BP	0.34 (2)
Hexyl	HP	0.41 (2)
Octyl	OP	0.57, 0.47 (2)
3,7-Dimethyl-6-octenyl	CP	0.62
Bicyclo[2,2,1]hetp-5-en-2-methyl	NMP	0.52
2-(Methacryloyoxy)isopropyl	MPP	0.51
6-Mercaptohexyl	MHP	0.53
6-Chlorohexyl	CHP	0.52

4'-(N,N'-Dicyclohexylureidocarbonyl)-
  2,2'-azobis(2-methylbutyronitrile)-
  4-carboxyethyl
	AZP	0.74

2,2'-Dimethyl-2,2'-azobisbutyronitrile- 4,4'-carboxyethyl	AZDP	0.07

(1) Thin layer chromatography was carried out on silica gel.
     Developing solvent system was chloroform-methanol-acetic acid-
     water (25:15:4:2, v/v).
(2) The values are those corresponding to the monoanilinium salts,
     respectively.

Table VI. Coupling Effects on Peroxide-Cured EPDM (1)

Phosphate Coupling Agent	Control, none	OP	CP	NMP	MPP		
Treated, wt%	0	0.92	0.92	0.92	0.46	0.92	1.83
Attached, wt%	0	0.73	0.75	0.53	0.20	0.30	0.90
Tensile strength, MPa	4.5	4.4	4.7	6.0	5.9	7.4	7.6
100% Modulus, MPa	2.2	1.9	2.8	3.1	2.9	3.4	3.7
200% Modulus, MPa	2.6	2.2	3.9	4.9	4.6	5.5	6.6
Elongation, %	370	370	290	280	290	310	260
Hardness (2) (JIS)	66	64	68	68	67	68	69
Tear (2) (JIS-B), kN/m	10	8	11	11	11	11	12
Set (2), %	8	8	4	2	4	4	3

(1) Recipe: EPDM (Sumitomo Esprene 502), 100; $CaCO_3$, 100; ZnO, 5.0; dicumyl peroxide (97%), 2.7
Cure: 165°C, 15 min.
(2) Tested by the method of Japanese Industrial Standards (JIS K 6301).

Table VII. Coupling Effects on Sulfur-Cured SBR (1)

Phosphate Coupling Agent	Control, none	OP	CP	MHP		
Treated, wt%	0	0.91	0.92	0.46	0.92	1.83
Attached, wt%	0	0.73	0.71	0.24	0.44	1.01
Tensile strength, MPa	3.6	4.8	4.8	5.7	5.9	5.8
100% Modulus, MPa	1.3	1.3	1.4	1.6	1.9	2.1
200% Modulus, MPa	1.5	1.4	1.5	1.8	2.2	2.5
300% Modulus, MPa	1.8	1.7	1.8	2.1	2.4	2.9
Elongation, %	480	530	570	530	530	510
Hardness (2) (JIS)	58	59	61	60	62	63
Tear (2) (JIS-B), kN/m	16	15	18	20	20	20
Set (2) %	7	9	11	8	8	7
Swelling, (3) %	327	328	324	294	270	261

(1) Recipe: SBR(Nipol 1502), 100; $CaCO_3$, 100; ZnO, 5.0; stearic acid, 1.0; sulfur 2.0; 2-benzo-thiazyl disulfide, 1.5; diphenyl guanidine, 0.5. Cure: 150°C, 15 min.
(2) Tested by the method of Japanese Industrial Standards (JIS K 6301).
(3) Weight increase after immersion in benzene for 48 h at room temperature.

Table VIII. End Group Effects of Polybutadiene on Vulcanizate Properties of $CaCO_3$-Loaded SBR (1)

End Group of Polybutadiene	Blank	$-OCH_3$	$-OH$	$-OPO_3H_2$
Tensile strength, MPa	7.3	8.1	11.1	12.4
300% Modulus, MPa	1.7	1.7	1.7	3.6
200% Modulus, MPa	1.3	1.4	1.4	3.0
100% Modulus, MPa	1.2	1.4	1.4	2.3
Elongation, %	610	650	680	680
Hardness (JIS)	52	52	53	61
Tear (JIS-B), kN/m	14	15	15	24
Swelling, (2) %	347	339	344	252

(1) Recipe: SBR (Nipol 1502) 100, filler 100, zinc oxide 5.0, stearic acid 1.0, sulfur 2.0, 2-benzothiazyl disulfide 1.5, diphenyl guanidine 0.5. Cure: 150°C, 7 min.
(2) Weight increase after immersion in benzene for 48 h at room temperature.

Titanate or Borate Modification

Patents and technical papers about titanate modification have been increasing in number rapidly. Monte and Sugerman [70] have been exploring its versatile utility for polymer composites. Han et al. [71] and Sharma et al. [72] studied the coupling effects on mineral filler-loaded polyolefins. The results of their application studies have been accumulated, however, as pointed out by Ishida [73], the data on the molecular structure at filler/matrix interface are quite limited.

Organic borate coupling agent (e.g. dicetyl isopropylborate [74]) was stated to be useful for a calcium carbonate-filled polyolefin. Real molecular structure at the composite interface has not been discussed yet.

Other Modifications

The adsorption of surfactants such as sodium dodecyl sulfate onto calcium carbonate has been frequently studied. Calcium salts of these surfactants have generally good water solubilities and are easily washed away from the solid surface. Saleeb [75] measured the adsorption of bis(2-ethylhexyl)calcium sulfosuccinate onto calcium carbonate and five other particulate adsorbates. At relatively low concentrations, the adsorbed molecules form closely packed double layers on the calcium carbonate surface.

Chibowski [76] used a radiotracer technique to show that the adsorptive behavior of sodium dodecyl sulfate on calcium carbonate was affected by the presence of polyacrylamide on the particle surface as well as in the liquid phase. The interaction between the sulfate and the polymer was discussed.

In the surface modification of calcium carbonate filler by the salt deposition method, it is critically important that the modified surface layer has a quite limited water solubility. Nakatsuka et al. [69] observed that the solubility of the modifying salt has significant effects on the physical property retention of vulcanizates under wet conditions (Figure 5). Sodium mercaptoacetate is readily soluble in water, whereas calcium mercaptohexyl phosphate is sparingly soluble.

Preparation of a sparingly soluble salt on the calcium carbonate surface was also attained by use of carbonate ion on the surface. Vogt [77] found that fatty amines reacted with calcium carbonate to afford fatty ammonium carbonate (Eq. 8).

$$CaCO_3 + 2RNH_3Cl \rightleftharpoons (RNH_3)_2 CO_3 + CaCl_2 \qquad (8)$$

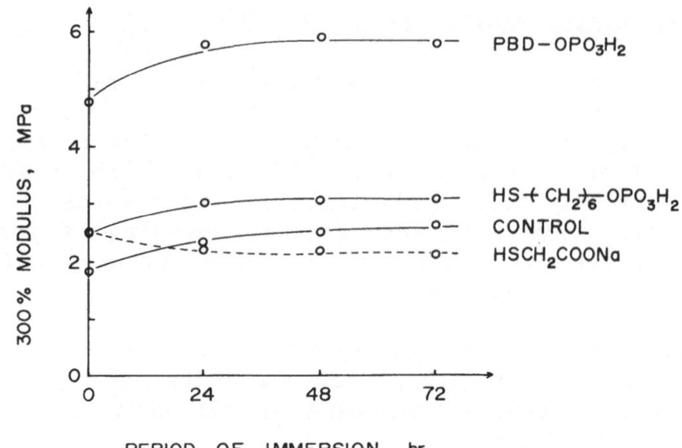

Fig. 5. Effects of modifier's solubility on preservation of SBR vulcanizate modulus (M_{300}) under wet conditions at 70°C.

The ammonium carbonate has a limited water solubility when R is alkyl functionality having at least one primary amine group.

CALCIUM CARBONATE-POLYMER INTERACTION

In the preceding section, the surface modifications of calcium carbonate with some surface modifiers were described. The interaction of the calcium carbonate with polymer molecules should also be addressed. Often matrix polymers are modified by copolymerization or graft polymerization of polar monomeric units to obtain favorable interactions with calcium carbonate filler.

Calcium Carbonate-Specific Polymer Interaction

Fowkes and Mostafa [78] showed that the adsorption of polymers onto fillers from organic solvents is dominated by the acid-base interactions among the solvents, polymers, and filler surfaces. As shown in Figure 6, the acidic polymer of post-chlorinated polyvinylchloride adsorbs most strongly from neutral solvents.

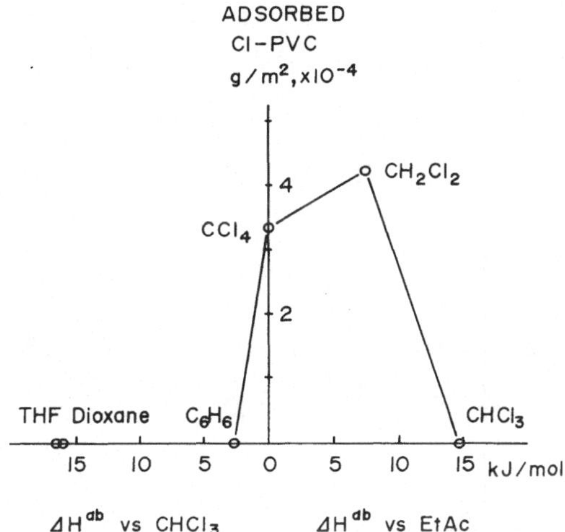

Fig. 6.   Interactions among post-chlorinated polyvinylchloride-
          organic solvent-calcium carbonate. The acidity or basi-
          city of the solvent (ΔH^{ab}) were calculated by Drago's
          parameters.

Nakagaki and Ninomiya [79] discussed the effects of sodium
polyacrylate in water on the aggregation and sedimentation state of
calcium carbonate particles. Kawalewski et al. [80] studied
ethyleneoxide oligomer-modified calcium carbonate in a polypropylene matrix. They ascribed the physical property improvements of
the composite to the easy translocation of the liquid-modified
filler in the polymer matrix during deformation.

For polymer matrix modification, it would be enough to give
an example. Sasaki et al. grafted maleic anhydride to polypropylene [81] and to polyethylene [82]. Calcium carbonate composites
with the grafted polymer exhibited much improved mechanical
properties compared with the non-grafted polymers.

Graft Polymerization Onto Calcium Carbonate

Graft polymerization onto inorganic mineral fillers is a
method to establish strong interactions between the filler surface
and the polymer matrix. However, the complicated procedures dis-
courage its practical use for filler surface modification.

Graft polymerization onto calcium carbonate particulates has
been accessible by different methods: mechanochemical method [83],
γ-ray preirradiation [84]. By chemical methods, two distinct
approaches are available; initiation from the solid surface and

Table IX. Graft Polymerization of Styrene on the Phosphate-Modified $CaCO_3$ (1)

Grafting System	AIBN mg	Attached Agent mg/g-$CaCO_3$	Polymerized Styrene g	Conv. %	Graftivity %	Grafted Polystyrene mg	Graft Efficiency %
AZP	0	1.8	0.076	2.8	4.1	21.2	28
	0	97.3	0.864	31.8	54	598	69
AZDP	0	1.5	0.110	4.1	7.1	38.4	35
	0	43.0	0.917	33.9	59	722	79
MPP-AIBN	15.0	0	0.923	33.9	0.03	0.13	0.02
	15.0	8.3	0.909	33.4	0.18	0.90	0.10
	15.0	38.9	1.000	36.8	0.21	1.04	0.11

(1) Polymerization conditions: $CaCO_3$ 0.50 g, styrene monomer 2.72 g, 80 $\pm$ 0.2°C, 60 min.

copolymerization of the comonomeric unit attached to the solid surface [85,86]. As comonomer units, acrylic or methacrylic acid are commonly used. It is of great interest to determine what molecular structural changes would appear for the monomeric unit by graft copolymerization.

Nakatsuka et al. [87] compared the graftivity and graft efficiency of the two chemical grafting approaches by using organo phosphates having methacryloyoxy or active azo groups (Table IX). The graftivity and graft efficiency stand for the weight percentages of graft polymer to the graft polymer-filler composite and of graft polymer to total polymer, respectively. The results in Table IX showed that the initiating graft is much preferable to the copolymerizing graft as a method of chemical graft polymerization. A scanning electron micrograph of the styrene-grafted calcium carbonate, obtained by the initiating graft, is shown in Figure 4.

The carboxylic acid having initiating ability was employed for chemical graft polymerization [88]. For the same purpose, calcium carbonate was covered with calcium sulfite [89]. Silane coupling agent was also used for graft polymerization onto calcium carbonate [36]. Such polymer-grafted filler particulates were generally much more hydrophobic than those prior to the polymerization. It has been shown that graft functional polymethacrylates have reactivities for organic agents as high as homopolymers in appropriate solvents [90].

It deserves notice that Howard et al. [91] found a practical application of graft polymerization. They prepared transition metal catalysts for ethylene polymerization on the mineral surfaces of calcium carbonate, clay, and aluminum trihydroxide. The homogeneous polyethylene-grafted filler had a fine impact toughness as well as reinforcement.

REFERENCES

1. D. L. Thomas and P. R. S. Gibson, Rubber and Plastic Age, 50, 888 (1969).
2. A. Bosshard and P. Delfosse, Ger. Offen., 2,309,516 (1974), Chem. Abstr., 82, 44403f (1975).
3. Kirk-Othmer, "Encyclopedia of Chemical Technology, 2nd Ed.," Vol. 4, p. 2, Wiley (1964).
4. T. Shiraishi, U. S. Pat., 1,863,945 (1932).
5. W. Johno, T. Watanabe, and M. Ashida, Kogyo Kagaku Zasshi, 60, 515 (1957).
6. J. W. Mellor, "A Comprehensive Treatise on Inorganic and Theoretical Chemistry," Vol. 3, p. 814, Longmans, Green and Co., London (1923).

7. J. L. Wray and F. Daniels, J. Amer. Chem. Soc., 79, 2031 (1957).
8. J. M. Criado and J. M. Trillo, J. Chem. Soc., Faraday Trans. 1,
    74 (4), 961 (1975).
9. O. Yamaguchi, N. Takashita, and K. Shimizu, Bull. Chem. Soc.
    Jpn., 52 (4), 1217 (1979).
10. G. Crowe and P. E. Kummer, Plastics Compoundings, 1978, 14.
11. Y. Shiraki, Nippon Gomu Kyokaishi, 53, 17 (1980).
12. T. Ikegami, Japan Plastics, 31, (8), 43 (1980).
13. H. Shibazaki and S. Edagawa, Jpn. Kokai Tokkyo Koho, 40,830
    (1979). Chem. Abstr., 91, 58796j (1979).
14. J. Ariya, K. Takitani, K. Mitarai, and K. Yamamoto, Jpn. Kokai
    Tokkyo Koho, 138,097 (1977). Chem. Abstr., 88, 154060y
    (1978).
15. S. Witekowa and W. Farbotko, Chem. Abstr., 78, 140701q (1973).
16. L. Domka, Chem. Abstr., 91, 177362z (1979).
17. M. Iovchev, Chem. Abstr., 70, 32033z (1969).
18. H. Miki and K. Matsui, Sekko to Sekkai, 94, 29 (1968).
19. T. Fujiwara, M. Takehashi, and T. Ikegami, Jpn. Kokai Tokkyo
    Koho, 17,925 (1981). Chem. Abstr., 95, 26706w (1981).
20. J. Ariya, K. Takitani, K. Mitarai, and K. Yamamoto, Ger. Offen.
    2,716,794 (1977). Chem. Abstr., 88, 90498n (1978).
21. Y. Nakahara, M. Mizuguchi, and K. Miyata, J. Colloid Interface
    Sci., 68, 401 (1979).
22. G. D. Parfitt and K. S. W. Sing eds., "Characterization of Pow-
    der Surfaces," Chap. 1 and 2, Academic Press (1976).
23. R. G. Gammage and S. J. Gregg, J. Colloid Interface Sci., 38,
    118 (1972).
24. T. Morimoto, J. Kishi, O. Okada, and T. Kadota, Bull. Chem. Soc.
    Jpn., 53, 1918 (1980).
25. G. Goujon and B. Mutaftscheiev, J. Colloid Interface Sci., 57,
    148 (1976).
26. M. Maruta and K. Yamada, Thermochim. Acta, 14, 245 (1976).
27. T. Hattori, K. Shirai, M. Niwa, and Y. Murakami, Bull. Chem.
    Soc. Jpn., 54, 1964 (1981).
28. T. Nakatsuka, H. Kawasaki, S. Yamashita, and S. Kohjiya, J.
    Colloid Interface Sci., in press.
29. H. Ishida and J. D. Miller, Proc. 38th Ann. Tech. Conf., Reinf.
    Plastics/Composites Inst., SPI, Sect. 4-E (1983).
30. V. Rives-Arnau, G. Munuera, and J. M. Driado, Spectrosc. Lett.,
    12, (10), 733 (1979).
31. C. D. Wagner, D. A. Zatko, and R. H. Raymond, Anal. Chem., 52,
    1445 (1980).
32. T. Yamanaka and K. Tanabe, J. Phys. Chem., 80, 1723 (1976).
33. J. Johnstone and E. D. Williamson, J. Amer. Chem. Soc., 38, 975
    (1916).
34. P. Somasundaran and G. E. Agar, J. Colloid Interface Sci., 24,
    433 (1967).
35. E. P. Plueddemann ed., "Interfaces in Polymer Matrix Composites",
    chap. 5 and 6, in L. J. Broutman and R. H. Krock eds, "Com-
    posite Materials," vol. 6, Academic Press (1979).

36. T. Nakatsuka, H. Kawasaki, K. Itadani, and S. Yamashita, J. Appl. Polym. Sci., 24, 1985 (1979).
37. M. R. Rosen and E. D. Goddard, Polym. Eng. Sci., 20, 413 (1980).
38. T. C. Williams, Ger. Offen., 2,524,863 (1975), 2,559,639 (1977).
39. Y. Furusawa, Y. Shiraki, K. Kondo, and H. Kodama, Jpn. Tokkyo Koho, 39,061 (1977).
40. T. Nakatsuka, H. Kawasaki, and K. Itadani, Jpn. Kokai Tokkyo Koho, 113,619 (1980). Che. Abstr., 67042z (1981).
41. S. E. Berger and G. A. Salensky, Ger. Offen., 2,743,682 (1978). Chem. Abstr., 89, 90668p (1978).
42. N. Ikeuchi and T. Ano, Jpn. Kokai Tokkyo Koho, 28,143 (1976). Chem. Abstr., 85, 6762b (1976).
43. H. J. Hass, H. Hanisch, K. M. Roedder, and R. Buening, Brit. Pat., 2,026,006 (1980). Chem. Abstr., 93, 48916a (1980).
44. T. Shiraki, T. Ibaragi, M. Ishihara, and M. Honda, Jpn. Kokai Tokkyo Koho, 109,545 (1978). Chem. Abstr., 90, 169615y (1979).
45. J. Szczypa, St. Chibowski, and K. Kuspit, Trans. Inst. Mining Met. Sect. C, 88 (March), 11 (1979).
46. E. Suito, Sekko to Sekkai, 94, 3 (1968).
47. O. I. Ivanishchenko and Y. P. Giadkikh, Kolloidnyi Ahurnal, 41, (4), 774 (1979).
48. Kh. A. Scarifkhodzhaeva, N. D. Ryabova, and E. A. Aripov, Deposited Doc. 1975, VINITI, p101. Chem. Abstr., 87, 4780y (1977).
49. G. Cruz-Romero and N. T. Coleman, Soil Sci. Soc. Amer. Prec., 38, 738 (1974).
50. J. Kasai, Sekko to Sekkai, 94, 54 (1968).
51. J. Hutchinson and J. D. Birchall, Elastomerics, 1980 (July), p. 17.
52. K. Yomo and K. Okamura, Jpn. Tokkyo Koho, 24,399 (1979). Chem. Abstr., 92, 23464t (1980).
53. R. H. Lalk and S. Evani, U.S. Pat., 4,151,341 (1979).
54. T. Takechi and T. Goto, Jpn. Kokai Tokkyo Koho, 72,793 (1977). Chem. Abstr., 87, 118843b (1977).
55. S. Yamashita, Y. Minoura, H. Okamoto, T. Nukui, and E. Morimoto, Nippon Gomu Kyokaishi, 51, (3), 194 (1978).
56. H. Shibazaki, S. Edagawa, H. Hasegawa, M. Funazu, N. Moriyama, and Y. Fukumoto, Jpn. Kokai Tokkyo Koho, 165,960 (1980). Chem. Abstr., 94, 193171d (1981).
57. H. Shibazaki, S. Edagawa, H. Hasegawa, M. Funazu, N. Moriyama, and Y. Fukumoto, Jpn. Kokai Tokkyo Koho, 165,961 (1980). Chem. Abstr., 94, 209747j (1981).
58. T. Ikegami, T. Takeuchi and T. Goto, Jpn Kokai Tokkyo Koho, 25, 646 (1978). Chem. Abstr., 89, 61298n (1978).
59. I. Sasaki and F. Ide, Kobunshi Ronbunshu, 38 (2), 67 (1981).
60. E. Oda, E. Saito, and S. Nishimura, Jpn. Tokkyo Koho, 35,978 (1978). Chem. Abstr., 89, 216379r (1978).
61. T. Abe, R. Tanizawa, and K. Kanda, Jpn Kokai Tokkyo Koho, 84, 045 (1978). Chem. Abstr., 89, 164497f (1978).

62. H. Matsushima and S. Miyata, Jpn. Kokai Tokkyo Koho, 56,174
     (1978). Chem. Abstr., 90, 25594t (1979).
63. Y. Avnimelech, Nature, 288, 255 (1980).
64. T. Nakatsuka, H. Kawasaki, K. Itadani, and S. Yamashita, J.
     Colloid Interface Sci., 82, 298 (1981).
65. T. Nakatsuka, H. Kawasaki, K. Itadani, and S. Yamashita, J.
     Appl. Polym. Sci., 27, 259 (1982).
66. T. Ohzeki, M. Akutsu, and J. Kawai, U.S. Pat., 4,258,142
     (1981).
67. S. Saeki, N. Inoue, and N. Suzuki, Jpn. Kokai Tokkyo Koho,
     128,728 (1982).
68. T. Nakatsuka, H. Kawasaki, and K. Itadani, Jpn. Kokai Tokkyo
     Koho, 168,954 (1982).
69. T. Nakatsuka, H. Kawasaki, and S. Yamashita, submitted.
70. S. J. Monte and G. Sugerman, Paper of the Meeting of Rubber
     Div., Amer. Chem. Soc., 110th, p. 43 (1976).
71. C. D. Han, C. Sandford, and H. J. Yoo, Polym. Eng. Sci., 18
     (11), 849(1978).
72. Y. N. Sharma, R. D. Patel, I. H. Dhimmar, and I. S. Bhardwaj,
     J. Appl. Polym. Sci., 27, 97 (1982).
73. H. Ishida, "Recent Progress in the Studies of Molecular and
     Microstructure of the Interfaces in Composite Materials,
     Adhesive Joints and Coatings," Proc. of the Symposium on
     Adhesive Aspects of Polymer Coatings, K. L. Mittal ed.,
     Plenum (1982).
74. M. M. Fein, B. K. Patnaik, and F. K. Y. Chu, U.S. Pat., 4,073,
     766 (1978). Chem. Abstr., 88, 171001x (1978).
75. F. Z. Saleeb, Kolloid Z. -Z. Polym., 239 (1), 602 (1970).
76. S. Chibowski, J. Colloid Interface Sci., 76, 371 (1980).
77. J. C. Vogt, Tenside Deterg., 13 (2), 90 (1976). Chem. Abstr.,
     85, 158932v (1976).
78. F. M. Fowkes and M. A. Mostafa, Ind. Eng. Chem. Prod. Res. Dev.,
     17, 3 (1978).
79. M. Nakagaki and H. Ninomiya, Yakugaku Zasshi, 96 (9), 1127
     (1976).
80. T. Kawalewski, R. Kalinski, and M. Kryszewski, Colloid Polym.
     Sci., 260, 652 (1982).
81. I. Sasaki, K. Ito, T. Kodama, and F. Ide, Kobunshi Ronbunshu,
     33 (2), 122 (1976).
82. I. Sasaki, T. Kodama, and F. Ide, Kobunshi Ronbunshu, 33 (3),
     162 (1976).
83. A. B. Taubman, L. P. Yanova, and G. S. Blyskosh, J. Polym. Sci.,
     A-1, 9, 27 (1971).
84. B. Dinh-Ngoc, J. G. Rabe, and W. Schnabel, Angew. Makromol.
     Chem., 46, 23 (1975).
85. R. Kroker, M. Schneider, and K. Hamann, Progr. Org. Coatings,
     1, 23 (1972).
86. R. Laible and K. Hamann, Adv. Colloid Interface Sci., 13, 65
     (1980).

87. T. Nakatsuka, H. Kawasaki, and S. Yamashita, J. Colloid Inter-
     face Sci., in press.
88. Yu. A. Avereva, V. A. Popov, V. V. Guzeev, E. P. Shvarev, S. S.
     Ivanchev, and G. P. Gladyshev, Dokl. Akad. Nauk, SSSR, 252
     (5), 1174 (1980).
89. T. Yamaguchi, H. Hoshi, M. Hirakawa, and I. Watanabe, Ger.
     Offen., 2,237,256 (1973). Chem. Abstr., 79, 5877g (1973).
90. T. Nakatsuka, H. Kawasaki, and S. Yamashita, submitted.
91. E. G. Howard, R. D. Lipscomb, R. N. MacDonald, B. L. Glazar,
     C.W. Tullock, and J. W. Collette, Ind. Eng. Chem. Prod. Res.
     Dev., 20, 421 (1981).

OSMOTIC PRESSURE-FILLED CRACKS

J. P. Sargent and K. H. G. Ashbee

H. H. Wills Physics Laboratory
University of Bristol
Tyndall Avenue
Bristol BS8 1TL, England

ABSTRACT

In-service loss of load transfer is a slow process and, since
it involves action by diffused water, the suggestion has been made
that it proceeds by way of an interfacial failure process akin to
stress corrosion in monolithic solids. It can occur without application of external stress. Debonded pockets in accelerated tests
are often osmotic pressure-filled and propagate as cracks into the
adjacent resin. It is demonstrated that osmotic pressure-filled
cracks in polyester resin are elastic in nature and evidence is
presented for concluding that their growth is by a slip/stick mechanism. An explanation involving crazing is offered to account for
the alternate periods of crack blunting and sharpening. The appearance of fracture surface markings delineating positions of
crack arrest are attributed to Griffith's [1] observation that in a
plate subjected to other than uniaxial tension perpendicular to the
crack, the maximum stress concentration is close to but not at the
crack tip. Measurement of crack profiles is used to determine the
rate dependence of Young's modulus and, by assuming that the
periods of crack growth are adequately described by linear elastic
fracture mechanics, an attempt is made to compare these periods for
three different test temperatures.

INTRODUCTION

The consequence of osmosis, due to solutes in the matrix
[2,3] or at the interface between matrix and fibers [3,4,5], is the
propagation of osmotic pressure-filled cracks within the resin.
Figure 1, reproduced from reference 4, shows examples of resin
cracks at fiber interfaces (a) in E-glass/polyester resin and (b)

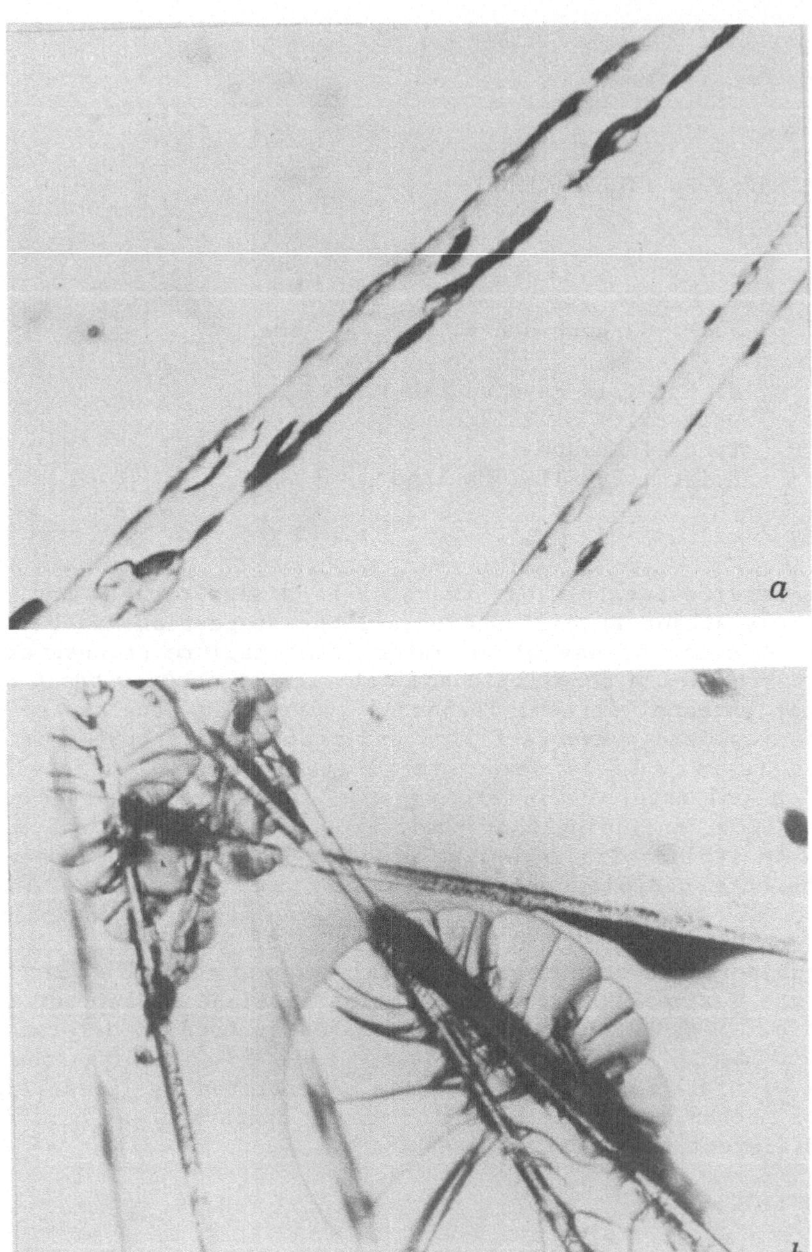

Fig. 1.   Resin cracks accompanying debonding of (a) E-glass and
          (b) C-glass fibers in polyester resin composites immersed
          in boiling water for 20 h (Reference 4).

in C-glass/polyester resin composites. To fully understand weathering caused by osmosis, it is evidently necessary to explore the
nature of cracks in resin matrix materials. Of particular interest
is the justification for using linear elastic fracture mechanics
(LEFM) to describe accelerated test observations and to interpolate
such descriptions to weathering in the field. With this in mind,
it is necessary first to establish that cracks of the kind described in references 2-5 are essentially elastic cracks, i.e.
cracks which can be treated by LEFM.

*Whether cross-linked, and therefore truly rubbery, or not
cross-linked, and therefore having only transient rubbery elasticity, all high polymers which do not crystallize become glasses
below some more or less well-defined transition temperature. In
truth, of course, that temperature is dependent on rate of change
of temperature, and a material may have several glass transition
temperatures at which various modes of motion within it freeze out.
Below this temperature, or range of temperatures, they are true
glasses, not very different in basic properties from any other
glasses. Of course a typical atom in the molecular chain is bound
to only two of its neighbors by strong covalent binding forces, and
to another eight or so neighbors through van der Waals forces,
which are weaker - but one ought not to exaggerate this factor of
weakness for the van der Waals forces: it is not much more than a
factor of 10. Boiling points are a rough measure of binding energies, and most substances boil between 100 and 3000K. The factors
in strength which come from failure to attain ideal strength are
more serious than that ratio, and an intrinsically weaker substance
may possibly compete very well if for any reason it can be made to
fall less short of its ideal strength. In fact, the polymeric
organic glasses are weaker than inorganic glasses, but scarcely
more so than in proportion to density, and at not too low a temperature some of them have the advantage over the inorganic glasses in
absorbing more energy at fracture in spite of lower absolute
strength - F. C. Frank [6].* The expectation that the polymeric
glasses are not very different in basic properties from any other
glasses, e.g. the inorganic glasses, anticipates the following experimental justification for use of LEFM to describe crack propagation in resins.

ELASTIC NATURE OF RESIN CRACKS OBSERVED IN ACCELERATED TESTS

To demonstrate the elastic nature of osmotic pressure filled
cracks in polyester resin, and in particular to demonstrate the
absence of any permanent (i.e. plastic) contribution to the crack
opening displacement, the crack shown in Figure 2(a) and others
have been re-photographed after drying by annealing the sample in
air for 1 hr at 100 °C. This treatment causes the crack faces to
come into contact with each other, Figure 2(b). Seen face-on, such
cracks in dried samples of the same resin generate interference

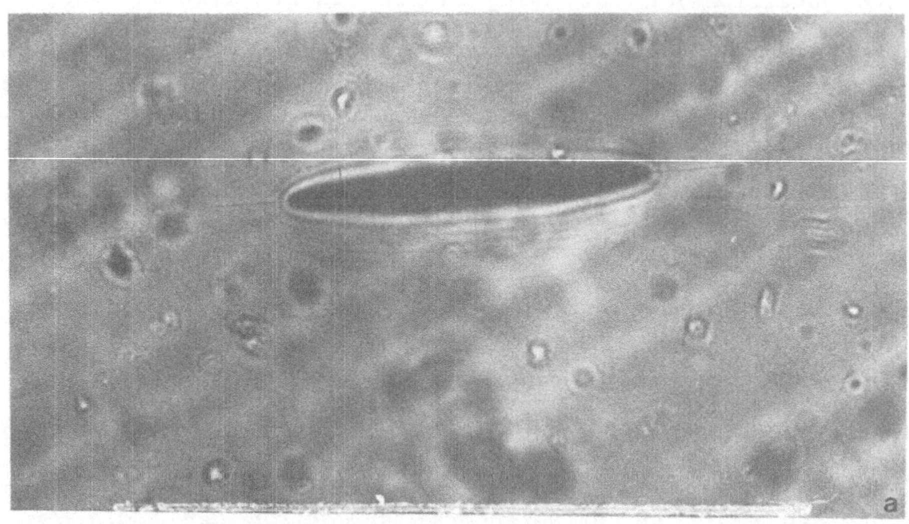

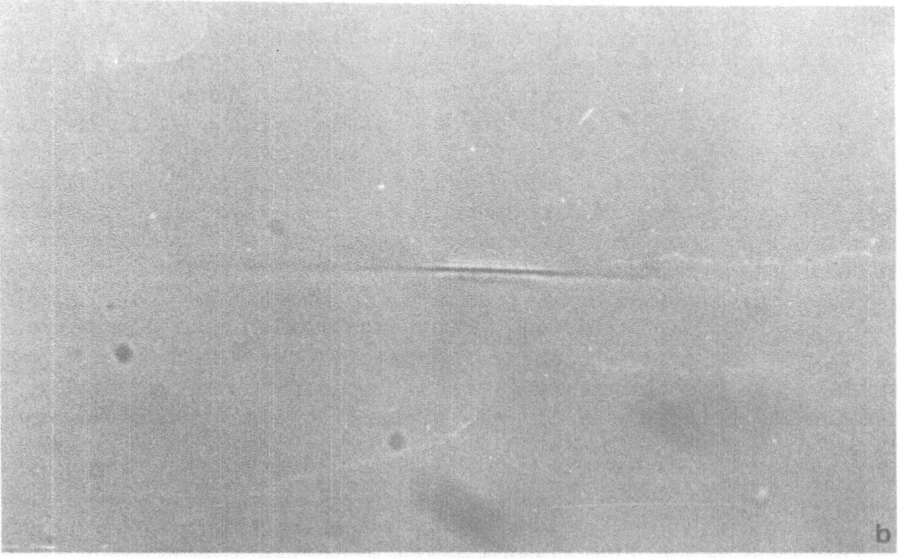

Fig. 2.   Edge-on view of (a) penny-shaped crack in polyester resin designated A in reference 2 and (b) edge-on view of penny-shaped crack after drying.   The marker is 0.1 mm.

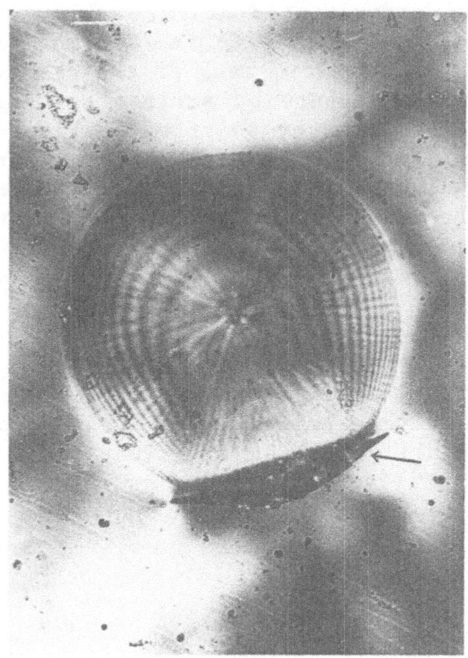
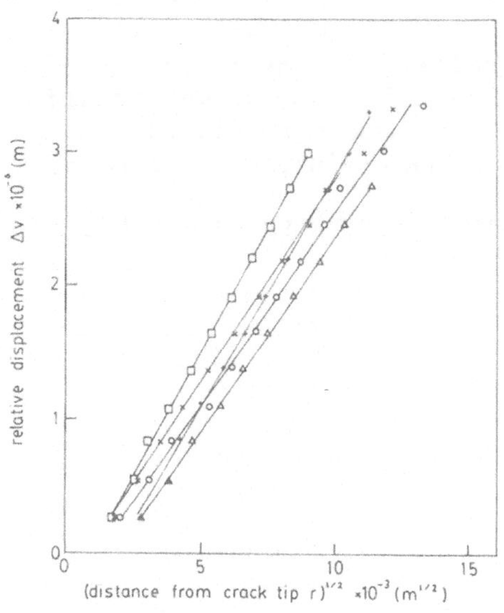

Fig. 3.   Face-on view of (a) penny-shaped crack after drying.
Notice the interference fringes. (b) Data from profiles
of penny-shaped cracks after drying.

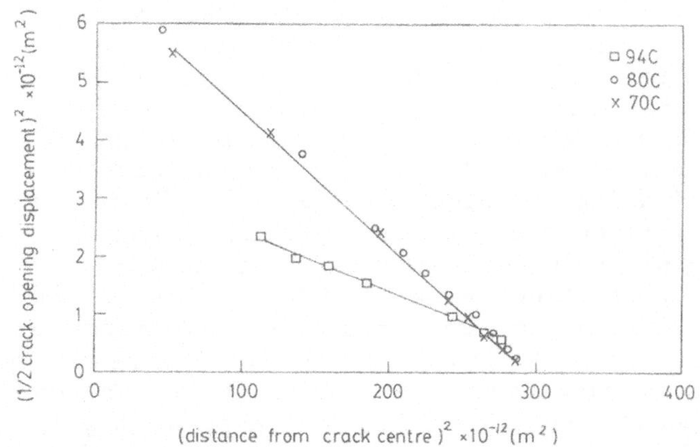

Fig. 4. $n_0{}^2$ versus x^2 data from profiles of penny-shaped cracks.
The values for E deduced from these graphs are 0.43 GPa
at 70°C, 0.43 GPa at 80°C, and 0.86 GPa at 94°C.

patterns, Figure 3(a), and, by counting the fringes, the very close proximity of the crack faces can be demonstrated, Figure 3(b). It could be argued that drying the resin might lead to the generation of negative pressure inside the crack. However, many of the cracks in dried resin are connected to the atmosphere by surface seeking cracks; a surface seeking crack is indicated by an arrow in Figure 3(a). In any case, creation of negative pressure would surely be relieved by the formation of bubbles containing occluded gas and/or low molecular weight material. The small residual space between the crack faces is attributed to obstruction by solute deposited during drying, see reference 2.

RATE DEPENDENCE OF YOUNG'S MODULUS

Elastic cracks, i.e. Griffith cracks, are elliptic in nature. This elliptic form has been re-examined by many authors including Sack [7], Sneddon [8], Elliott [9], and Westergaard [10]. The specific case of pressure-filled elastic cracks was considered by both Sneddon [7] and Westergaard [10]. Expansion of Westergaard's equation [4.6] gives:

$$\eta^2_{o} = \frac{-4\,(1-\nu^2)^2\,p^2\,x^2}{E^2} + \frac{4(1-\nu^2)^2\,p^2\,a^2}{E^2} \qquad [1]$$

where η_0 = half crack opening displacement, x = distance measured from the crack centre, a = length of semi-minor axis of the crack, ν = Poisson's ratio, E = Young's modulus.

When formed, osmotic pressure-filled cracks often contain resolvable undissolved solute. Undissolved solute can be seen at the centre of the crack shown in Figure 4(a) and (b) of reference 2 for example. The water solution within such cracks is evidently saturated with solute. The osmotic pressure, estimated in reference 2 for the saturated hot water solution identified in cracks inside the polyester resin system designated system A in that reference, is 473.3 bars. The present accelerated test data were obtained using the same resin in water immersion tests at 70, 80 and 94°C. Since osmotic pressures for saturated solutions of common inorganic solutes increase only marginally in this temperature range, a constant value of 500 bars was assumed when using equation (1) to compute the respective magnitudes of E from the data shown in Figure 4. Static values for E were determined on the same samples by simultaneous measurements of the longitudinal and transverse sound wave velocities at 5Mhz; no significant differences from the room temperature value of 4.0 GPa for the dry resin were found. It is therefore concluded that the variations in E obtained from Figure 4 are attributable solely to its time dependence.

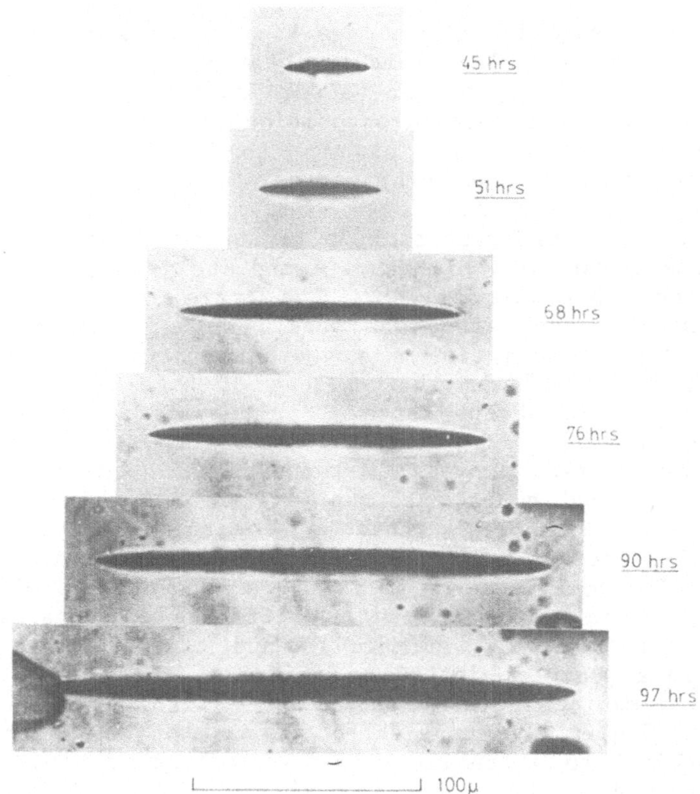

Fig. 5.   Edge-on view of a single penny-shaped crack 0.2 mm below
the surface, photographed six times during its growth.

MEASUREMENT OF CRACK GROWTH RATES

Figure 5 shows an edge-on view of a penny-shaped crack photographed after several times of immersion in water at 94°C. Measurements of the crack diameter reveal that the crack growth rate is
constant at 1.1×10^{-9} ms^{-1}. The same experiment on cracks in specimens immersed in water at 70°C and 80°C reveal constant crack
growth rates of 5×10^{-12} ms^{-1} and 7×10^{-11} ms^{-1}, respectively.

CRACK SURFACE TOPOGRAPHY

Close inspection of the crack surfaces sometimes reveals the presence of more or less concentric steps, i.e. abrupt changes in fracture path. Examples can be seen on the crack surface shown in Figure 6. The steps delineate positions at which the crack was presumably at rest and are assumed to be direct evidence for a slip/stick mode of crack propagation.

Crazing is the precursor to fracture in many polymeric materials, the occurrence of which immediately suggests a model for slip/stick crack propagation. The craze is comprised of highly drawn material and, since the necking-down phenomenon which gives rise to high draw ratio is a characteristic of high temperature drawing, it is evident that a zone of high temperature material must exist ahead of the craze. For polymers, room temperature is near the melting temperature so high temperature probably means just a few degrees above room temperature. Intermittent falls in temperature, and hence interruptions of the necking-down process associated with crazing, would be expected to cause crack arrests until the stress is raised sufficiently to reheat the craze leader. The small changes in fracture path, each time crack growth resumes, is in accordance with Griffith's [1] evaluation of the tensile stress at the surface of and tangential to an elliptic cavity in a plate subjected to principal stresses σ_1 and σ_2, respectively making angles Φ and $\pi/2 - \Phi$ with the major axis of the ellipse. Inglis [11] considered the case of an elliptic hole in a plate subject to a tensile stress σ applied in a direction making an angle Φ with the major axis of the ellipse, Figure 7(a), and found that the tensile stress at the surface of and tangential to the hole is:

$$\sigma_{\beta\beta} = \sigma \frac{\left[\sinh 2\alpha_o + \cos 2\Phi - e^{2\alpha_o} \cos 2(\Phi - \beta) \right]}{\cosh 2\alpha_o - \cos 2\beta} \qquad [2]$$

Referring to Figure 7(b) for $\sigma_2 = 0$ the Inglis solution, with the sense of increasing β reversed gives:

$$\sigma_{\beta\beta} = \sigma_2 \frac{\left[\sinh 2\alpha_o + \cos 2\Phi - e^{2\alpha_o} \cos 2(\Phi - \beta) \right]}{\cosh 2\alpha_o - \cos 2\beta} \qquad [3]$$

and for $\sigma_1 = 0$ it gives:

$$\sigma_{\beta\beta} = \sigma_1 \frac{\left[\sinh 2\alpha_o + \cos 2(\pi/2 - \Phi) - e^{2\alpha_o} \cos 2(\pi/2 - \Phi + \beta) \right]}{\cosh 2\alpha_o - \cos 2(-\beta)} \qquad [4]$$

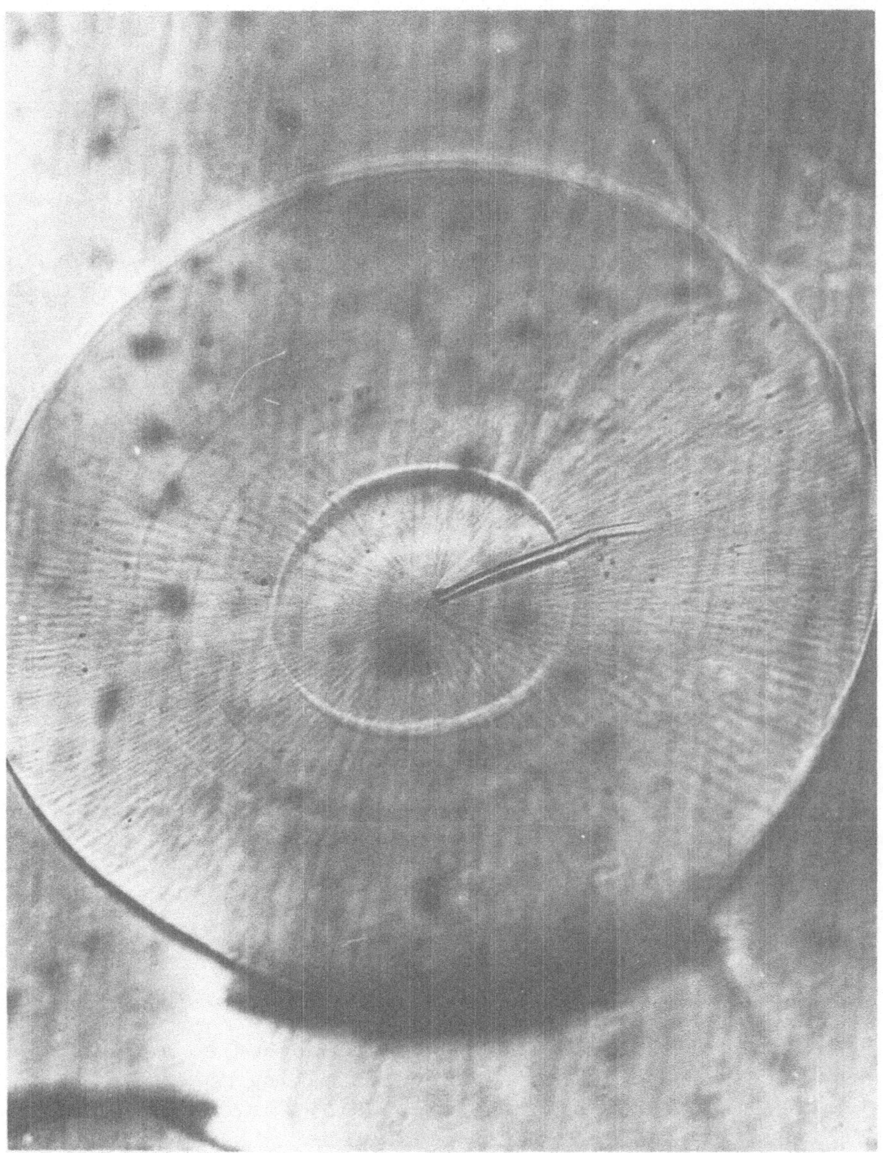

Fig. 6. Crack showing concentric growth ring structure.

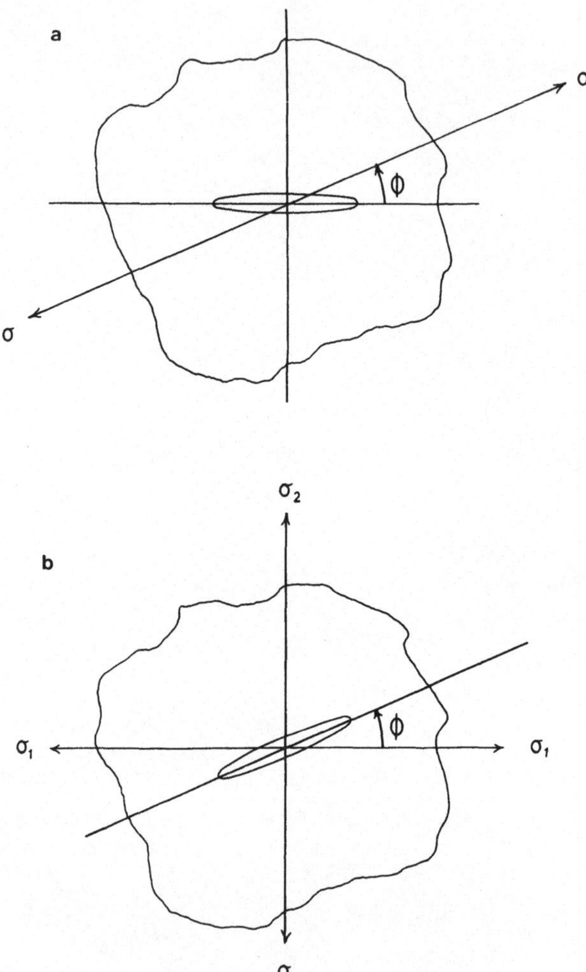

Fig. 7. (a) An elliptic hole in a plate subject to a tensile stress σ applied in a direction making an angle Φ with the major axis of the ellipse. (b) For $\sigma_2 = 0$ the Inglis solution with the sense of increasing β reversed.

Hence by superposition,

$$\sigma_{\beta\beta} = (\sigma_1 + \sigma_2)\ \frac{\sinh 2\alpha_o + (\sigma_1 - \sigma_2)\ \left[e^{2\alpha_o}\cos 2(\Phi-\beta)-\cos 2\Phi\right]}{\cosh 2\alpha_o - \cos 2\beta}$$

[5]

The values of Φ and β for which $\sigma_{\beta\beta}$ is a maximum are found by differentiation. Putting $\partial\sigma_{\beta\beta}/\partial\beta = 0$ and taking as solution $\sin 2\beta = A\alpha_o + 0(\alpha_o^2)$, it can be demonstrated that $\sigma_{\beta\beta}$ is a maximum at two pairs of points on each crack. If $\Phi = 0$ or $\pi/2$, these points are at the ends of the major and minor axes respectively. For all other values of Φ both pairs of points are very near the ends of the major axis as sketched schematically in Figure 8(a). The two extremal values of $\sigma_{\beta\beta}$ always have opposite sign. Continuation of fracture from one such point, thereby leaving a step on the fracture surface, is illustrated in Figure 8(b).

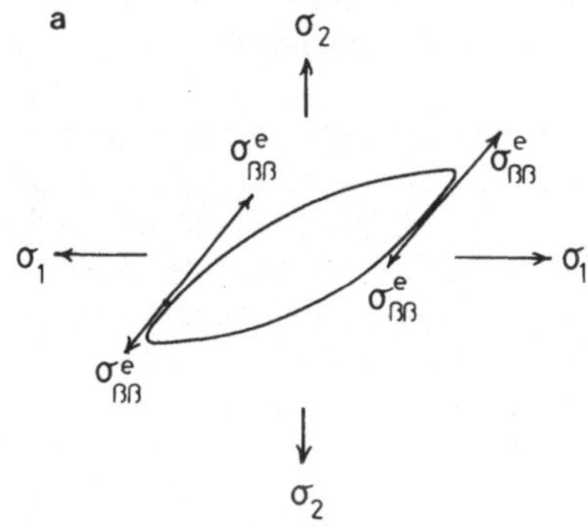

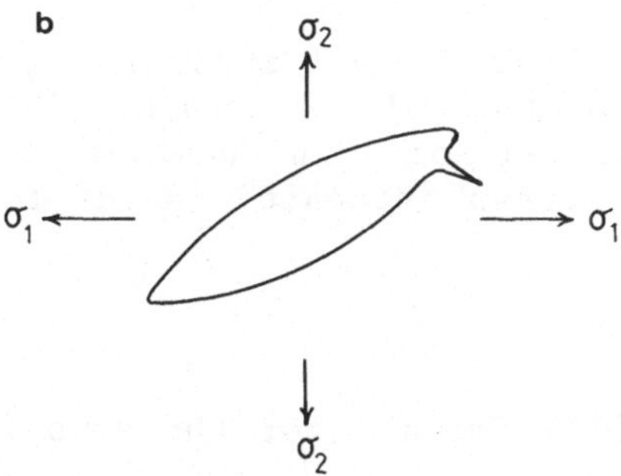

Fig. 8. Locations of maximum tensile stress (a) on an elliptic cavity in a plate subjected to biaxial tension. (b)Crack propagation from one such location.

DISCUSSION

Fracture mechanics is concerned with the propagation of atomically sharp cracks. It may well be that, when propagating, osmotic pressure filled cracks in polyester resin are indeed atomically sharp. It may also be the case that the periods of propagation are short in which case the time dependence of elastic modulus may be ignored and *linear* elastic fracture mechanics (LEFM) may be used to describe the cracks. In the following discussion, both these assumptions are made.

Crack Growth Times

The usual starting point for analysis of crack growth in inorganic glasses [12] is the parabolic relationship between stress intensity factor (K) and crack length (a):

$$K = \sigma Y \sqrt{a} \qquad [6]$$

where σ is the tensile stress normal to the crack and Y is a constant determined by the crack geometry.

In polymer matrix composite materials, cracks can nucleate and grow in the absence of any externally applied stress; the driving force for fracture is derived from osmosis [2] with each crack containing an internal osmotic pressure (p). The Griffith energy (U_G), the elastic free energy released on creation of a penny--shaped crack inflated by an internal pressure (p), is identical to that for a crack of the same lateral dimensions formed in the presence of a previously uniform uniaxial stress $\sigma = p$ applied normal to its plane [8].

$$U_G = \frac{8(1 - \nu^2) a^3 p^2}{2E} = \frac{1}{2} PV \qquad [7]$$

ν is Poisson's ratio, E is Young's modulus and V is the elastic expansion, i.e. the volume, of the crack. Hence it is concluded that p may be substituted for σ in equation (6) in order to obtain an equation for the stress intensity factor for a pressure filled crack

$$K = pY\sqrt{a} \qquad [8]$$

Using van't Hoff's [13] formula for the osmotic pressure of a dilute solution:

$$p = \frac{mRT}{V} \qquad [9]$$

where m is the number of moles of dissolved solute. It will be assumed that m remains unchanged during crack growth.

Substituting V from equation (7) we get:

$$p = ka^{-3/2} \tag{10}$$

where the constant k is given by:

$$k^2 = \frac{mRTE}{8(1-v^2)} \tag{11}$$

Hence, equation 8 becomes:

$$K = \frac{Yk}{a} \tag{12}$$

It should be noted that, whereas for crack growth in a monolithic solid subjected to an applied stress which does not vary with time, K increases with crack growth (equation (6)), the case for pressure-filled cracks concerns a K which decreases with crack growth (equation (12)).

Re-arranging equation (12) and differentiating we get:

$$v = \frac{da}{dt} = f(K) = \frac{-Yk}{K^2} \cdot \frac{dK}{dt} \tag{13}$$

Hence the time required for a crack inflated by osmotic pressure to propagate under the action of that pressure is:

$$t = -Yk \int_{K_{initial}}^{K_{Ic}} \frac{dK}{K^2 v} \tag{14}$$

$$= \frac{Yk}{v} \left[\frac{1}{K_{I\ initial}} - \frac{1}{K_{Ic}} \right] \quad \text{if v is constant} \tag{15}$$

Stress Intensity Factors

The cracks described here are elastic cracks and, if they can be described by linear elastic fracture mechanics, meassurements of their profiles can be used to obtain values for stress intensity

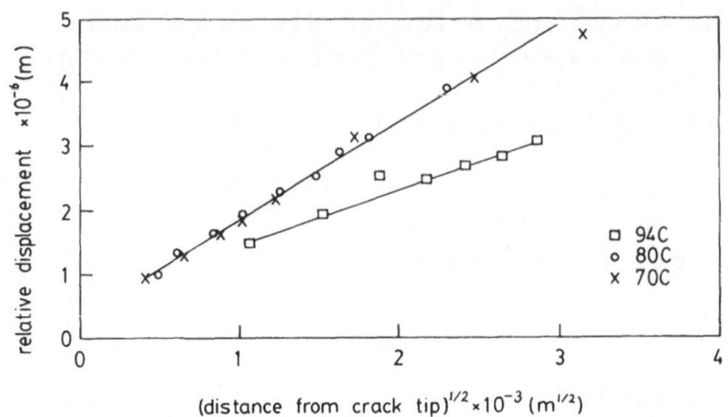

Fig. 9. Δv versus $r^{1/2}$ data for cracks grown at 70°C, 80°C and 94°C.

factor (K_I). Eshelby's [14] equation (2.10) for the parabolic relationship between crack opening displacement (ΔV) and distance from the crack tip (r) is:

$$\Delta V = \frac{K_I}{(2\pi)^{1/2}} \cdot r^{1/2} \cdot \frac{8(1-\nu^2)}{E} \qquad [16]$$

Figure 9 shows plots of ΔV versus $r^{1/2}$ for cracks at 70, 80 and 94°C from the slopes of which it is deduced that the corresponding stress intensity factors are 0.26, 0.28 and 0.28 MPam$^{1/2}$ respectively.

If the crack velocity is always the same during periods of crack growth, the relative times of overall crack growth at 70, 80 and 94°C are in the ratios of the respective times given by equation (15).

ACKNOWLEDGEMENTS

This work was supported by U.S. Army Grant DAJA-81-C-0214.

REFERENCES

1. A. A. Griffith, 1st Intl. Conf. Appl. Mech. (Delft 1924) 55.
2. K. H. G. Ashbee, F. C. Frank and R. C. Wyatt, Proc. Roy. Soc. (1967) A300, 415.
3. N. R. Farrar and K. H. G. Ashbee, J. Phys. D. (1978) 11, 1009.
4. K. H. G. Ashbee and R. C. Wyatt, Proc. Roy. Soc. (1969) A312, 553.
5. Elizabeth Walter and K. H. G. Ashbee, Composites (1982) 13, 365.

6. F. C. Frank, Discussion Meeting on New Materials held in The
    Lecture Theatre at the Royal Institution, June 1963 and
    published in Proc. Roy. Soc. (1964) A282, 9-16.
7. R. A. Sack, Proc. Phys. Soc., (1946) 58, 729.
8. I. N. Sneddon, Proc. Roy. Soc., (1946) 58, 729.
9. H. A. Elliott, Proc. Phys. Soc., (1947) 59, 208.
10. H. M. Westergaard, J. Appl. Mech., (1939) A49.
11. C. E. Inglis, Trans. Inst. Naval Architects (1913) 55, 219.
12. A. G. Evans, J. Mater. Sci. (1972) 7, 1137.
13. J. van't Hoff, Phil. Mag. (1888) 26:81, 2662.
14. J. D. Eshelby, Sci. Prog., Oxf. (1971) 59, 161.

ARAMID/EPOXY VS. GRAPHITE/EPOXY: ORIGIN OF THE DIFFERENCE IN
STRENGTH AT THE INTERFACE

L. Penn*, F. Bystry, W. Karp and S. Lee

Composite Materials Department
Ciba-Geigy Corporation
Ardsley, New York 10502

*Midwest Research Institute
Kansas City, Missouri 64110

ABSTRACT

In single filament pull-out tests, the graphite/epoxy interfacial bond strength was found to be twice as large as the
aramid/epoxy interfacial bond strength. This difference carried
over to unidirectional filamentary composites made with the same
resin system. Short beam shear test results for graphite/epoxy
were nearly twice as high as those for aramid/epoxy. Graphite/-
epoxy similarly exceeded aramid/epoxy in the Iosipescu shear tests
used as a cross check to the short beam shear test.

To determine the reasons for this observed difference between
aramid and graphite systems in interfacial adhesion and related
tests, three factors were considdered: intermolecular interactions,
chemical bonding, and mechanical interference. Evidence is
presented to show that mechanical interference caused by thermal
mismatch plays an important role in explaining the difference
between aramid/epoxy and graphite/epoxy at the interface.

INTRODUCTION

The three main factors that can influence adhesion and
therefore, adhesive performance at the interface in composites are
intermolecular interactions, chemical bonds, and mechanical interference. We have been investigating these factors in aramid/epoxy
composites and in graphite/epoxy composites. Intermolecular inter-

actions, electrostatic in origin, are sometimes known as secondary
bonds. They include dipole interactions, dispersion (London)
interactions, and hydrogen bonds. These intermolecular interactions require from 2-6 kcal/mole for rupture. In contrast,
chemical bonds, the second factor, require 60-100 kcal/mole for
rupture. Chemical bonds across the interface would obviously
increase the interfacial strength. The last of the three factors,
mechanical interference, can act in several ways. A lock and key
fit at an interface due to surface roughness can greatly increase
the force required to separate the adhering materials. More
importantly, for composites made with smooth fibers, pressure
exerted by the matrix on the fiber due to thermal mismatch can
affect the adhesive performance of the interface.

In our laboratory, a single filament pull-out test has been
successfully developed for aramid/epoxy systems. Because a single
filament imbedded in resin is the basic building block of the
composite, its use may be expected to shed light on some of the
processes that occur during loading of a composite. Our findings
were that the graphite/epoxy interface seemed to be about twice as
strong as the aramid/epoxy interface. Furthermore, allegedly
interface-sensitive tests on composites such as short beam shear
and Iosipescu shear tests [1,2] echoed the same greater strength of
the graphite over the aramid. In this paper, we discuss the
contribution of intermolecular interactions and of chemical bonding
to the performance difference between aramid and graphite composites. Furthermore, we examine the proposition that mechanical
interference from thermal mismatch is the major cause of performance difference and we describe experiments designed to test
this.

EXPERIMENTAL

<u>Materials</u>

The graphite fiber used in this study was Hercules AS4
(modulus 28,000,000 psi) obtained from the manufacturer without
sizing. The aramid fiber used was Du Pont's Kevlar 29 (modulus
8,500,000 psi) also obtained without sizing. For a supplementary
set of experiments to check the effect of modulus alone, Kevlar 49
aramid with a modulus of 18,000,000 psi (twice that of Kevlar 29)
was used.

Two different matrix resin systems were used in this work.
The first system was used in making the laminates and in some of
the single filament pull-out work. Designated epoxy A, the first
system was N,N,N´,N´-tetraglycidyl methylene dianiline
(CIBA-GEIGY's MY 720) cured with 4,4´-diamino-diphenyl sulfone

(CIBA-GEIGY's HT 976) and a small amount of BF .MEA catalyst. This
is a strong but brittle tetrafunctional epoxy currently used in
high performance composites. Designated epoxy B, the second system
was diglycidyl ether of bisphenol A (CIBA-GEIGY's 6010) cured with
modified triethylenetetramine (CIBA-GEIGY's 956). This is a difunc-
tional epoxy system less brittle than the first.

Unidirectional Laminate Preparation

Preparation of unidirectional preimpregnated tape ("prepreg")
was carried out by passing the multifilament yarn (graphite or
aramid) through a solution of the epoxy A resin components and
winding it onto a rotating drum winder in a single layer. The
resultant prepreg was formed into a unidirectional lay-up for processing into a laminate of the desired thickness. A press-clave
consolidated the lay-up into a laminate of 65% fiber volume
according to the following cure schedule: 1/2 h at 120°C under
vacuum and 1 h at 177°C under 85 psi pressure. This was followed
by a post cure of 4 h at 150°C and 7 h at 200°C.

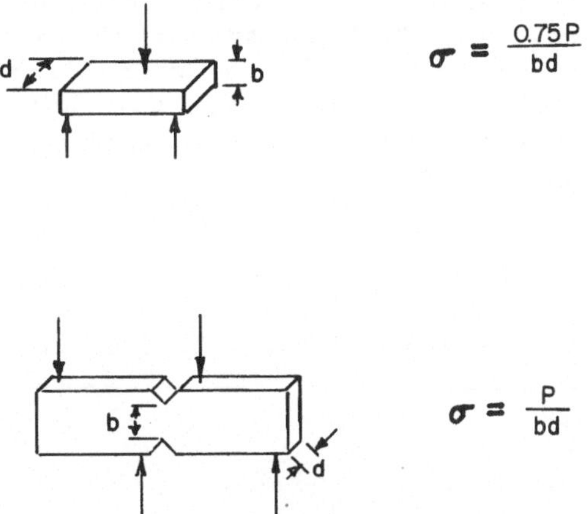

Fig. 1. Short beam shear (top) and Iosipescu (bottom) test speci-
         mens and loading patterns. Computations for failure
         strength show factor of 0.75 to adjust for parabolic stress
         distribution in SBS specimen as compared to assumed uniform
         stress distribution in Iosipescu specimens. P_B is the
         breaking load.

Laminate Testing

For the short beam shear test, ASTM-2344, ten specimens 0.75 in x 0.25 in x 0.125 in were cut from each laminate. Short beam shear tests were carried out on the universal testing machine at a cross-head speed of 0.5 in/min.

The Iosipescu test is a double-notched four-point bend test for shear strength [1,2]. Ten specimens 2.0 in x 0.50 in x 0.125 in were cut from each laminate for this test, with the fibers parallel to the long axis of the specimen. Figure 1 shows the loading pattern of the test specimens and the equation used to calculate failure strength for each test.

Surface Characterization of Fiber

The surfaces of the fibers and of the cured resins were characterized by contact-angle measurements. Contact angles were obtained for a series of probe liquids on each type of solid surface being investigated. The liquids, a nonhomologous series with a range of surface tensions, are given in Table 1. Liquids were purified before used until their surface tensions were within 1 dyne/cm of accepted literature values.

Contact angles were calculated from wetting force measurements carried out on a sensitive electrobalance [3]. The fiber or cured resin specimen, hung from an arm of the elctrobalance, was slowly immersed and then emerged in the probe liquid at a rate of 0.4 mm/min, low enough for contact-angle values to be independent of rate. This gave the steady state advancing and steady state receding contact angles from:

Table 1. Probe Liquids and Their Surface Tensions (γ_L)

Liquid	γ_L, dyne/cm
Water	72.8
Formamide	58.4
Methylene iodide	50.8
Ethylene glycol	48.2
Bromonaphthalene	43.9
Dimethyl acetamide	35.4
Hexadecane	26.7

$$F = \gamma_L P \cos\theta_{a.r}$$

where F is the wetting force (from the electrobalance chart recording) in dynes, P is the specimen perimeter in cm, γ_L is the liquid surface tension (measured independently) in dynes/cm and $\cos\theta_{a,r}$ is the cosine of the contact angle, advancing or receding. To obtain a representative value for the $\cos\theta_a$ (or $\cos\theta_r$) of any probe liquid on a given solid, values from at least 10 replicate solid specimens were averaged.

<u>Single Filament Pull-Out Tests</u>

The bond strength at the interface was evaluated directly by a single filament pull-out test. Specimens were prepared so that a single filament passed perpendicularly through a film of resin which was then cured while the filament was suspended in place. Curing was done in an oven at atmospheric pressure. Figure 2 shows the specimen configurations for aramid (left) and for graphite (right).

Figure 3 shows a fixture for preparation and test of several specimens. Note the cardboard end tabs on the top of each filament. After cure, the apparatus containing the specimens was placed on the crosshead of the testing machine, the filaments were attached to the load cell, and were pulled out of the cured resin film one by one.

The greatest difficulty in preparing and setting up this type of test was to keep the resin film thin enough so that the filament would pull out rather than simply break in tension. By trial and error, we found that for the aramid fiber the resin film had to be less than 0.3 mm thick, and for the graphite fiber the resin film had to be less than 0.05 mm thick. The configuration on the left of Figure 2 where the resin adhered to the two blades was easier and faster to set up than the configuration on the right, where the resin for each filament was in a separate hole. The configuration on the left was perfectly adequate for the aramid filaments but did not produce a thin enough film for the graphite filaments. The configuration with the round holes for each filament was suitable for graphite because the holes were drilled in shim stock only 0.025 mm thick which enabled formation of a resin film below 0.050 mm thick.

Typically, during the pull-out test, the load increased linearly on the load versus time trace and then suddenly dropped when debonding occurred [4]. After the tests were complete, the

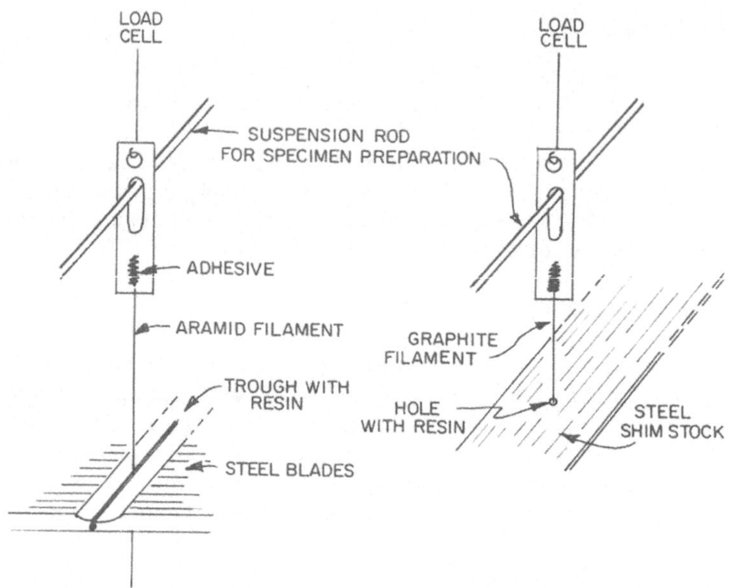

Fig. 2.   Single filament pull-out test configurations for aramid
           (left) and graphite (right). The one on the right provides
           the thinner film required by the small diameter graphite
           fiber. The single filament passes down through the resin
           film and is cured in place in an oven at atmospheric
           pressure.

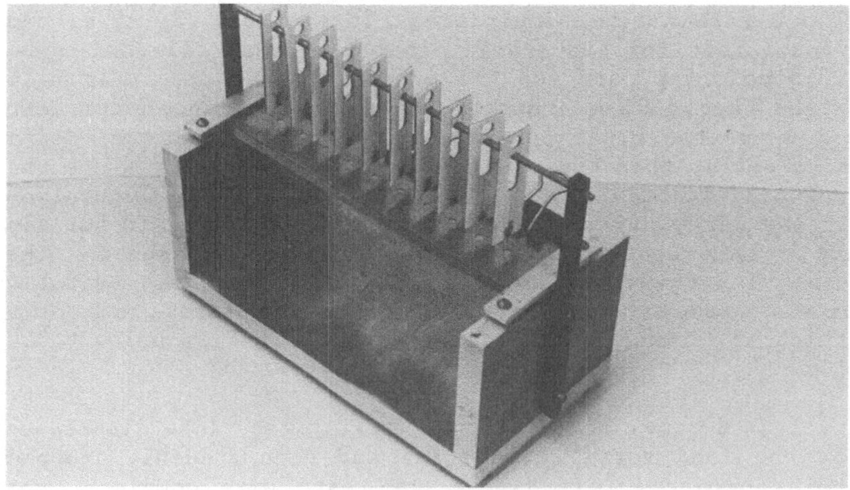

Fig. 3.   Example of fixture used for preparation of a group of
           single filament pull-out specimens. The whole apparatus
           is placed on the testing machine crosshead and filaments
           are pulled out of cured resin film one by one.

embedment depth was determined with a microscope and a micrometer and the bond strength was calculated by dividing peak load by interfacial area. The nominal filament diameters of 12 μm for aramid and 8 μm for graphite were used in calculating interfacial area. In the case of aramid, using the nominal rather than the actual diameter for each filament was no problem since the manufacturer produces a uniform product with very little diameter variation. In contrast, graphite fiber is manufactured with much greater diameter variation (filament diameters ranging from 4 μm to 12 μm in one spool). Therefore, for graphite, when the nominal diameter value was used in the calculation of interfacial area, the actual variation caused additional scatter in the interfacial strength values. After calculation of the interfacial bond strength for each specimen, statistical decision theory was applied to the data. The z-scores were calculated and used to test the significance of sample differences.

After pull-out, many of the filaments were examined in the scanning electron microscope (SEM) to check for adhering matrix material or for fiber damage. For the aramid/epoxy single filament pull-out tests which were carried out in the SEM, each imbedded filament was separately mounted. Each specimen was coated with conductive material prior to placement in SEM. A small drilled metal sheet containing the resin film and its imbedded filament was affixed to the SEM specimen stage and the free end of the filament was attached to an externally controlled screw. As the screw was turned to apply tension to the filament, the pull-out process was observed on the display screen of the SEM.

Studies on Degree of Cure

For the experiments where test groups were processed at different cure temperatures, it was necessary to make sure that the only variable was thermal mismatch and to eliminate extraneous variables. It was required that the degree of cure in the resin be a constant, no matter what the cure temperature, so that the resin's viscoelastic properties would be constant. Torsional braid analysis (TBA), a dynamic mechanical test method, was used to check the viscoelastic properties of the epoxy resin after various cure times at selected temperatures.

TBA provides a measure of the relative rigidity and damping of the test polymer as a function of temperature, and in doing so, displays the glass transition of the polymer. When the TBA trace of the resin cured at one condition was exactly equivalent to the TBA trace of the resin cured at another condition, the degrees of cure were considered to be equivalent.

                                                L. PENN ET AL.

Table 2. Failure Strength Results for Aramid/Epoxy and Graphite Epoxy

System	Single Filament Pull-out, psi	Short Beam Shear, psi	Iosipescu Shear, psi
Aramid/Epoxy A	4,800 ± 1,100	7,800 ± 500	12,900 ± 700
Graphite/Epoxy A	12,300 ± 3,200	13,600 ± 700	16,700 ± 600

RESULTS AND DISCUSSION

Comparison of Aramid and Graphite Composites

The single filament pull-out results for aramid fiber and for graphite fiber in the tetrafunctional epoxy system are shown in Table 2. The large scatter in these data is typical of single filament test data [5,6] and probably reflects the natural variation of interfacial strength from spot to spot. In such small specimens, averaging effects would be reduced. Also shown in Table 2 are the failure strength results of short beam shear and Iosipescu shear tests in unidirectional laminates. The key point from the Table is that the graphite/epoxy strength is 1-1/3 to 2 times larger than the aramid/epoxy strength for all three types of test, the single filament pull-out, the short beam shear, and the Iosipescu shear.

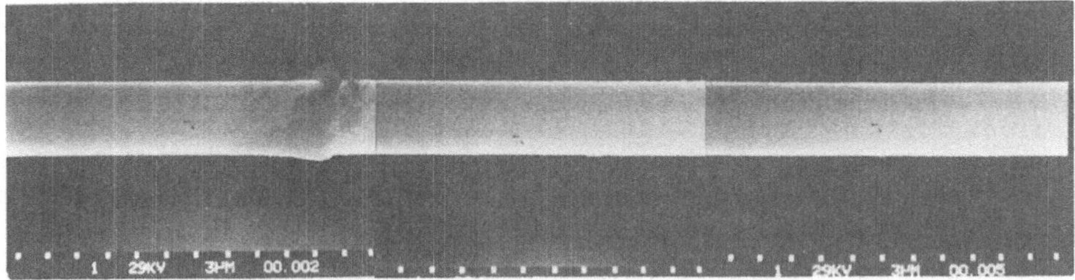

Fig. 4. Scanning electron micrograph of debonded area of graphite single fiber pulled from cured resin film. Although the meniscus formed by the resin before it cured still adheres to the fiber, the debonded region is clean, indicating adhesive failure at the interface rather than the cohesive failure in the resin film. The cracking of the meniscus during test is observed as a small discontinuity on the rising force portion of the load-deflection curve.

From single filament tests actually carried out in the SEM and from post-failure micrographs, it was clear that the aramid/epoxy failures were interfacial [4]. The graphite/epoxy failures also were interfacial. Figure 4 shows a scanning electron micrograph of the clean debonded area of a graphite fiber pulled out from the resin film. Thus, for the materials studied here, the single filament pull-out test is indeed a test of interfacial bond strength. The fact that the results for shear strength of the unidirectional composites parallel the single filament bond strength suggests that the shear strength in the composite is interface-dependent.

One explanation that would come to mind for the difference in adhesive performance would be the modulus difference between the two fibers. Would the graphite modulus (28,000,000 psi) being three times higher than that of aramid (8,500,000 psi), bring about a tow-fold higher adhesive performance for the graphite? Perhaps the more compliant aramid, contracting radially as the fiber is loaded longitudinally, would peel away from the interface. The stiffer graphite would not be expected to contract as much radially, and would not experience as much peeling load. If this Poisson effect were the explanation for the interfacial strength difference observed between graphite/epoxy and aramid/epoxy, then we would expect the failure load in the single filament test to be independent of imbedment depth. However, the failure load goes up directly with imbedment depth until the fiber failure strength itself is reached [4]. Furthermore, were fiber modulus difference to play an important role in determining interface strength, one would expect a difference between the commercially available high and low modulus versions of the aramid fiber in single filament pull-out tests and in laminate shear tests. Although the modulus of the Kevlar 49 aramid is twice as high as the modulus of the Kevlar 29 aramid, no such differences were noticed when a separate study of the high modulus fiber in single filament and in laminate tests was carried out.

It will be of obvious value to discover the specific reasons for the higher performance of the graphite/epoxy interface. This understanding could lead to a way to improve certain modes of composite performance.

Fiber Surface Analysis

To determine if the intermolecular (secondary bonding) inter-actions at the aramid/epoxy interface differed much from those at the graphite/epoxy interface, the surfaces of both fibers were characterized by contact angle analysis. The results are shown in Figure 5, where the x-axis is an ordered list of the probe liquids' surface tensions and the y-axis is a contact angle cosine scale.

There is one bar for each probe liquid on the solid: the bottom of
each bar marks the average $\cos\theta_a$ value made by the probe liquid and
the top of each bar marks the average $\cos\theta_r$ made by the probe
liquid. (Typical scatter around each average $\cos\theta_a$ or $\cos\theta_r$ value
is ± 0.03 cosine units). It is clear from Figure 5 that the
surfaces of aramid fiber and of graphite fiber are physicochemi-
cally very similar. The differences in surface analysis shown here
are not large enough to cause a two-fold difference in interfacial
bond strength. The surface differences must be much greater than
those shown in Figure 5 to exert a significant influence on adhesive bond strength [7,8]. Therefore, we must look elsewhere for
the cause of the interfacial strength difference.

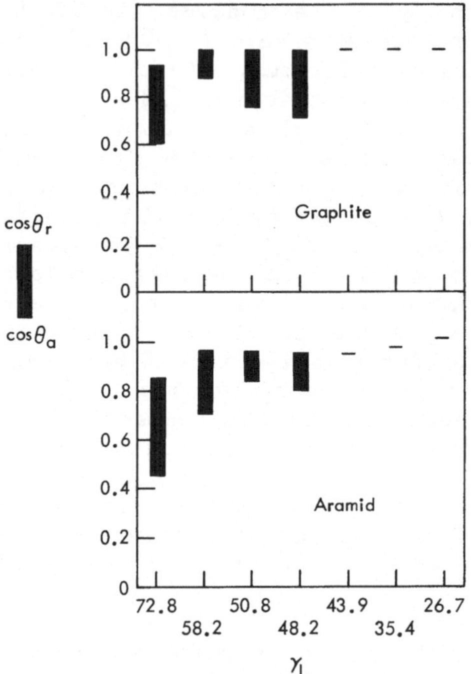

Fig. 5.   Contact angle bar graphs for aramid fiber and graphite
          fiber. The y-axis is the scale for contact angle cosine
          values. The x-axis is an ordered list of the surface ten-
          sions of the probe liquids. There is one bar for each con-
          tacting liquid on each type of fiber: the bottom of each
          bar marks the $\cos\theta_a$ value while the top of each bar marks
          the $\cos\theta_r$ value. Typical scatter (not shown) at the top
          and bottom of each bar is ± 0.03 cosine units. The two
          bar graphs should be compared like fingerprints.

Possible Chemical Bonding at Interface

There is no spectroscopic way to investigate the aramid/epoxy or graphite/epoxy interface structure as has been done with glass composites because the interface to volume ratio in filamentary composites is not large enough to give a strong spectroscopic signal. This ratio could be increased by using small particles, but in the case of aramid, they are not available and in the case of graphite, they are opaque to light. In the absence of ability to directly measure the interfacial structure, a discussion of the possibilities can be held.

It has been shown that the surface of aramid fiber is oxidized [9,10]. From the chemical structure of the aramid, the ESCA data, and the mass spectral decomposition data, it is likely that the surface oxidation is in the form of carbonyl, carboxylate, and hydroxyl groups. The surface of graphite fiber is also oxidized. Studies have shown that the functional groups also include carbonyl, carboxylate (carboxylic acid), and hydroxyl groups [11,12]. Although a quantitative comparison cannot be made, the possibilities for chemical bonding between fiber functional groups and matrix resins molecules may be similar for both aramid and graphite. However, it is not fair to draw a conclusion at this time, and it must be emphasized that chemical bonding as a reason for interface differences cannot be ruled out completely.

Investigation of Mechanical Interference

The final factor to be considered as affecting interfacial strength is mechanical interference. A significant lock and key configuration may be ruled out by the fact that under the SEM at 800X both the aramid and graphite fibers are relatively smooth. (Shallow longitudinal grooves shown in some graphites at much higher magnification would not be expected to influence shear strength because they do not increase the surface area much above the nominal value and they are parallel to the fiber direction). Another kind of mechanical interference is proposed here as an explanation for the difference in performance between aramid/epoxy and graphite/epoxy systems. This is the radial compression or tension (tight or loose fit) exerted by the matrix on the fiber as a result of thermal mismatch during cool-down from the high curing temperature. Recognition of residual thermal stresses in composites is not new [13]. However, we propose here to use them to explain a specific result and then to test that explanation. Table 3 contains the basis for our explanation.

Table 3.   Literature Values of Room Temperature Thermal Expansion
           or Contraction Coefficients for Matrices and Fibers

Material	α_L, x 10^{-6}/°C	α_T, x 10^{-6}/°C	Ref.
Aramid	−2	59	(14)
Graphite	−2	22	(15)
Epoxy A	45	45	(15)
Epoxy B	60–70	60–70	(16,17)

The thermal expansion coefficients listed in Table 3 were
used to estimate the coefficients for the materials in our study.
(It should be remembered that thermal expansion coefficients are a
function of temperature and generally increase slightly with temperature for the range −200°C to 200°C). The fibers are anisotropic
so there is a large difference between α_L and α_T . The fibers
shrink in length but increase in diameter as the temperature
rises. Because the aramid used by us was the lower modulus and
less anisotropic form than the one in Table 3, its α_L would be a
bit higher and its α_T would be a bit lower than in the Table. The
epoxies are isotropic and so there is no difference in their radial
(α_T) and axial (α_L) expansion coefficients. However, the epoxies
are different from each other because of different degrees of
cross-linking: epoxy B has a higher α than does the stiffer, more
highly cross-linked epoxy A.

Considering only α_T of the fibers along with the α of the
cured resins, we can assume the following relations:

Graphite < Epoxy A < Aramid < Epoxy B

These relations lead to the prediction of a radial pressure
build-up at the interface in a filamentary composite. At the
elevated cure temperature, the cross-linked resin network is formed
around the fiber in a nearly stress-free manner. When the composite is cooled down to room temperature, contraction strain
occurs,and it occurs to a different extent in each material component, according to each value of α. (It is helpful to recall that
when an annulus or ring cools down from a high temperature, both
inner and outer diameters decrease, i.e., the hole gets smaller).
Thus, when a fiber imbedded in an annulus of cured resin cools
down, the fiber diameter will decrease and the resin will also
shrink down around the fiber. From the relations of α given in the
previous paragraph, the diameter decrease of the epoxy A resin
annulus would be expected to be much greater than the diameter
decrease of the graphite fiber over a given temperature drop.

Thus, the resin would exert a radial compression on the fiber at room temperature. This compression would produce better contact at the interface and would also hold the fiber tighter during a shear debond test. On the other hand, the α relations show that for the aramid imbedded in epoxy A, the diameter decrease in the fiber would be more than the diameter decrease of the resin annulus during cool-down, giving a looser fit at the interface at room temperature.

The equation for pressure at the interface of a cylindrical fiber in an infinite medium is helpful to the discussion, although it cannot be assumed to accurately represent the experimental configuration used in our laboratory. The equation is:

$$P = \frac{(\alpha_m - \alpha_f)\ \Delta T + \varepsilon_c}{\dfrac{(1 + \upsilon_m)}{E_m} + \dfrac{(1 - \upsilon_f)}{E_f}}$$

where α_m = expansion coefficient of matrix, α_f = transverse expansion coefficient of fiber, υ_m = Poisson ratio of matrix, υ_f = transverse (in-plane) Poisson ratio of fiber, E_m = matrix elastic modulus, E_f = transverse elastic modulus of fiber, and ε_c = linear curing shrinkage in matrix above glass transition (this would only be a small part of the total volumetric curing shrinkage).

For both types of fiber, the second term in the denominator is an order of magnitude less than the first term so it can be ignored. This simplifies the comparison since now the denominator is constant. The numerator term ε_c is estimated to be very small, about 3×10^{-3} at the maximum. Below are rough estimates of the numerator terms for cooling from 177°C to room temperature, a ΔT of 150°C. Since the glass transition of epoxy A is 250°C, the whole cooling process occurs while the resin is in the glassy state.

System	$(\alpha_m - \alpha_f)\Delta T$	ε_c
Aramid/epoxy A	−.0023	+.003
Graphite/epoxy A	+.0057	+.003

This very rough estimate shows that it is possible that for aramid/epoxy A, the thermal mismatch counteracts and alleviates most of the curing stress, while for graphite/epoxy A, the thermal mismatch adds to the curing stress, producing an even greater radial compression.

To test for evidence of a mechanical interference effect due to thermal mismatch, experiments were carried out to compare the interfacial strength of a fiber/epoxy system processed at two different temperatures. For two temperatures widely apart, epoxy A did not yield equivalent degrees of cure, no matter how much adjustment in cure time was made, so it had to be eliminated from this part of the study. Epoxy B did yield equivalent degrees of cure (as determined by TBA) at several different time/temperature conditions. The final two curing conditions selected were: (1) 120°C for 3 h; and (2) 70°C for 64 h.

These two conditions both gave a resin system with a glass transition temperature (T_g) of 95°C. The fact that this T_g is above the cure temperature in one case and below the cure temperature in the other case deserves comment. It is generally assumed that some stress relaxation can occur in a cured resin at temperatures above its T_g, although this is not usually determined quantitatively. In the case of the specimens cured at 120°C, some stress relaxation during cool-down may occur above the T_g and this may diminish the stress buildup due to thermal contraction. However, below the T_g, with continued cooling, no further relaxation would occur and thermal contraction stresses would grow until room temperature was reached. Ideally the experimenter would like as much difference as possible between the cool-down stresses achieved for the two curing conditions. For the specimens cured at 70°C, thermal contraction stresses build up over the full cool-down range from 70°C to 20°C, at ΔT of 50°C. For the specimens cured at 120°C, the thermal contraction stresses build from an unknown starting temperature somewhere between 95°C and 120°C. Thus for these specimens the effective ΔT value would be between 75°C and 100°C.

From the α relations presented earlier and the ΔT estimates discussed above, one would expect the group of specimens of the aramid/epoxy B processed at the higher temperature to have the higher interfacial shear strengths. The results of single filament pull-out tests on aramid/epoxy B specimens processed at the two temperatures are shown in Table 4.

Table 4. Single Filament Pull-Out Test Results for Aramid/Epoxy B Specimens Processed at Two Different Temperatures

System	Cure Temperature °C	No. Specimens	Interfacial Shear Strength, psi
Aramid/Epoxy B	70	59	4000 ± 1100
Aramid/Epoxy B	120	105	4700 ± 1600

Although the difference is small, it is statistically significant at the 0.01 level. This verifies the existence of a noticeable mechanical interference effect.

The only question about this experiment is the effect of thermal mismatch between the metal frame and the epoxy resin film. The α values for metals are much smaller than for epxoy resins. This would be expected to put the resin film in radial tension as it adheres to the metal. However, since the geometry of the metal/epoxy interface in the apparatus used for the aramid/epoxy studies is not radially symmetric, the stress distribution in the cooled specimen is difficult to estimate, even qualitatively. Even so, the key finding is that a strength difference at the fiber/epoxy interface does arise from the summation of mechanical interferences due to thermal mismatches.

The geometry in the case of the apparatus used for graphite/-epoxy single filament tests is more clear cut. Here the resin film is a true disc and its outer rim, the epoxy/metal interface, is circular. The thermal mismatch at the epoxy/metal interface would produce a radial tension in the resin film during cool down. The higher the processing temperature, the greater would be the resultant radial tension in the resin film on cool down. At the fiber/resin interface, the opposite would be true: the higher the processing temperature, the higher the radial compression exerted on the fiber by the resin. If the fiber/resin interface were the most influential one, then a higher processing temperature would give a higher interfacial strength. The actual results of the single filament pull-out tests on graphite/epoxy B specimens processed at the two temperatures are shown in Table 5.

Table 5. Single Filament Pull-Out Test Results for Graphite/Epoxy
          B Specimens Processed at Two Different Temperatures

System	Cure Temperature °C	No. Specimens	Interfacial Shear Strength, psi
Graphite/Epoxy B	70	80	11,400 $\pm$ 3,940
Graphite/Epoxy B	120	82	14,200 $\pm$ 7,470

The table shows that the specimens processed at the higher temperature exhibit the higher interfacial shear strength. Although the scatter appears large, the difference in interfacial shear strength is statistically significant at the 0.01 level.* These results also show that the thermal mismatch effect at the fiber/matrix interface overrules the thermal mismatch effects farther away.

Although it is difficult to model the situation at the interface in any of the single filament pull-out tests, it does seem evident that interfacial adhesive performance is affected by mechanical interference from thermal mismatch.

Conclusions

The factors that influence interfacial strength (adhesive performance) have been separated into three types for consideration. The three types are: (1) intermolecular interactions, (2) chemical bonds, and (3) mechanical interference. The role of each of these in explaining the difference between aramid/epoxy and graphite/epoxy interfacial bond strength was examined. Through surface contact angle analysis, it was determined that the intermolecular interactions at the graphite/epoxy interface were not sufficiently different from those at the aramid/epoxy interface to cause a two-fold strength difference. Consideration of literature data on the functional groups at the surface of aramid and of graphite fibers suggest that these may be quite similar. If so, then chemical bonding would not explain the two-fold strength difference. Mechanical interference arising from thermal contraction mismatch was experimentally investigated. The results suggested that the mechanical interference stresses could indeed explain the difference between the graphite/epoxy and aramid/epoxy interfaces. The tighter fit at the graphite/epoxy interface could be the source of the greater bond strength of the graphite fiber with the matrix.

---

*A further indication of the superior interfacial strength of the population cured at 120°C compared to that cured at 70°C is the fact that many 120°C – cure specimens failed in fiber tension rather than pulling out of the resin. Thus, the average value for interfacial strength of the population cured at 120°C was calculated with the strongest interfaces missing from the data set.

Generally, people have assumed that residual thermal stresses are
undesirable in composites. Although this may often be true, in
some situations residual stress may have a constructive effect.
Thermal stress that gives a mechanical interference which causes a
tighter fit or better contact for adhesion at the interface is constructive rather than detrimental for interfacial bond strength.
Perhaps in the future, workers can use matrix materials with
coefficients of thermal expansion that lead to stronger interfaces
and better shear strength.

REFERENCES

1. N. Iosipescu, J. Materials, 2, 537 (1967).
2. J. M. Slepetz, T. F. Zagaeski, and R. F. Novello, Technical
    Report AMMRC TR 78-30, Army Matemials and Mechanics Re-
    search Center, Watertown, Massachusetts (July 1978).
3. G. Mozzo and R. Chabord, 23rd Annual Technical Conference,
    Reinforced Plastics/Composites Institute, SPI, Section 9-C
    (1968).
4. L. S. Penn and S. M. Lee, Fibre Science Technol., 17, 91
    (1982).
5. L. Drzal, M. J. Rich, J. D. Camping, and W. J. Parks, 35th
    Annual Technical Conference Reinforced Plastics/Composites
    Institute, SPI, Section 20-C (1980).
6. L. J. Broutman and F. J. McGarry, Mod. Plast., 40, 161 (1962).
7. L. S. Penn and E. R. Bowler, Surf. Interface Anal., 3, 161
    (1981).
8. L. S. Penn, F. A. Bystry, and H. J. Marchionni, Polymer Com-
    posites, 4, 26 (1983).
9. L. Penn and F. Larsen, J. Appl. Polym. Sci., 23, 59 (1979).
10. R. Allred, personal communication of ESCA results, Massachu-
    setts Institute of Technology, 1982.
11. G. Gynn, R. N. King, S. F. Chappell, and M. L. Deviney,
    "Improved Graphite Fiber Adhesion," Air Force Wright Aero-
    nautical Laboratories, Report No. AFWAL-TR-81-4096, Wright
    Patterson Air Force Base, Ohio 45433, September 1981.
12. E. Fitzer, K. H. Giegl, and L. M. Manocha, Proceedings of the
    5th London International Conference on Industrial Carbon
    and Graphite (1978), p. 405.
13. H. T. Hahn, J. Composite Materials, 10, 266 (1976).
14. C. C. Chiao and T. T. Chiao, in "Handbook of Composites,"
    George Lubin, Editor, Van Nostrand Reinhold Company, New
    York, 1982, p. 272.
15. A. K. Miller and D. F. Adams, "Advanced Composite Materials -
    Environmental Effects," ASTM-STP 658, J. R. Vinson, Editor,
    American Society for Testing and Materials, Philadelphia,
    1978, p. 121.
16. G. Marom and B. Gershon, J. Adhesion, 7, 195 (1975).
17. H. Lee and K. Neville, "Handbook of Epoxy Resin, McGraw-Hill
    Book Company, New York, p. 17-16.

FRACTO-EMISSION FROM FIBER-REINFORCED AND PARTICULATE FILLED

COMPOSITES

J. T. Dickinson, A. Jahan-Latibari, and L. C. Jensen

Department of Physics
Washington State University
Pullman, WA 99164-2814

ABSTRACT

Fracto-emission (FE) is the emission of particles and photons
during and following crack propagation. The types of particles we
have observed include electrons (EE), positive ions (PIE), photons
(phE), and excited and ground state neutral emission (NE). In this
paper we present our work on the characterization of the various FE
components and measurements relating FE to the fracture events and
material properties involved. FE characteristics measured include
total emission, time dependence relative to crack propagation,
species of neutral and ionic components, energy of charged species,
and time correlations between pairs of FE components. Experiments
on fracture of epoxy, single fibers, fiber/epoxy strands, particulate filled epoxy, and multi-ply fiber/epoxy systems will be
presented.

INTRODUCTION

When a crack propagates through a material, the crack walls
are left in a highly excited, non-equilibrium state. For non-
metals, this departure from equillibrium involves: (1) broken
bonds, (2) liberated fragments (e.g., free radicals, atoms, molecules), (3) defects (e.g., in crystals, point defects), (4) charge
separation often involving production of charged species, a variety
of types of electron traps and associated electric fields, and (5)
a localized rise in temperature. All of these factors represent
concentrated energy which can contribute to the ejection or emission of charged particles, neutral particles, and photons from the
fracture surfaces. We refer to all forms of such emission accompanying fracture as "fracto-emission" (FE). Our experimental

studies of the characteristics of FE from a wide range of materials are presented in references 1-13. A review of our work on FE accompanying adhesive failure can be found in reference 3. The basic behavior we have observed can be summarized as follows:

(1) Some form of crack propagation appears to be a necessary prerequisite for the occurrence of FE.

(2) FE is a wide-ranging phenomenon. We have observed electron (EE) and positive ion emission (PIE) from all materials tested including inorganic crystalline materials, ceramics, glasses, glassy polymers, elastomers filled and unfilled, fiber-reinforced composite, and molecular crystals.

(3) The few systems we have studied to date emit photons (phE) in air and in a vacuum environment.

(4) Interfacial failure between epoxies, polymers, glasses, graphite, and metals produces very intense, long-lasting energetic EE and PIE. This is believed to be due to the production of high concentrations of surface free radicals and surface charge due to charge separation. The EE and PIE energy distributions which we have measured for these systems are broad, slowly decreasing functions peaking near 0 eV but extending to > 1 keV. These energies are believed to be due to charged particles being accelerated in the presence of the electric fields due to charge patches on the fracture surface.

(5) Polymeric systems have a strong dependence of EE intensity on crack velocity (V_c). Presumably, this is due to a higher density of free radicals and trapped electrons produced by more primary bond scissions at higher V_c. At lower V_c, the polymer has time for slipping and unraveling of chains allowing it to deform and tear with less "damage".

(6) In support of this, more highly cross-linked polymers produce higher intensity and longer-lasting FE (for the same reasons).

(7) The measurements we have made on the mass of PIE produced during fracture indicate that the masses are chain fragments; this implies a sensitivity to where the fracture has occurred on an atomic scale.

Our initial work on FE from composites has concentrated on fracture of individual fibers, unfilled resins, and unidirectional fiber/epoxy systems. A few studies of multi-directional fiber/epoxy systems have also been carried out. In addition, we have recently examined FE from particulate-filled epoxy. The results of these studies will be presented here.

II. EXPERIMENTAL PROCEDURE

Details of our experimental procedure are given in references 1-13. In brief, experiments were performed in vacuum at pressures ranging from 10^{-8} to 10^{-6} torr. Our vacuum systems are equipped with devices to stress samples in various ways including tension, flex, and compression, while measuring stress and/or strain. The detectors used for charged particles are Channeltron Electron Multipliers (CEM) which produce fast (10 ns) pulses with approximately 90% absolute detection efficiency for electrons and nearly 100% efficiency for positive ions. The gains of the CEM´s used were typically 10^6 to 10^8 electrons/incident particle. The detectors were positioned 1 to 4 cm away from the sample with a bias voltage on the front cone of the CEM to attract the charged particles of interest. Background noise counts ranged from 1 to 10 counts/sec. Standard nuclear physics data acquisition techniques were employed to count and store pulses, normally as functions of time. The time scales of interest are submicrosecond to several second intervals, which we can easily cover with commercial electronics. Single fibers and epoxy-filament strands were tested in tension at a rate of 1% per second. Fiber samples consisted of 5 to 20 fibers adhesively bonded to Al sheet metal shaped to fit into clamps in the vacuum system. To reduce the probability of fiber pull-out, the fibers were stretched across a sharp Al edge, where approximately 90% of them fractured.

Kevlar, E-glass, and S-glass epoxy strands and unidirectional graphite-epoxy composites made from Union Carbide Thornel 300 graphite fibers and NARMCO 5208 epoxy resin were also fractured in tension. A sharp notch was made in the center of the tension sample to control the fracture initiation.

Graphite-epoxy composites made from Union Carbide Thornel 300 fibers and various NARMCO epoxy resins were tested in flex as well. The fiber directions in these composites were (0), (45), and (0,90,90,0) degree to the long axis. Samples were tested with a span-to-depth ratio of 30:1 and a strain rate of 0.064 mm/sec. Acoustic emission (AE) and EE were detected from graphite-epoxy composite fractured in flex. AE was detected with a PZT transducer with a resonant frequency of 175 KHz (Acoustic Emission Technology Corporation AC175L). The bursts were typically 500 μsec in duration. The filtered and amplified signal was fed into a discriminator to eliminate background noise, and the resulting pulses were counted on a multi-channel scaler. Thus the count rate displayed is determined by both the number and size of AE bursts (the number of "rings" that trigger the discriminator). To reduce the influence of mechanical AE in our experiments, the mechanical supports were covered with teflon tape. Fracture of a uniform material (PMMA), which will have no interlaminar shear or delamination, showed no prefracture AE in our system. Figure 1 shows schematically the electron multiplier and AE transducer arrangement which simultaneously detect AE and EE from the sample. Load and deflection were also measured.

Another composite structure we have investigated recently is a particulate-filled epoxy. The epoxy was EPON-828 (Z-hardener) filled with irregularly shaped alumina particles with an average diameter of approximately 10 μm. This brittle material was broken in flex.

III. RESULTS AND DISCUSSION

<u>Filament-Epoxy Strands:</u> An early observation we made involving adhesive failure and its effect on charged particle emission concerned the fracture of composites. Starting with the constituents of a composite, the EE time distributions of the fracture of individual 10 to 20 μm filaments of Kevlar[TM], Thornel 300 graphite, E-glass and S-glass, as measured with a CEM 1 cm from the sample, are shown in Figure 2. Also shown is the EE from the fracture of unfilled DER 332/T403, a bisphenol type A resin. With the exception of Kevlar filaments, repeated experiments showed no evidence of a measurable rise time to the peak emission. The brittle fibers with small cross section break on a nanosecond time scale. The peak emission occurs during fracture and decays rapidly away, typically 10 to 100 μsec, as shown in Figure 2.

When these fibers are placed in epoxy resin and fractured, the results are significantly different. Figure 3 shows the EE and PIE resulting from the failure of a strand containing Kevlar fibers in DOW DER 332 epoxy. These curves were taken simultaneously with two detectors. In general, EE exceeds PIE in terms of total emission by 10 to 40%; in Figure 3, PIE has been normalized to the EE at a single point. On the time scale shown, the time required for fracture was less than one channel. Thus, the signal rises from a noise count of 0.1 to 10 per second to peaks of 10^4 to 10^5 counts per second. Note that in this case the decay from the peak lasts for many seconds. Intense emitters such as these give detectable emission for as long as two hours after fracture. Also, we note that the decay kinetics for both EE and PIE are essentially identical, suggesting that a common rate-limiting step is shared by the two types of emission.

Examination of the fracture surface on a number of systems involving adhesive failure or delamination with a SEM indicates that the production of interfaces is responsible for the considerable differences between Figure 2 and Figure 3. This feature of

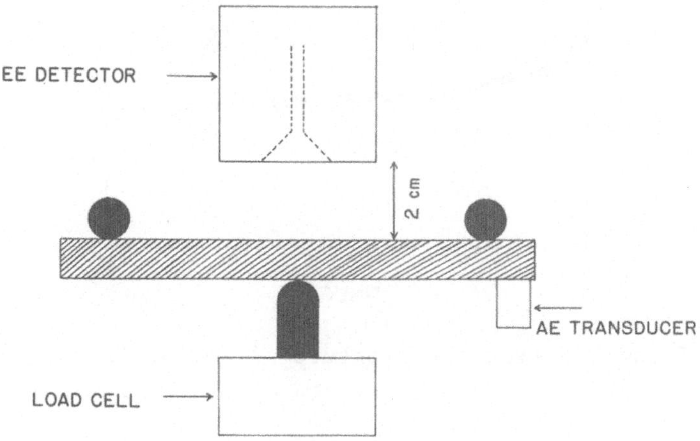

Fig. 1. Schematic diagram of experimental arrangement for EE, AE, and load measurements on composite materials in flex.

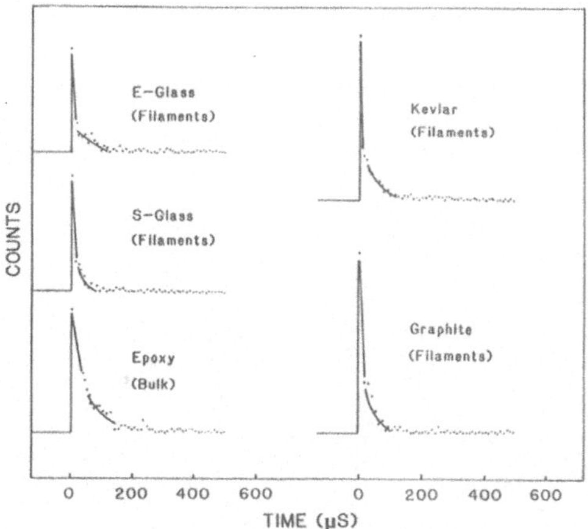

Fig. 2. The time distribution of EE due to the fracture of graphite, E-glass, Kevlar filaments and bulk epoxy (Dow DER 332/T304). Note the fast time scale.

                    J. T. DICKINSON ET AL.

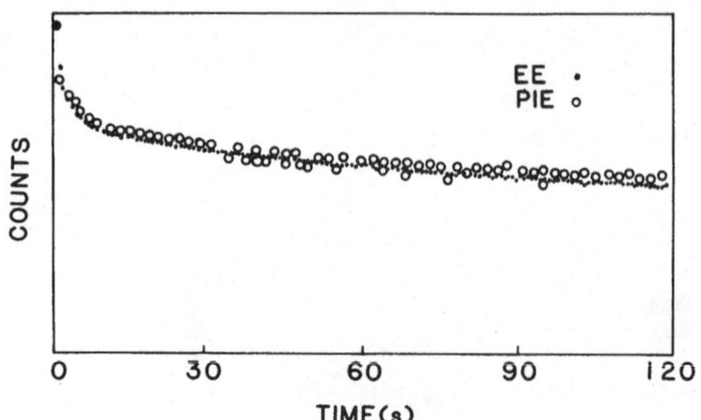

Fig. 3. EE and PIE from the fracture of Kevlar/Epoxy strands.

ELECTRON EMISSION FROM FRACTURE OF
FIBRE-REINFORCED EPOXY UNDER TENSILE STRAIN

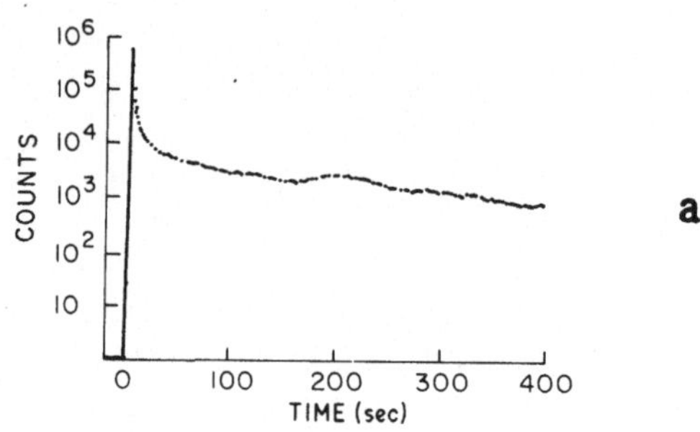

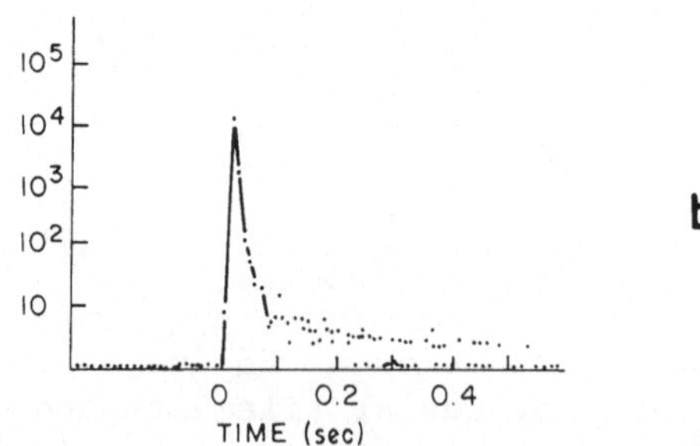

Fig. 4. Electron Emission during and following fracture of a)E-
         glass and b)S-Glass-epoxy strands. Note the different
         time scales.

intense, long-lasting emission may serve as a measure of the extent of delamination that has occurred. In support of this, in Figures 4a and 4b we compare the EE for two types of epoxy strands made from 20 μm diameter E-glass filaments and 10 μm S-glass filaments embedded in DOW DER 332 epoxy. (Note the different time scales for the two different materials). Examination under the SEM shows that there is considerably more delamination and separation of the filaments in the case of E-glass than for S-glass epoxy strands, which apparently results in considerably higher count rates and longer lasting emission. By far the predominant emission is coming from the surfaces created by the separation of the filaments from the matrix.

The results of experiments on unidirectional graphite-epoxy composites (Thornel 300/5208) are shown in Figure 5. The samples were 0.25 mm thick and 2.4 mm wide. The EE and PIE were measured from two separate samples. The resulting emission plotted on a log scale shows the rapid rise during fracture and slow decay following fracture. Examination of the fracture surfaces shows extensive delamination and interfacial-like failure, consistent with the results on DOW DER 332 Kevlar strands.

<u>Energy Distributions:</u> Because the EE and PIE from systems involving interfacial failure frequently were intense and long lasting, we were able to take measurements of the energy distributions, n(E), using retarding potential analysis. The curves in Figure 6 represent n(E)dE, where dE is 2 eV, plotted on a log scale and normalized to unity at the peak. Both curves are nearly identical, showing a peak near 0 eV and a significant number of higher energy particles in the tail. It is well known that charge separation is a common occurrence with adhesive failure and can leave the surfaces in a highly charged state. Thus, the probable cause of the high energy particles is the release of the charges in the physical proximity of charge patches of the same sign, yielding an acceleration of the particles to the observed energies.

Preliminary experiments involving fracture of the filaments and neat resin alone do not seem to yield emission at such high energies. Thus, we appear to have a distinct indicator of interfacial failure in a composite system:

a) intense, long-lasting EE and PIE

b) the presence of high energy EE and PIE.

Retarding grids could easily reject the low energy particles and thus obtain a signal which is entirely due to interfacial failure. Proper steps to quantify these measurements could allow a precise determination of the degree of delamination/interfacial

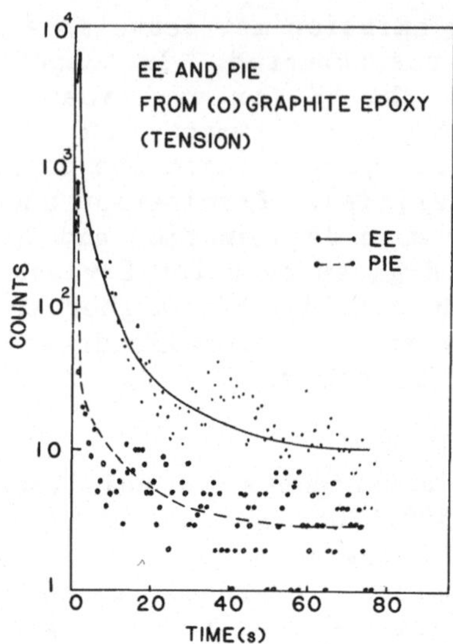

Fig. 5.   EE and PIE from the tensile failure of unidirectional graphite/epoxy composite (Union Carbide Thornel 300 graphite fiber and NARMCO 5208 epoxy resin).

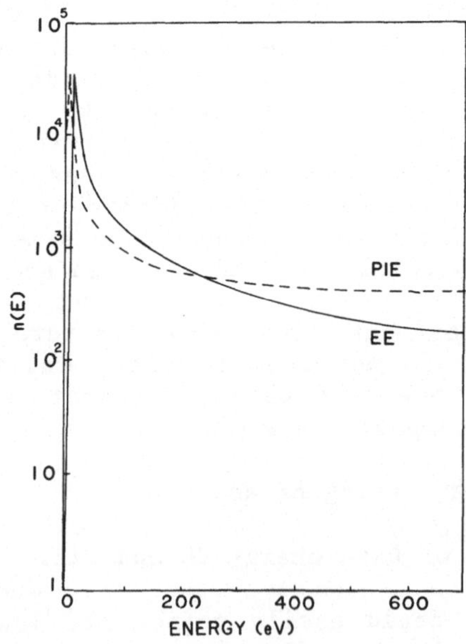

Fig. 6.   Energy distribution on a log scale for EE and PIE from Kevlar/Epoxy strands.

failure that has occurred during a fracture event.

TOF Measurements of PIE Mass from Fracture of Filaments: To determine the masses of the positive ions emitted from Kevlar and E-glass fibers, we have devised a time-of-flight (TOF) technique [14] (shown schematically in Figure 7). The fibers, upon fracture, emit bursts of EE and PIE simultaneously, as determined by a number of experiments using two detectors close to the sample. By triggering a multichannel analyzer (MCA) with the EE burst, we can record the time of arrival of the corresponding PIE burst down a 25 cm drift tube. From the leading edge of these TOF distributions, one can measure a TOF corresponding to the fastest (presumably the lightest) positive ions emitted.

Figure 8 shows the leading edges of TOF distributions obtained for PIE from Kevlar and E-glass fibers for 500 V drift tube potential (-V). Five to ten curves like these were acquired and an average time obtained for each value of -V. The average time at each voltage was plotted vs. the inverse square root of the potential, and the slopes of these lines were used to calculate M/q. M/q values of 60 $\pm$ 20 a.m.u. and 48 $\pm$ 12 a.m.u. were obtained for Kevlar and E-glass, respectively.

Although the uncertainties in these values of M/q are relatively large, we emphasize that the M/q of the PIE accompanying fracture of materials has previously been totally unknown. The technique we have employed here favors the detection of the lightest masses if more than one mass is emitted. Nevertheless, we should have some sensitivity to the presence of heavier masses which should show up as a shoulder on the leading edge of the TOF distribution at longer times. Careful examination of a number of TOF distributions both for Kevlar and E-glass show no clear evidence of heavier masses. Therefore, our present results indicate that PIE accompanying fracture from these materials consists of relatively light ions.

For each material our uncertainties do not allow a unique value of mass to be assigned to the observed PIE; and, therefore, a number of candidates have been examined. For Kevlar, we can rule out absorbed H_2O and ions of common background and atmospheric gases. If we assume q = e, the PIE mass from Kevlar is considerably smaller than a monomer. Likely candidates are:

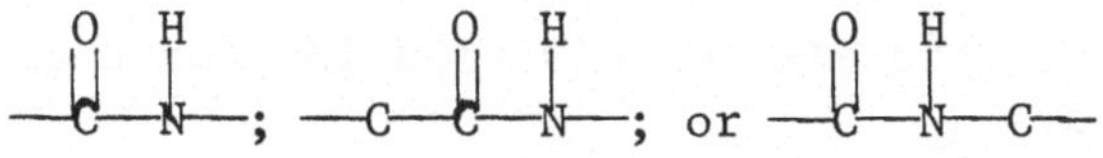

$$
\begin{array}{ccc}
\mathrm{O} & \mathrm{H} & \\
\| & | & \\
-\mathrm{C} - \mathrm{N} - ; & -\mathrm{C} - \overset{\mathrm{O}}{\underset{\|}{\mathrm{C}}} - \overset{\mathrm{H}}{\underset{|}{\mathrm{N}}} - ; & \text{or} \quad -\overset{\mathrm{O}}{\underset{\|}{\mathrm{C}}} - \overset{\mathrm{H}}{\underset{|}{\mathrm{N}}} - \mathrm{C} -
\end{array}
$$

all of which could be produced by a chain bond cleavage. For

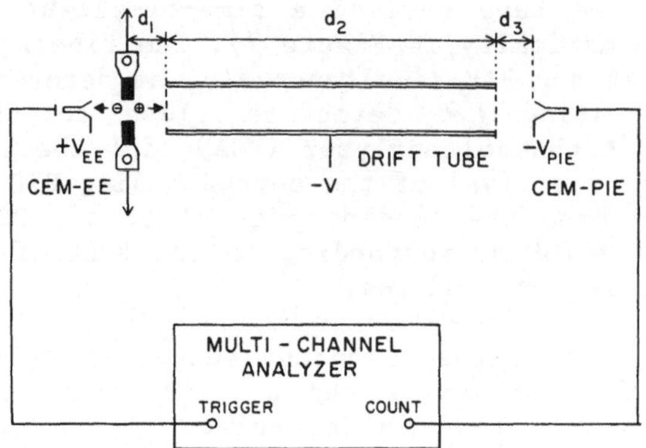

Fig. 7.   The experimental arrangement for use in the time-of-flight technique.   The distances are $d_1=d_3=1$ cm, $d_2=25$ cm.

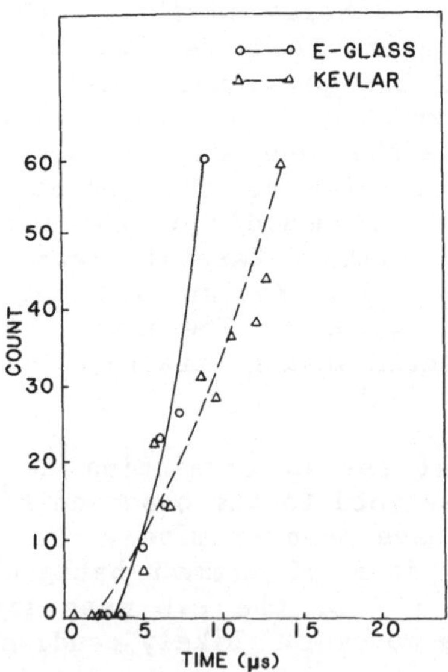

Fig. 8.   The leading edge of the PIE TOF distribution for E-glass and Kevlar fibers.

E-glass, if we assume q = e, we can rule out H_2O^+ and O_2^+. Fragments of the constituent atoms/molecules of E-glass which are possible candidates include Ca^+, K^+, and possibly SiO^+. Reference 11 discusses a TOF technique that takes advantage of the coincidence between individual electrons and positive ions that we have observed. This method has been applied to the measurements of PIE masses from the fracture of filled polybutadiene, an elastomer. We also applied the method to Kevlar-epoxy stands.

Figure 9 is the resulting TOF curve from fracture of a Kevlar-epoxy strand for the same drift tube length of 25 cm and a tube voltage of -2kV. The peaks over the time interval of 1 to 5 μsec are due to the heavier ions; we are still in the process of trying to identify them. The large feature in the first channel (0-0.25 μsec) is also a positive ion (It can be shifted slightly with a different voltage on the tube.). For reasonable initial kinetic energies (less than a few keV), the only masses capable of reaching the detector that fast are mass 1 or 2, i.e., hydrogen. We conclude, therefore, H^+ or H_2^+ is a predominant component of the PIE from this particular material. Of course, both the epoxy resin and the Kevlar filaments contain abundant quantities of hydrogen.

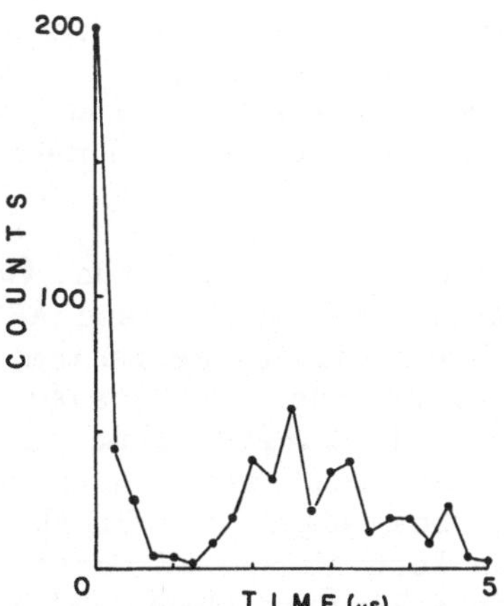

Fig. 9. The TOF for PIE from the fracture of Kevlar/Epoxy strands. The major peak near 0 μs is attributed to H^+ or H_2^+.

AE and EE from Flexural Testing of Fiber-Reinforced Composites: To further explore FE from composites, we simultaneously examined the AE and EE accompanying flexural failure. Figures 10 through 12 show the results of AE and EE measurements from $(0°)_{16}$, $(45°)_{16}$, and $(0,90,90,0)_{16}$ graphite-epoxy composites. Load vs. deflection curves are also included in Figures 10 through 12 to better understand the dependence of AE and EE on the deformation and failure of composite materials. In general, the AE data obtained from these experiments can be characterized as follows: first, an inital rapid rise from zero due to the initial load applied to the specimen; second, the steady buildup of the AE count rate prior to failure. Finally, a large burst followed by a drop in AE count rate at catastrophic failure.

Concerning AE only, our result differs somewhat from those of Barnby and Parry [15]. Barnby and Parry observed no acoustic activity prior to fracture for unidirectional fiberglass-epoxy composite notched flex samples. In their experiment, the onset of failure and large load drop was indicated by the onset of AE. However, their result on cross-ply $(0/90)°$ material showed the AE buildup immediately following the application of load. Fitz-Randolph, et al. [16] have shown the steady increase of AE with deflection for unidirectional boron-epoxy composite.

Composite materials generally exhibit a variety of failure modes including matrix crazing or microcracking, debonding, fiber failures resulting from statistically distributed flaw strength, delamination and void growth [16]. Some of these events, prior to failure, will be clearly detectable in both EE and AE.

The basic requirement for detecting fracture events with EE (or PIE) is that the newly created fracture surfaces are in some manner in communication with the vacuum so that the particles can escape from the sample and be detected. Thus, correlations or lack of correlations provide information on the mechanisms leading up to failure.

For example, in Figure 10, for the $(0°)_{16}$ graphite-epoxy system at the early stages of loading and AE buildup, shear and delamination is the dominant failure mechanism; and the main source of AE is from interlaminar shear. Because of the statistical nature of the strength of the fibers, some may fracture at a very low stress in tension which will also contribute to the AE count rate. Loose fibers at the edges may break at any time during loading and will produce both AE and EE bursts simultaneously. As the loading advances, interlaminar shear and internal delamination will lead to AE. Matrix crazing, in the tension side of the sample

and separation of tiny bundles of fibers from the tension side of
the sample will all contribute to simultaneous AE and EE. Thus, the
slow buildup of EE prior to failure is attributed to small microcracks formed on the surface air tension. The bursts of EE prior
to failure are considered to be due to "larger" events such as edge
cracking or bundles fracturing on the front surface. Finally, the
test specimen fails catastrophically (where the load drops)
accompanied by large bursts of AE and EE occurring together. One
can frequently see several plys failing successively.

Even though some of the composite failure mechanisms
described above also apply to angle ply laminates, transverse
cracking and interfacial failure will predominate. For ($45°$)$_{16}$
interfacial failure is the main failure mechanism. This is seen in
Figure 11 where we observe large bursts of AE and EE simultaneously. Presumably, interlaminar shear contributes to the
continuous AE buildup. One interesting feature of AE and EE data
from $(0,90,90,0)°_{16}$ samples is the AE buildup without any
appearance of EE prior to failure. Large interlaminar shear
deformation and failure will occur in $90°$ (interior) laminates
prior to the failure of $0°$ (exterior) laminates (Figure 12). These
internal events apparently cannot be seen with EE due to their
being internal to the sample.

The results of these experiments indicate that with EE it is
possible to detect micro-fracture events such as microscopic separation of tiny bundles of fibers, interfacial failure and matrix
crazing in fiber-reinforced composites. Even though the EE technique is not able to detect internal failure such as interlaminar
shear failure, it provides evidence of failure at early stages of
fracture. Also, it clarifies the source of AE as a function of
strain by the presence or absence of AE-EE correlations. Finally,
comparisons of the techniques tell precisely the onset times for
internal and external failure.

<u>EE From Particulate-Filled Epoxy:</u> Another form of reinforced
plastics which have gained popularity are particulate filled
plastics. Particles of silica or alumina are incorporated into
plastics primarily because of their low cost. In addition, some
material properties may improve to some extent. In our studies, we
examined EPON-828 epoxy (Z-hardener) filled with irregularly shaped
alumina particles. This material is quite strong and brittle, so
we fractured most of the samples in a three-point flexure mode.
The cross-section of the sample was 2 mm x 6 mm. A typical EE
curve plotted on a log scale is shown in Figure 13, where t = 0
corresponds to the instant of failure. The material for this emission curve is filled at an Al_2O_3/epoxy ratio (α) of 3 to 1 by
weight.

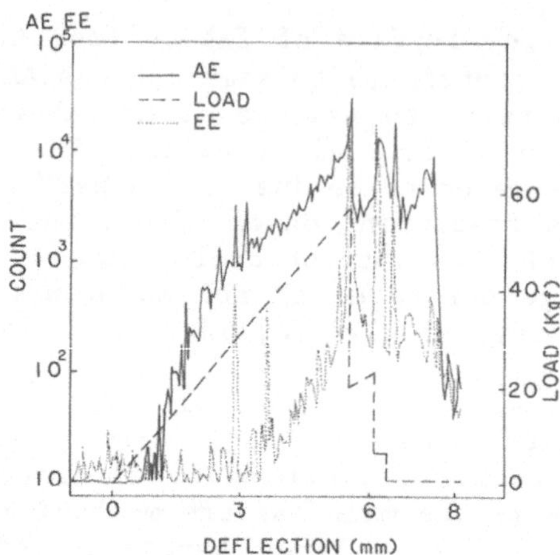

Fig. 10.　The EE, AE, and load accompanying the flexural straining of 16 layer, unidirectional graphite-epoxy composite. (Union Carbide Thornel 300 graphite fiber and NARMCO 5209 epoxy resin).

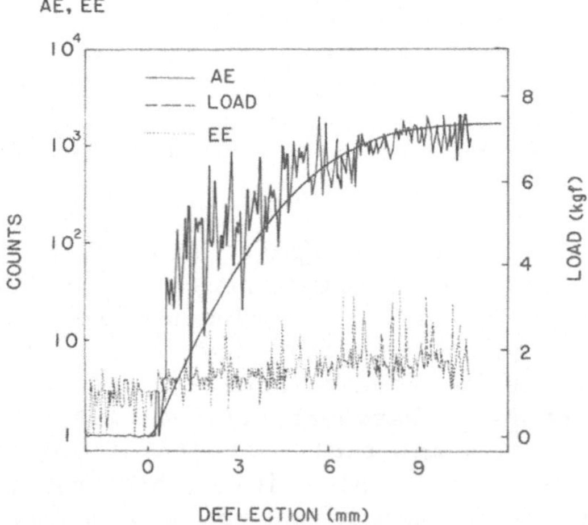

Fig. 11.　The EE, AE, and load accompanying the flexural straining of 16 layer (±45°), graphite-epoxy composite. (Union Carbide Thornel 300 and NARMCO 5208 epoxy resin).

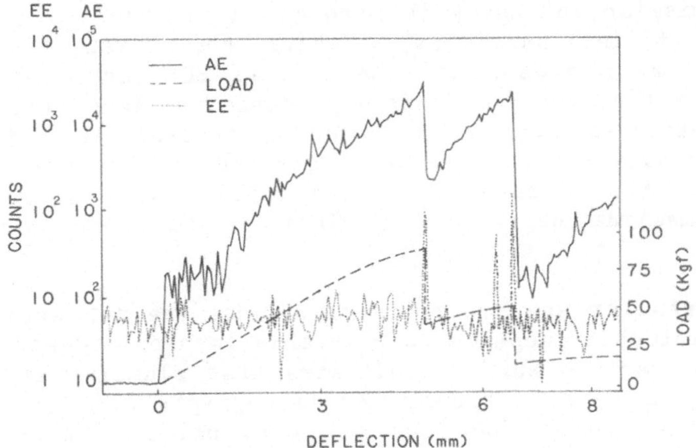

Fig. 12. The EE, AE, and load accompanying the flexural straining
of 16 layer, cross ply (0, 90, 90, 0)° graphite-epoxy
composite. (Union Carbide Thornel 300 graphite fiber and
Fiberite 934 epoxy resin).

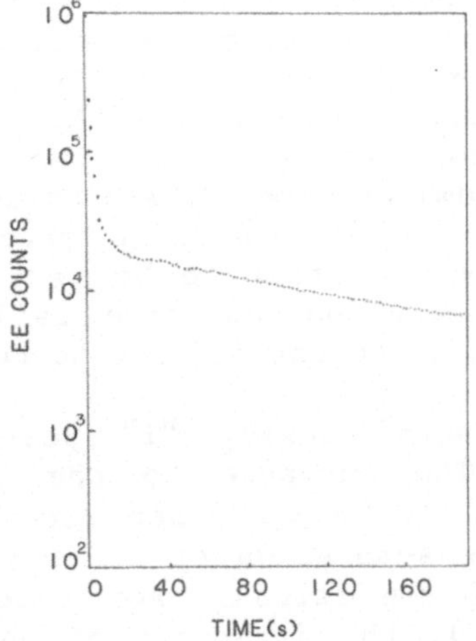

Fig. 13. Typical EE curve plotted on a log scale from the fracture
of an alumina particle filled epoxy.

The emission intensity is strongly influenced by the concentration of filler particles. Taking the first channel (0.8 sec/channel) as a measure of the initial EE count rate vs. the Al_2O_3/epoxy ratio, one sees this dependence in Figure 14. The total emission (measured over a duration of several hundred seconds) follows basically the same curve. Compared to the unfilled material ($\alpha = 0$), the EE intensity rises rapidly as α increases, and reaches a maximum near $\alpha = 1$. This is followed by a slower decline.

These results are preliminary, and we are not entirely sure why the EE intensity depends on α in this fashion. Optical inspection of the fracture surface indicates that alumnia particles are indeed being exposed, although SEM micrographs are far less convincing. Thus, we are not sure at this point of the degree of interfacial failure that is occurring. Secondly, as α increases, the mechanical parameters such as fracture energy, surface energy, and the degree of interfacial failure are bound to change. We are obviously interested in correlating these mechanical properties with the resulting EE.

<u>Photon-Emission Measurement From Filament-Epoxy Strands</u>: We have performed in air a number of experiments on the phE from epoxy strands of filaments with a strand cross-section of 0.5 mm . Figure 15 shows the visible phE vs. time during the straining and failure of epoxy strands of Kevlar, E-glass, and graphite. Several show phE prior to failure, possibly due to crack formation on a surface visible to the photomultiplier, or to chemiluminescence as observed by George and Pinkerton [18], and Fanter and Levy [19]. The decay that we observe in these curves for the phE following fracture is within the time constant of the electrometer used to measure the photon detector current. Although the cause of the major burst of phE during fracture is unknown, we suspect, as with many cases of tribo-luminescence, that breakdown is occurring between charge patches due to the high potentials produced by charge separation. We expect this to be particularly intense at instances when delamination and adhesive failure are occurring. Further experiments need to be carried out to confirm this.

PhE was also measured during "T" peel tests of two-ply Kevlar-epoxy panels. The entrance to the photomultiplier was approximately 2 cm from the "crack", and directed toward it. PhE was observed only during separation of the plys and decayed immediately upon release of the stress. For a constant area of new fracture surface (5 cm), the intensity of phE per unit area of fracture surface was found to depend strongly on the crack velocity, defined as the linear rate of the creation of new surface (cm/sec). Figure 16 shows the phE for a typical delamination. Figure 17 shows this dependence where the ordinate represents the

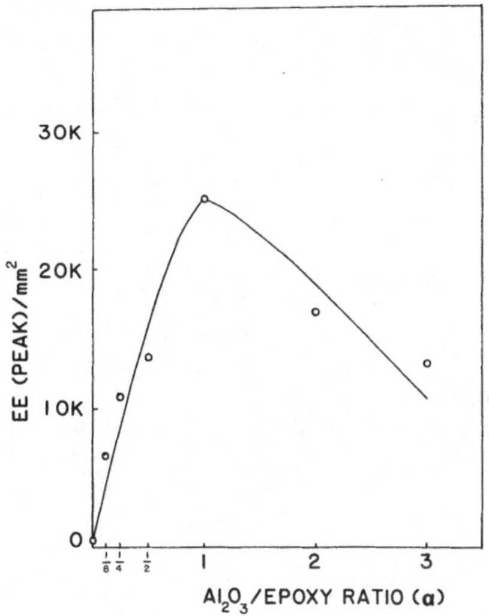

Fig. 14. Peak EE as a function of the Al_2O_3/epoxy ratio, α.

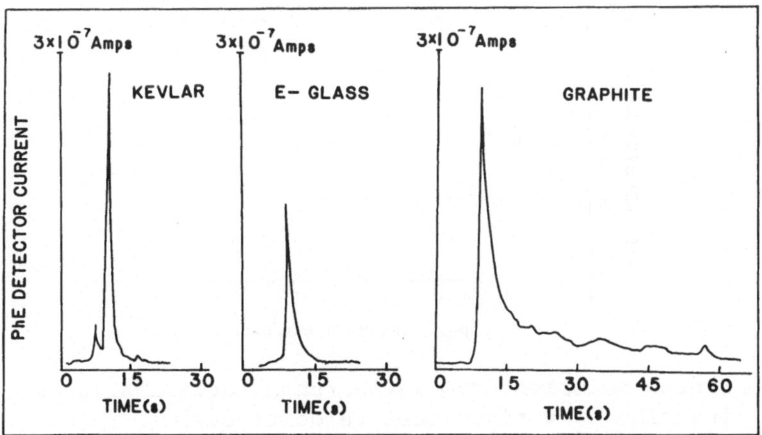

Fig. 15. Photon-emission accompanying the fracture of Kevlar, E-Glass, and Graphite epoxy strands.

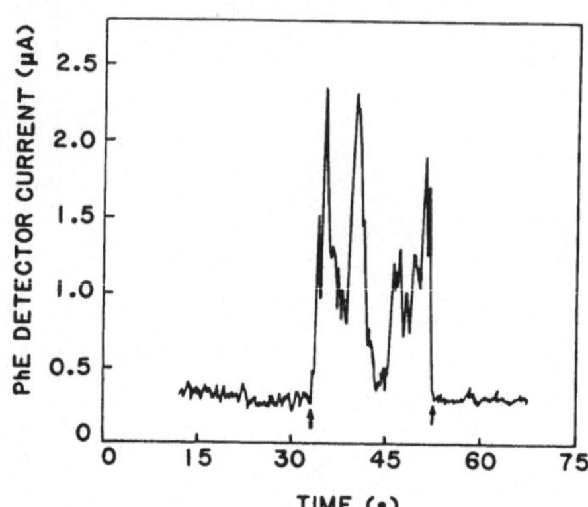

Fig. 16.   Photon-emission from the delamination of a Kevlar/epoxy
           Composite.

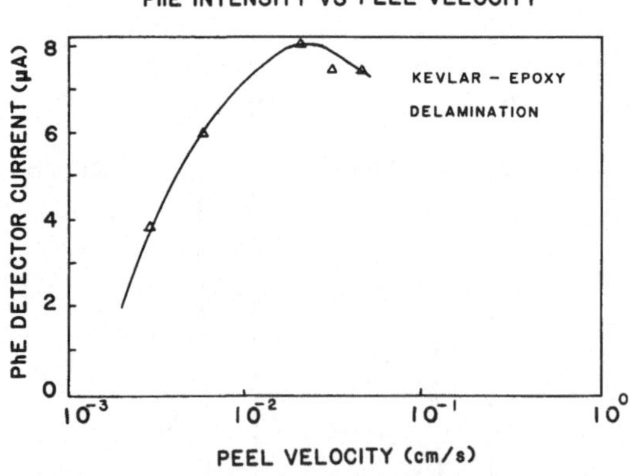

Fig. 17.   Photon-emission from Kevlar/epoxy delamination as a
           function of various peel velocities.

area under the emission curves for various velocities. The light intensity tends to increase for more rapid separation of the two surfaces, with a saturation occurring at a velocity of 10^{-1} cm/s.

CONCLUSIONS

We have tried to show a variety of FE results on a number of systems involving composites, in particular where adhesive failure at interfaces is occurring, and we have tried to indicate some of the parameters that are influencing this emission. The need for careful studies of the physics and chemistry of these phenomena is obvious. The usefulness of FE as a tool for NDT or for investigation of failure mechanisms requires a broad based attack combining fracture mechanics, materials science, and fundamental fractoemission studies on materials of mutual interest. Potential areas of usefulness of FE in studying composite failure include the following:

1. Using FE as a probe of crack growth on an extremely wide range of time scales. This need not be catastrophic fracture and might involve crazing, micro-cracking, linking of microcracks, and other pre-failure events.

2. The energies of the FE components may serve as a measure of the density of the charge distributionsd created on the fracture surface and relate to debonding parameters between fiber and resin.

3. FE may serve as a way to measure the surface temperature at the crack tip by careful modeling of the emission curves at short times after fracture. Our modeling to date has required an elevated temperature of fracture that decays quickly away [14].

4. FE may serve as a means of measuring instantaneous crack velocity. Certainly the instant of crack formation, and the onset and duration of dynamic crack growth can be measured readily.

5. FE may serve as a probe of the locus of fracture in composite materials and in illuminating failure mechanisms.

6. FE may serve as an NDT tool, perhaps in conjunction with acoustic emission. FE should be particularly useful when sensitivity to events near the surface is desired.

7. FE may be related in important ways to fracture mechanics parameters such as surface energy, fracture strength, or fracture toughness. If reliable connections could be made to such parameters, FE might be used to measure them.

ACKNOWLEDGEMENTS

First we wish to thank our Washington State University colleague, Ed Donaldson, for his helpful discussions and contributions. We also wish to thank those people who have contributed specimens used in these studies, particularly R. L. Moore, Lawrence Livermore Laboratory, for samples of filaments and fiber/epoxy strands. We are also appreciative of interest and advice from O. Ishai, A. Gray, L. C. Clements, and H. Nelson of the NASA-Ames Research Center, and W. D. Williams, Sandia National Laboratories.

This work was supported by the Office of Naval Research contract N00014-80-C-0213, National Science Foundation Grant DMR-8210406, Sandia National Laboratories, NASA-Ames Research Center, and a grant from the M. J. Murdock Charitable Trust.

REFERENCES

1. J. T. Dickinson, P. F. Braunlich, L. Larson, and A. Marceau, Appl. Surf. Sci. 1, 515 (1978).
2. D. L. Doering, T. Oda, J. T. Dickinson, and P. F. Braunlich, Appl. Surf. Sci. 3, 196 (1979).
3. L. A. Larson, J. T. Dickinson, P. F. Braunlich, and D. B. Snyder, J. Vac. Sci. Technol. 16, 590 (1979).
4. J. T. Dickinson, D. B. Snyder, and E. E. Donaldson, J. Vac. Sci. Technol. 17, 429 (1980).
5. J. T. Dickinson, D. B. Snyder, and E. E. Donaldson, Thin Solid Films 72, 225 (1980).
6. J. T. Dickinson, E. E. Donaldson, and D. B. Snyder, J. Vac. Sci. Technol. 18, 238 (1981).
7. J. T. Dickinson, E. E. Donaldson, and M. K. Park, J. Mat. Sci. 16, 2897 (1981).
8. J. T. Dickinson and L. C. Jensen, J. Polymer Sci. Polymer Physics Ed. 20, 1925 (1982).
9. J. T. Dickinson, M. K. Park, E. E. Donaldson, and L. C. Jensen, J. Vac. Sci. Technol. 20, 436 (1982).
10. J. T. Dickinson, L. C. Jensen, and M. K. Park, J. Mat. Sci., 17, 3173 (1982).
11. J. T. Dickinson, L. C. Jensen, and M. K. Park, Appl. Phys. Letters 41, 443 (1982).
12. J. T. Dickinson, L. C. Jensen, and M. K. Park, Appl. Phys. Letters 41, 827 (1982).
13. H. Miles and J. T. Dickinson, Appl Phys. Letters 41, 924 (1982).
14. J. T. Dickinson, to appear in Proceedings of the Symposium on Recent Developments in Adhesive Chemistry, ACS Seattle, 1983.
15. J. T. Barnby and T. Parry, J. Phys. D: Appl. Phys. 9, 1919
16. J. Fitz-Randolph, D. C. Phillips, P. W. R. Beaument, and A. S. Tetelman, J. of Mat. Sci. 7, 289 (1972).
17. C. K. H. Oharan, J. Eng. Mat. & Tech. 100, 233 (1978).

18. G. A. George and D. M. Pinkerton, Proceedings of a Critical
     Review of Characterization of Composites, June 8-10, 1981,
     Massachusetts Institute of Technology, (Office of Naval
     Research, 666 Summer St., Boston, MA, 1981).
19. D. L. Fantor and R. L. Levy, in ACS Symposium Series No. 95,
     Durability of Macromolecular Materials, R. K. Iby, Editor,
     p. 211 (American Chemical Society, Washington, D.C. 1979).

SURFACE TREATMENT OF CONDUCTIVE CARBON BLACKS AND THEIR EFFECT

ON THE PROPERTIES OF CONDUCTIVE COMPOSITES

P. Datta and R. N. Friel

RCA Laboratories
Princeton, NJ 08540

ABSTRACT

Methods have been developed for surface treatment of
conductive carbon blacks. The surface function groups in
conductive carbon blacks were analyzed using Fourier transform IR
spectroscopy. The surface treatment methods have been developed on
the basis of the observed surface functional groups. In this
method, the surface functional group of the conductive carbon
blacks are reacted with fatty acid chlorides in a solvent
suspension. Qualitative observation indicated improved compatibility between the surface treated carbon black and PVC matrix which
resulted in improvement in melt flow properties, dispersion,
electrical conductivity, and mechanical properties. A comparison
of properties between chemical treated and untreated carbon black
composites is discussed.

INTRODUCTION

Conductive carbons have been used successfully for many years
to increase the electrical conductivity of thermoplastics. However, their incorporation results in an increase in melt viscosity
and degradation of some mechanical properties of the
thermoplastics. The fabrication of moldable electronic devices
requires thermoplastics filled with a well-dispersed conductive
carbon black and having low melt viscosity with good mechanical
properties [1,2]. Plueddemann and Stark [3] have shown that
surface modification of inorganic fillers such as clay silicas and
silicates with silane coupling agents improved melt viscosity,
dispersion, and flow during molding of compounds formulated from
them. Plueddemann et al.[4] and Collins [5] described various
types of organosilane coupling agents that are effective for
inorganic fillers.

Monte and Sugarman [6] described the effect of organotitanate coupling agents on filler-polymer interactions and claim that titanates greatly improve melt viscosity and dispersion of inorganic filler-polymer composites. Burrell [7] predicted dispersion of fillers in inks from the filler wetting properties and pigment dispersion parameters. Seymour [1] has shown that surface treatment of calcium carbonate with stearic acid considerably improves the dispersion, flow, and elongation in the polyvinylchloride (PVC) systems formulated with this filler.

The surface chemistry of the carbon black particles also influences the conductive properties of the plastic composites. Carbon black is comprised of spherical particles that tend to cluster by the so-called grouping effect [1,7]. Studebaker [8] reviewed the surface chemistry of carbon black in detail. He considered carbon blacks as a series of imperfect polycyclic aromatic hydrocarbon in various states of oxidation. A large number of papers [9-11] have been published on the surface chemistry of carbon black. Spackman and Charlesby [12] reviewed the literature on the free radical nature of carbon black surfaces.

The purpose of this study is to determine the nature of the surface functional groups on the conductive carbon black and to modify the surface by reacting these functional groups with surface coupling agents. We have successfully treated Ketjen black surfaces with fatty acid chlorides. The surface treated Ketjen blacks produce reduced melt viscosity, improved dispersion, and electrical conductivity when formulated with a standard PVC compound.

EXPERIMENTAL RESULTS AND DISCUSSION

A. Surface Treatment Procedure

Ketjen black EC (KB) (a product of AKZO Chemie, Netherlands), a conductive carbon black, was treated in solvent suspension with three generic types of coupling agents. One percent by weight of coupling agent was dissolved in heptane. One-hundred fifty grams of KB pellets were placed in a high speed Waring Blender for two minutes to disaggregate the KB pellets. Three-hundred grams of 0.5% coupling agent solution were added slowly to the carbon black, and the mixture was blended for five minutes. The surface treated samples were stored in plastic bags for at least 48 h and then vacuum dried at 120°C for 16 h in a vacuum oven in an open pan. The dried materials were stored in polyethylene bags.

The coupling agents used for this study were:

a) <u>Silane Coupling Agents</u>: Dow Corning Z-6020 (N-2-amino-ethyl-3-aminopropyltrimethoxysilane), Z-6030 (γ-methacryloxypropyl-trimethoxysilane), Z-6040 (γ-glycidoxypropyltrimethoxysilane), Z-6075 (vinyltriacetoxysilane), Z-6076 (γ-chloropropyltrimethoxy-silane), and methyloctadecyldichlorosilane (MODS). Silane coupling agents are obtained from Dow Corning Corp., Midland, MI and Petrarch Systems, Inc., Levittown, PA.

b) <u>Titanates</u>: Isopropyl-tri(dioctylphosphato) titanate (KR-12), isopropyltri(dioctylpyrophosphato) titanate (KR-38S), te-tra(2,2 diallyloxymethyl-1) butoxy titanium di(di-tridecyl)phos-phite (KR-55). All organotitanates are a product of Kenrich Petro-chemicals, Inc., Bayonne, NJ.

c) <u>Organic Functional Groups</u>: Stearoyl chloride (StCl), deca-noyl chloride (DCl), stearic acid, octadecyl alcohol (ODAL), and octadecylamine. These organic chemicals are obtained from Fisher Scientific Company, Springfield, NJ and used without purification.

Fourier transform IR spectra of the pellets were taken with a Digilab FTS-14 spectrometer. Spectra were taken using a TGS detec-tor with 2000 scans at a resolution 8 cm^{-1}.

A C. W. Brabender plastography dynamometer was used for the evaluation of the melt properties of surface treated KB-PVC compo-sites. Each compound was weighed according to the given formu-lation. The compound was then blended in a Waring Blender for four one minute periods.

Rheology measurements were performed with an Instron capillary extrusion rheometer using a die of 0.050 inch diameter and 2.0089 inch length with an L/D ratio of 40. The included entry angle of the die is 90°. All measurements were performed at 200°C using plunger velocities of 0.030-10.0 in/min. Generally, two separate runs were made, and the measured values were averaged. The apparent melt viscosities were calculated from shear stress and shear rate data.

Transmission electron micrographs of the thin section of the compounds were taken using Philips-300 at 100 kV.

A button-shaped sample was cut out of the conductive plate and then samples were metallized with 200 nm gold on both sides and connected to an automatic network analyzer interfaces with a desktop computer. The network analyzer measured the reflection co-efficient which was immediately converted by the computer to the resistivity and dielectric constant.

Dynamic mechanical properties of the carbon black filled PVC compound and surface treated carbon black filled compound were determined using a Rheometrics dynamic mechanical spectrometer. The measurements were performed at 10°C intervals from −120°C to 100°C at a frequency of 1 Hz.

B. Confirmation of Surface Treatment

The surface compositions of the untreated and surface treated Ketjen blacks were studied using Fourier transform IR spectroscopy (FTIR). The untreated and surface treated Ketjen blacks (KB) were vacuum dried at 120°C for four hours and then mixed with KBr to make a pellet. Figure 1 shows the spectra of stearoyl chloride treated Ketjen black. The spectrum of the untreated KB showed strong bands at 3950 cm^{-1}, 1715 cm^{-1}, and weak bands at 1360 and 1225 cm^{-1}. The band at 3950 cm^{-1} suggested the presence of either a phenol hydroxyl or inorganic hydroxyl. The band at 1715 cm^{-1} suggested the presence of carbonyl functional groups. The bands at 1360 and 1225 cm^{-1} are not assigned at present. The stearoyl chloride treated KB showed a relatively featureless broad band at 4000-3200 cm^{-1} and strong bands at 3000-2800 cm^{-1} and 1750 cm^{-1}. The strong bands at 3000-2800 cm^{-1} and 1750 cm^{-1}, respectively, are associated with aliphatic hydrocarbon and carbonyl ester functional groups. The characteristic band of the acid chloride group were not observed. This suggests that the acid chloride groups and the KB hydroxyl groups reacted with subsequent formation of ester functional groups.

C. Brabender Plastograph Study of Surface Treated KB in Polyvinyl-
chloride (PVC)

The compositions of standard PVC compound and surface treated KB in PVC compound are summarized in Table 1. Fifty gram samples were weighed out and fed into the Brabender using the ram chute assembly. The bowl temperature was kept at 150°C and all experiments were run at 33 rpm. The torque on the motors was plotted against time.

Fusion torque, equilibrium torque, material temperature, and bandwidths of the surface treated KB in PVC compound and conventional KB filled PVC and composition of unfilled PVC compounds are also summarized in Table 1. The equilibrium torque of the various surface treated KB's in the PVC compound was compared with conventional KB filled and unfilled PVC compound. The equilibrium torques of all the PVC compounds containing KB treated with organosilanes and organotitanates with the exception of KR 55 were found to be similar to that of untreated KB in the PVC. The PVC containing KB treated with stearic acid, stearic amine, and octadecyl alcohol has 200 to 400 m/gm lower equilibrium torque than untreated KB in the PVC. These polar fatty acid derivatives are

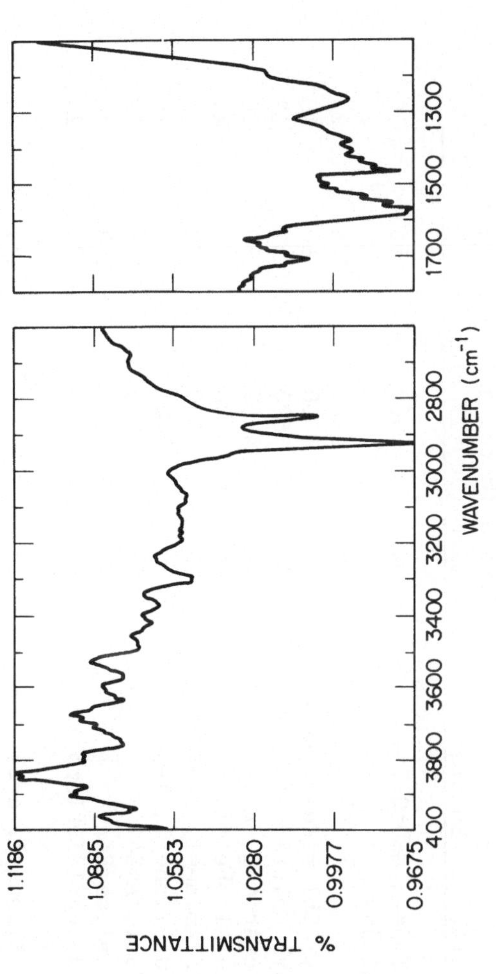

Fig. 1. A Fourier transform IR spectrum of KB treated with stearoyl chloride.

Table 1. Brabender Mixing Head Torque Data on Solvent Suspension Surface Treated Ketjen Black in a Standard PVC Compound A.

| Composition Parts per Hundreds | Torque (meter/gm) | | Material Temp. | | Bandwidth |
	Fusion	Equilibrium	Equil. °C	$\Delta T = T_E - T_B$	(m/gm)
A, AP480=95, T35=2, G30=0.75, G70=0.25, K175=2	6500	3500	156	6	100
KB=15, A=85	7800	5200	176	26	350
KB=15, Stearoyl Chloride=1, A=84	5300	3400	160	10	225
KB=15, Decanoyl Chloride=1, A=84	5800	3600	162	12	230
KB=15, Stearic Acid=1, A=84	6700	4900	170	20	275
KB=15, Stearic Alcohol=1, A=84	6900	5000	175	25	295
KB=15, Stearic Amine=1, A=84	6850	4850	173	23	300
KB=15, DC-6020 Silane-1, A=84	8400	5000	178	28	230
KB=15, DC-6075=1, A=84	8200	5200	178	28	375
KB=15, DC-6040=1, A=84	800	500	170	20	300
KB=15, Methyldodecyldichlorosilane =1, A=84	8500	5100	175	25	275
KB=15, UC-174 Silane=1, A=85	8100	5000	175	25	280
KB=15, KR-12 Titanate=1, A=84	8000	4900	170	20	280
KB=15, KR-34S Titanate=1, A=84	8000	5100	170	20	300
KB=15, KR-55 Titanate=1, A=84	5800	3600	164	14	250

good internal lubricants and reduce the melt viscosity of the carbon filled composites. KB treated with fatty acid chloride showed considerable activity. Stearoyl chloride and decanoyl chloride treated KB in the PVC have 1000 and 1500 m/gm lower equilibrium torque than conventional KB in the PVC compound. The equilibrium material temperature for the PVC containing the acid chloride treated KB was found to be 15°C lower than the same compound formulated from the other surface treated KB´s and untreated KB. These results indicate that KB treated with a fatty acid chloride produces lower frictional heat and reduces the melt viscosity of the filled PVC composite.

D. Instron Capillary Rheometric Melt Viscosity Results

Figure 2 shows the apparent melt viscosity as a function of shear rate for PVC compounds containing stearoyl and decanoyl chloride treated KB at a temperature of 200°C. The melt viscosity curves of the conventional KB filled and unfilled PVC compounds are included for comparison. The melt viscosity of all the PVC compounds containing KB treated with organosilanes was found to be equivalent to that of conventional KB in the PVC compound and were not included in the figure. Organotitanates with long hydrocarbon chain slightly lower the melt viscosity of the composites. The melt viscosity of organotitanate (KR 38S) treated KB composites is also shown in Figure 3. The compounds, containing KB treated with fatty acid, amine and alcohol, have similar melt viscosities to that of untreated KB containing compounds. The melt viscosity of the compound containing KB with fatty acid chloride is significantly lower than the melt viscosity of the conventional KB filled compound.

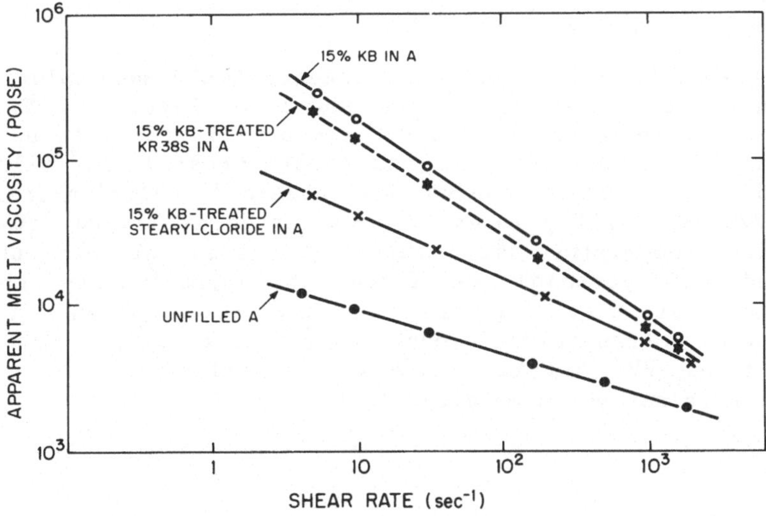

Fig. 2. Apparent melt viscosity vs. shear rate at 200°C for Compound A containing KB treated with stearoyl chloride.

E. Transmission Electron Micrographic Study of Dispersion

Transmission electron micrographs of thin sections (100 nm) of PVC compound, containing 15% conventional KB and the PVC compound containing 15% stearoyl chloride treated KB, were compared. Figure 3 shows transmission electron micrographs of these materials at 2800X. The stearoyl chloride treated KB displays better dispersion. The compound derived from 15% conventional KB displays large carbon-free insulating regions which account for the poor conductivity. The surface treated KB seems to display better dispersion due to displacement of hydrophilic polar hydroxyl groups from the surface with a hydrophobic hydrocarbon. The hydrophobic groups could reduce the amount of water which glues the carbon particles together and reduces the grouping effect [1,7].

F. Electrical Resistivity

AC and DC resistivities were determined for the composites containing 15% surface treated KB and conventional KB. The resistivity of carbon loaded PVC's has been found to exhibit a strong frequency dependence at high resistivities for composites with lower percent carbon. It has been found that the frequency dependence becomes less pronounced at high (15%) loading of KB. The resistivities of conductive composite at 15% KB loading are reported in Table 2. The resistivity of compounds derived from KB treated with fatty acid chloride is lower than those from untreated KB. All compounds from KB treated with organosilane, organotitanates, fatty acid, amine and alcohol have similar resistivity to that of the compound with untreated KB.

G. Dynamic Mechanical Properties

The shear modulus and loss tangent (G'G") were plotted as a function of temperature and are shown in Figure 4. The shear modulus of untreated KB filled PVC compound is found to be similar to that of silane and organotitanate-treated KB filled PVC compound. The shear modulus for stearoyl chloride-treated KB filled PVC is slightly lower than untreated KB filled PVC. Glass transition temperature for surface treated KB and untreated KB-filled PVC compound is found at approximately the same temperature as that for the unfilled PVC matrix. A slight variation of β-transition is observed for treated KB, untreated KB, and unfilled PVC matrix. However, detailed analysis of this transition is not yet completed.

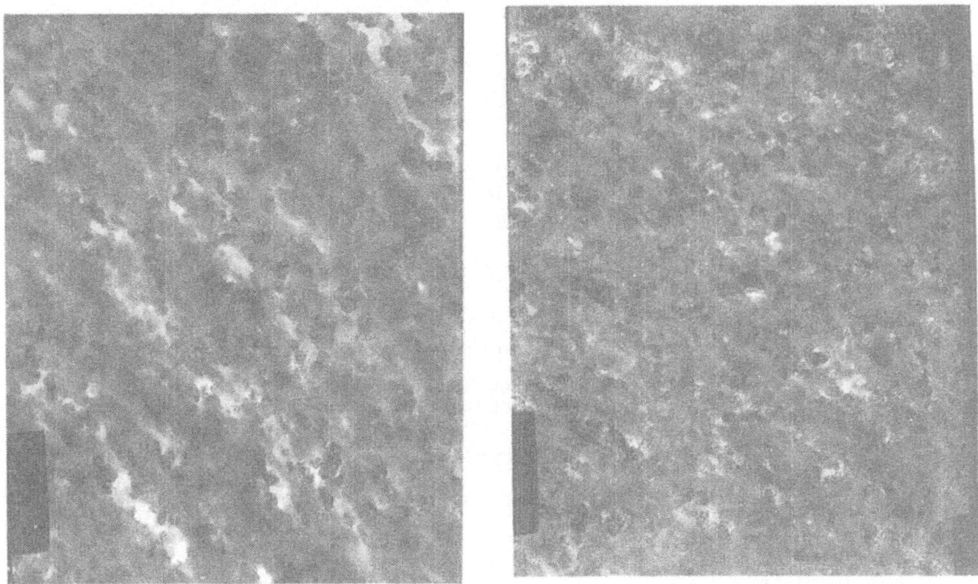

Fig. 3.   Transmission electron micrograph at 2800X of 15% weight
          of (A) conventional KB in A, and (B) stearoyl chloride
          treated KB in A.

Table 2. Resistivity Data on Solvent Suspension Surface-Treated
Ketjen Black in a Standard PVC Compound A

Compositions	Resistivity Ω-cm 900 MHz	DC
KB=15, A=85	4-6	4-6
KB=15, Stearoyl Chloride=1, A=84	0.4-1.5	0.3
KB=15, Decanoyl Chloride=1, A=84	1.5-2.5	2.5
KB=15, Stearic Acid=1, A=84	4.5	4.5
KB=15, Stearic Alcohol=1, A=84	5.6	6.2
KB=15, Stearic Amine=1, A=84	5	5
KB=15, DC-6020 Silane=1, A=84	3.5	3.4
KB=15, DC-6075 Silane=1, A=84	5.2	5.2
KB=15, UC-174 Silane=1, A=84	4.8	4.8
KB=15, KR-12 Titanate=1.0, A=84	5.0	5.1
KB=15, KR-38S Titanate=1.0, A=84	2.5	2.5
KB=15, KR-55 Titanate=1.0, A=84	3.5	4

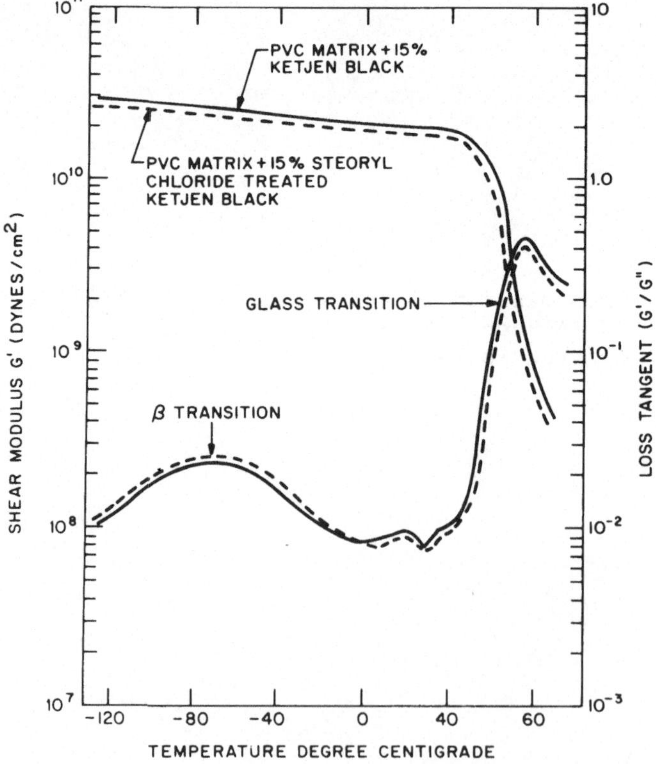

Fig. 4. Dynamic shear modulus and tan δ for 15% weight of (A) conventional KB in A, and (B) stearoyl chloride treated KB in A.

CONCLUSIONS

1. Fourier transform infrared spectroscopy is a feasible means for identification of surface functional groups on carbon blacks. Phenolic hydroxyl and carbonyl groups are identified on Ketjen black.

2. Fatty acid chlorides seem to react with hydrophilic surface groups on Ketjen black. The appearance of ester bands in the infrared spectra suggests that the acid chloride reacts with oxygen-containing functional groups on the KB surface.

3. The use of KB treated with fatty acid chlorides significantly improves the melt flow properties of carbon black filled PVC compound. KB treated with fatty acids and their other derivatives does not have a significant effect on the melt viscosity of the filled PVC compound.

4. KB surface treated with organosilanes and organotitanates did not produce any effect on the melt viscosity of the PVC compound except for the organotitanate with phosphite groups which reduces melt viscosity and marginal improvement of thermal stability of PVC compound.

ACKNOWLEDGMENTS

The authors wish to express appreciation to P. J. Zanzucchi, M. D. Coutts, H. H. Kawamoto, E. L. Allen, and N. A. Arroyo for their technical assistance.

REFERENCES

1. R. B. Seymour, Plastic Design and Processing 16, 7-11, July (1976).
2. L. P. Fox, RCA Review 39, 116 (1978).
3. E. P. Plueddemann and G. L. Stark, Modern Plastics 102, Sept. (1977).
4. E. P. Plueddemann and G. L. Stark, Proc. 28th Ann. Tech. Conf., Reinforced Plastics/Composites Inst., SPI, Section 21-E (1973).
5. W. Collins, The Modern Plastics Encyclopedia, McGraw Hill Co., NY, (1977-78).
6. S. G. Monte and G. Sugarman, Plastic Compounding 1, 56, June (1978).
7. H. Burrell, Polymer Preprints, ACS 35, 18-30 (1975).
8. M. L. Studebaker, Rubb. Chem. Tech. Note 30, 1400 (1957).
9. A. E. Austin and E. Adelson, Proc. Fifth Conf. on Carbon 1, 485 (1962).
10. M. L. Studebaker, Proc. Fifth Conf. on Carbon 2, 189 (1962).

11. A. V. Kiselev, G. A. Kozlov, and V. I. Lygin, Colloid J. USSR
     $\underline{28}$, 329 (1966).
12. J. W. C. Spackman and A. Charlesby, Proc. Fourth Rubber Tech.
     Conf., Inst. of the Rubber Industry, London, 274 (1963).

SURFACE MODIFIED ALUMINUM-POLY(VINYL ACETATE) INTERACTION IN THE
PRESENCE OF WATER

Kenneth E. Nietering and Wilmer G. Miller

Department of Chemistry
University of Minnesota
Minneapolis, MN 55455

ABSTRACT

The interaction of poly(vinyl acetate) with modified aluminum
surfaces has been investigated using the spin label technique to
monitor motion at the polymer-metal interface as a function of
temperature, coating thickness and exposure to water. The modified
aluminum surfaces used for adhesive bonding, designated CAA and
PAA, are prepared by anodization which gives an amorphous aluminum
oxide to which the polymer binds. A monolayer of adsorbed polymer
on either surface stays bound in the dry state to at least 150°C.
Upon exposure to water, the debonding was observed to depend on the
surface employed, and the method and time of exposure. The effect
of water is far less pronounced in the PAA than in the CAA
material. We conclude that debonding at room temperature precedes
bulk hydration of the oxide surface.

INTRODUCTION

The adhesive bonding of aluminum is a well-established technology. In preparation for bonding, the aluminum is cleaned and
surface modified by one of several techniques. Three of the more
commonly used techniques are the Forest Products Laboratory (FPL)
process, which involves an acid etch in dichromate [1], the chromic
acid anodization (CAA) process, which involves anodization in
aqueous CrO_3 after FPL treatment [2], and the phosphoric acid
anodization (PAA) process, which involves anodization in aqueous
phosphoric acid after FPL treatment [3]. Each of these treatments
leaves an amorphous aluminum oxide coating on the aluminum. The
morphology [4] and the resistance to attack by water [5,6], however, depend on the method of treatment.

The failure of adhesively bonded aluminum in the presence of water can occur cohesively in the oxide layer or in the organic adhesive; alternatively, it can occur adhesively at the adhesive-oxide or at the oxide-metal interface. The adhesive-oxide interaction is typically monitored by bulk property measurement, e.g., the peel test. We wish to report a study on adhesive-oxide interaction in the presence and absence of water, using primarily a technique which monitors polymer motion at the submolecular level at the interface. Only the PAA and CAA oxides were investigated. Although the typical adhesive in aluminum bonding is an epoxy, our studies are focused on linear poly(vinyl acetate). The technique employed to monitor motion is the spin-labeling technique, whereby a low level of a stable free radical (1-10 per polymer molecule) is covalently incorporated in the adhesive. The line shape of the electron spin resonance (ESR) spectrum is sensitive to the rotational motion of the label over a wide variety of rotational correlation times. Thus, if a monolayer or less of labeled polymer is deposited on the oxide surface, the effect of the surface on the polymer segmental dynamics can be monitored as a function of temperature, overlay of unlabeled adhesive, and the presence of environmental variables such as water. We have used the technique previously in a variety of other applications [7-11].

SAMPLE PREPARATION

The experimental approach is summarized in Figure 1. Aluminum was surface modified by anodization using either the CAA or PAA process. Electron microscopy, x-ray diffraction, ESCA and Auger electron spectroscopy analysis showed the resulting oxide surface to be amorphous with a morphology consistent with that published previously [4-6]. In the case of the PAA process the presence of $AlPO_4$ on the surface of the oxide [12] was detected as a phosphorous peak by Auger. In most cases the base aluminum was removed by dissolution in a bromine-methanol solution [13]. Nitroxide spin labeled poly(vinyl acetate), PVAc, of molecular weight 194,000, prepared as previously described [10,11], was adsorbed onto the PAA or CAA oxide from a dilute PVAc solution in CCl_4 to monolayer coverage over a period of 24 hours or longer. After exhaustive washing with fresh solvent the sample was vacuum dried. The oxide at this point was coated with less than 5 nm and perhaps as low as 1 nm of labeled polymer. These samples are designated "thinly coated oxides." In some instances the samples were overcoated with unlabeled PVAc by compression molding at 70-80°C with a three to one PVAc to oxide weight ratio. The mean thickness of the unlabeled PVAc overcoat was many micrometers. These samples were designated "thickly coated oxides." Details of the sample preparation and characterization can be found elsewhere [14].

RESULTS

Temperature Dependence-Coated Dry Oxides

The temperature dependence of the ESR spectrum from thinly-coated PAA is shown in Figure 2, where little observable change in ESR line shape, hence, nitroxide motion, is discernible to 142°C. Similar results were obtained with thinly-coated CAA. These results are similar to analogous studies on thinly-coated α-alumina and boehmite [15], and in contrast to studies on fumed silica [11], which binds the PVAc much more weakly than the various aluminas.

In thickly coated CAA samples heated to 90°C for approximately 30 min, the labeled polymer did not diffuse from the alumina surface into the unlabeled polymer matrix. This result was also observed in thickly coated α-alumina [15], again in contrast to studies on fumed silica [11], where diffusion from the oxide surface into the bulk polymer was observed to occur rapidly (less than 4 min). These results show that all aluminas which we have studied tightly bind PVAc to rather high temperatures (to at least 150°C), and that the diffusion of the surface attached polymer into a bulk polymer overlay is slow.

Thinly Coated Oxides - 100% Relative Humidity

Thinly-coated PAA was exposed to 100% relative humidity at ambient temperature. After 112 h no observable change had occurred in the ESR lineshape, even after heating to 140°C, Figure 3a. X-ray diffraction studies showed no change in the oxide. The sample was followed for 33 days at ambient temperature with no evidence of spectral lineshape changes.

Similar studies were carried out on thinly-coated CAA. After 50 h, a small fraction of the spin labels, less that 10%, showed rapid motion, Figure 4b. The fraction of rapidly moving nitroxides increased with time. After 73 days, Figure 4c, a mixed population of slow and fast moving nitroxides were observed. Longer term studies are in progress.

Thinly-Coated Oxides-Liquid Water

The ESR spectrum of thinly coated PAA, Figure 3b, showed a small percentage of fast motion nitroxides within 5 min after the addition of liquid water. The fraction of debonded nitroxides increased with time, Figures 3c,d. Even after 129 days, however, a substantial fraction remain adherent, as can be seen from the expanded scale spectrum shown in Figure 3d.

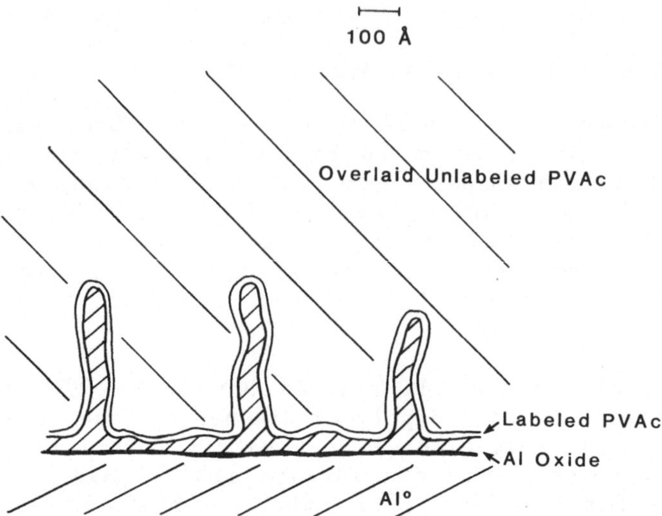

Fig. 1. Schematic representation of the systems studied, where a
monolayer of spin labeled polymer was coated onto an aluminum oxide coating, in some instances overcoated with
unlabeled polymer, and exposed to water.

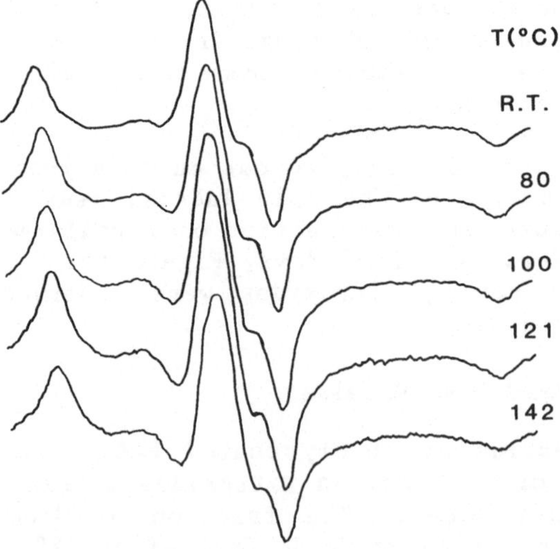

Fig. 2. The temperature dependence of the ESR spectrum of thinly-
coated PAA in the dry state at the indicated temperatures.

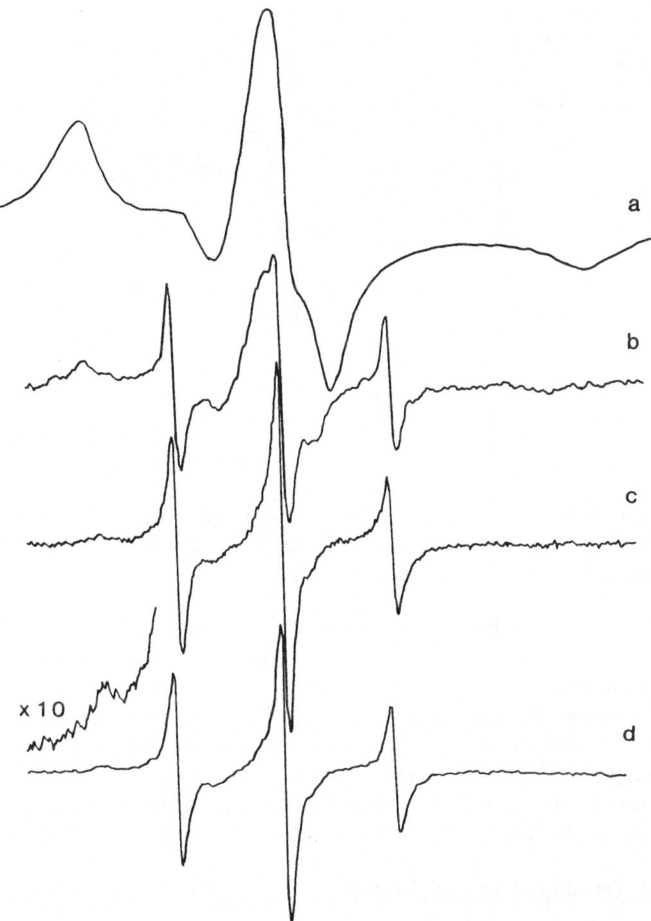

Fig. 3. ESR spectra of thinly coated PAA: (a) at 140°C after exposure to 100% relative humidity for 112 hours; at ambient temperature after (b) 5 minutes, (c) 18 days, or (d) 129 days exposure to liquid water.

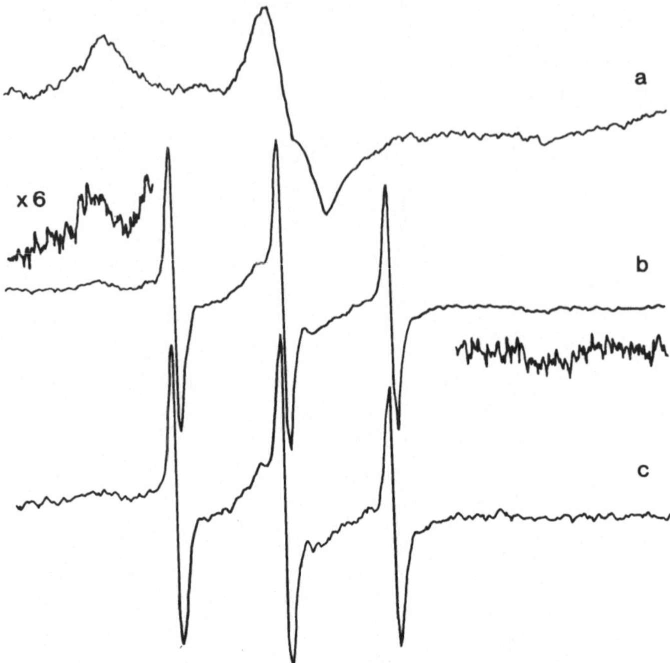

Fig. 4.   ESR spectra of thinly coated CAA at ambient temperature in
the dry state (a), or after 50 hours (b) or after 73 days
(c) at 100% relative humidity.

Similar studies with thinly coated CAA show the delamination at
ambient temperature of the polymer from the oxide was essentially
complete in less than 4 min after the addition of water, as shown
in Figure 5. X-ray diffraction indicated no measurable change in
the oxide when exposed to liquid water at ambient temperature for
one month, Figure 6, whereas a sealed sample exposed to 100%
relative humidity at ambient temperature, after heating to 160°C
for approximately 30 min, showed growth of pseudoboehmite.

<u>Thickly-Coated Oxides-Liquid Water</u>

A thinly coated CAA sample, compression molded with unlabeled
PVAc in 3 to 1 polymer to oxide weight ratio, was exposed to liquid
water at ambient temperature. After 5 days exposure the ESR spectrum, Figure 7b, showed that substantial amounts of the polymer
were still adhering to the oxide. After 17 days, Figure 7c, much
of the labeled polymer is no longer adhering to the oxide surface.
It is difficult to quantify the fraction which remains bonded due
to the noise level in this dilute spin system. It is clear,
however, that the thicker polymer coat slows down but does not stop
delamination from the oxide surface. No studies have been
completed with thickly coated PAA samples.

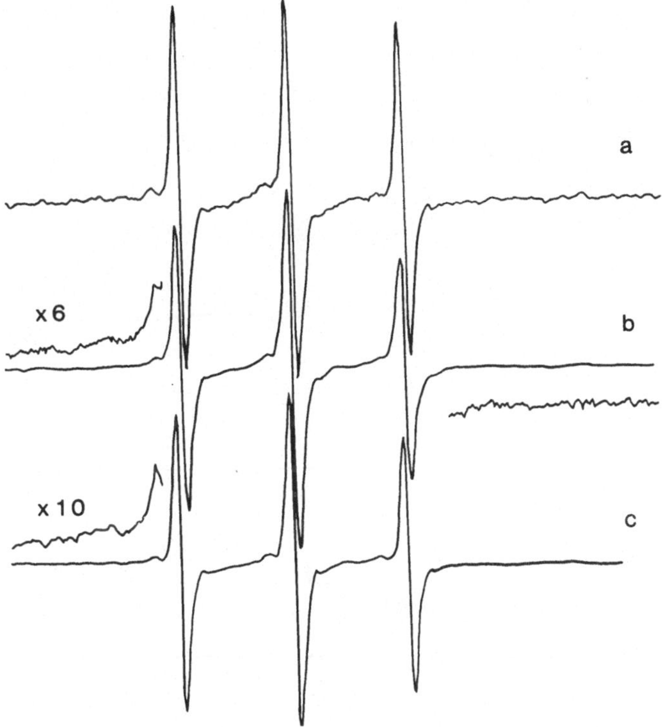

Fig. 5.   ESR spectra of thinly coated CAA upon exposure to liquid
water at ambient temperature for 4 minutes (a), 12 days
(b) or 123 days (c).

DISCUSSION

Superficially it is easy to deduce that the debonding of PVAc
from amorphous aluminum oxide depends on the preparation technique
as well as the manner of exposure to water. Closer scrutiny,
however, reveals a more complex situation. The reaction of bare
PAA and FPL prepared oxides to humid and to aqueous environments
has been studied extensively [5,6]. Hydration of the PAA oxide did
not occur over the time of study at room temperature in either a
humid or an aqueous environment. At elevated temperature (60°C)
in contact with liquid water the PAA oxide hydrates to
pseudoboehmite, and eventually aluminum hydroxide is formed on the
surface. By contrast, in 80% relative humidity at 50°C, no hydration was observable within 210 h; at 100% relative humidity, hydration did occur in a manner similar to that on contact with liquid
water, but on a longer time scale. It was concluded that the 100%
relative humidity results were caused by condensation to form bulk
water on the oxide. It has furthermore been concluded that the
degradation mechanism causing bond failure in PAA-treated adherends

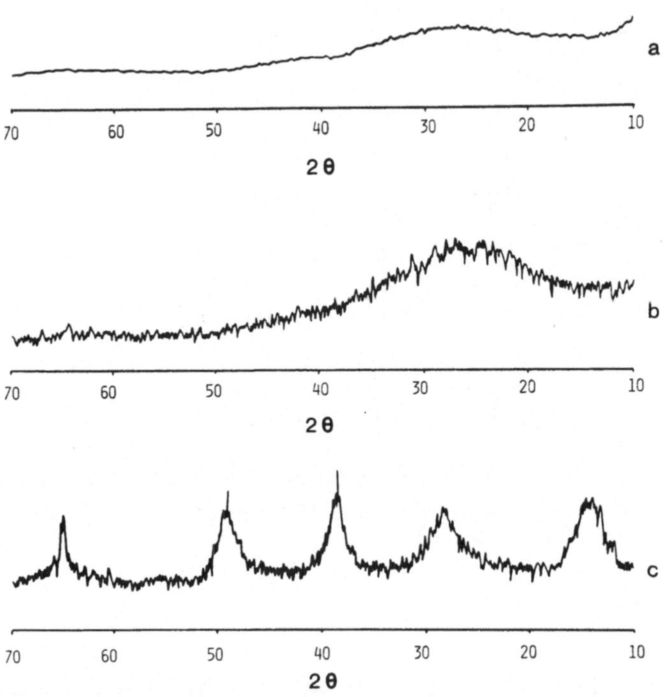

Fig. 6. X-ray diffraction of CAA before (a) and after (b) exposure
to liquid water for one month at ambient temperature, or
after (c) one day at room temperature and 100% relative
humidity, then heating to 160°C for one hour.

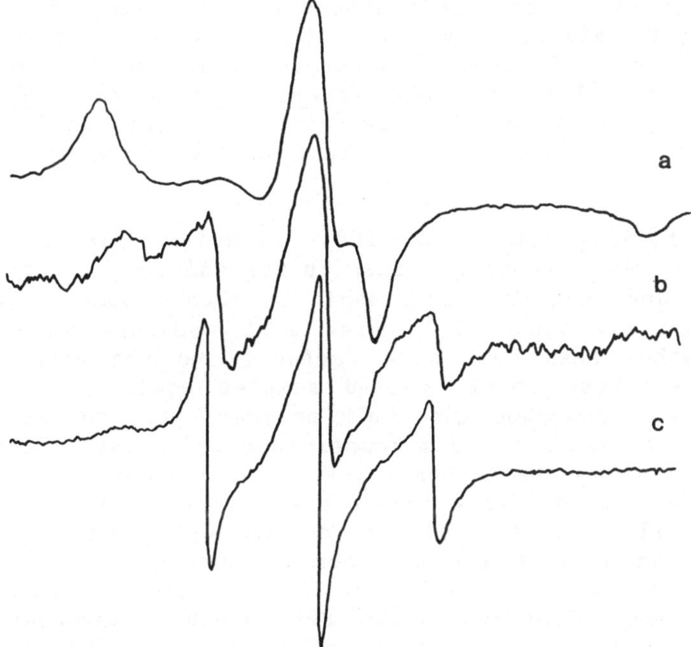

Fig. 7. ESR spectra of thickly coated CAA in the dry state (a), or after 5 days (b) or 17 days (c) exposure to liquid water at ambient temperature.

in a humid environment is the conversion of the oxide on the
aluminum to a hydroxide.

If bulk water is essential for the hydration of the oxide, it
is possible that an integral polymeric coating which does not take
up much water could slow down or prevent hydration. Bulk PVAc is
reported to absorb only 3-6 wt% water at room temperature [16].
Our sample at 100% relative humidity absorbed 4 wt%, or one water
molecule per five monomeric units, sufficient to lower the glass
transition temperature 25° as measured by scanning calorimetry.
Solid state ^{13}C NMR showed a slight sharpening of the spectrum with
the carbonyl carbon sharpening the most. Deuterium NMR at 46.06
MHz of a D_2O saturated sample showed a single line of 700 Hz linewidth at half-height, many times the linewidth in bulk D_2O. Thus
the small amount of water in saturated PVAc barely plasticizes it
and does not behave as bulk water. The water, which may be
temporally hydrogen-bonded to the carbonyl group, is in such short
supply that it could tie-up no more than 20% of the carbonyl groups
in the PVAc.

Our finding that in 100% relative humidity at room
temperature some debonding occurs in the CAA sample but not in the
PAA sample does not seem ascribable to local condensation. Local
condensation may occur, but water must condense on the polymer
surface rather than the oxide surface, and the amount of water
taken up by these thinly coated samples would have to be much
larger than is absorbed into bulk polymer. Our results on immersion in liquid water at room temperature reinforce this view. The
thinly and thickly coated CAA and thinly coated PAA show debonding,
although the rate of debonding in the thickly coated CAA sample is
much less than in the thinly coated CAA. We conclude that
debonding can take place at room temperature. This could be
ascribed to hydration of the oxide, or hydration (hydrogen bonding)
of the polymer. Inasmuch as the PVAc probably interacts with the
oxide surface through the side chain carbonyl groups, one could
envision debonding as a result of water competing with the oxide
for the carbonyls. However, with the CAA oxide sample much more
than 20% is debonded, hydration of the oxide seems more likely.
Although previous studies [6] found no hydration at room
temperature, a very thin hydration layer would not have been
detected. It is unknown if the initial hydration product is
pseudoboehmite, as is observed at later times. In previous studies
on crystalline aluminas [15] we have found that PVAc debonds at
room temperature from α-alumina when the oxide surface hydrates to
amorphous aluminum hydroxide, but does not debond when activated
alumina, a mixture of boehmite and γ-alumina, hydrates to form a
crystalline hydroxide. Although we do not understand these differences, we believe that hydration of the oxide surface occurs at
room temperature and without the presence of liquid water at the

oxide surface. The large difference between the CAA and PAA sample
results from the difference in surface composition of this oxide.
The hydration protection resulting from the PAA process has been
shown to be a result of the PAA surface containing aluminum
phosphate which may simply hydrate at a much slower rate,
particularly if the water which arrives at the surface is not
liquid-like.

In the polymer-surface systems we have studied debonding
appears to precede bulk hydration of the oxide at the interface.
If bulk water can then arrive by capillarity or by a break in the
polymer adhesive in coating, further hydration may rapidly occur.
It is thus possible that debonding precedes failure brought about
by volume changes due to massive oxide hydration, as has been
suggested [5,6].

ACKNOWLEDGEMENTS

This work was supported by the Department of Energy through
the Corrosion Research Center, University of Minnesota.

REFERENCES

1. H. W. Eichner and W. E. Schowalter, Forest Products Laboratory,
   Report No. 1813 (1950).
2. N. L. Rogers, U.S. Patent No. 3, (1950) 414,489.
3. G. S. Kabayashi and D. J. Donnelly, Boeing Co. Report No. DG-
   41517 (February, 1974).
4. J. D. Venables, D. K. McNamara, J. M. Chen, T. S. Sun, and R.
   L. Hopping, Appl. of Surf. Sci., 3, 88 (1979).
5. J. D. Venables, D. K. McNamara, J. M. Chen, B. M. Ditchek, T.
   I. Morgenthaler, T. S. Sun, and R. L. Hopping, 12th Nat.
   SAMPE Technical Conferences, Oct. 7-9, 1980, Seattle, Wash.
6. T. Sun, J. D. Venables and J. M. Chen, Martin Marietta Technical Report 80-34c, 1980.
7. W. G. Miller, Chapter 4 in "Spin Labeling," L. Berliner, ed.,
   Academic Press, 1979.
8. Z. Veksli, W. G. Miller and E. L. Thomas, J. Polym. Sci.,
   Polym. Symposium 54, 299 (1976).
9. Z. Veksli and W. G. Miller, Macromolecules, 10, 686 (1977).
10. W. G. Miller, W. T. Rudolph, Z. Veksli, D. L. Coon, C. C. Wu
    and T. M. Liang, in "Molecular Motion in Polymers by ESR,"
    R. F. Boyer and S. E. Keinath, eds., Harwood Academic Pub-
    lishers, New York, 1979, p. 145.
11. T. M. Liang, P. N. Dickson, and W. G. Miller, in "Characteri-
    zation of Macromolecules by ESR and NMR," ASC Symposium
    142, A. E. Woodward and F. A. Bovey, eds., American Chemi-
    cal Society, Washington, D.C., 1980, p. 1.
12. K. K. Knock and M. C. Locke, 13th National SAMPE Technical
    Conference October 13-15, 1981.

13. R. S. Alwitt, C. K. Dyer, and B. Noble, J. Electrochem. Soc.,
        129, 711 (1982).
14. K. E. Nietering, M. S. Thesis, University of Minnesota, 1983.
15. K. E. Nietering, J. E. Dickson and W. G. Miller, in preparation.
16. H. Burrell and B. Immergut, in "Polymer Handbook," 2nd Ed.,
        J. Brandrup and H. Immergut, eds., Interscience, New York,
        1975.

KINETICS AND MECHANISM OF AQUEOUS HYDROLYSIS AND CONDENSATION
OF ALKYLTRIALKOXYSILANES

E. R. Pohl and F. D. Osterholtz

Union Carbide Corporation
Tarrytown Technical Center
Tarrytown, NY 10591

ABSTRACT

The specific acid- and specific base-catalyzed hydrolysis rates of alkyltrialkoxysilanes have been measured in dilute aqueous solution at 25°C. The rates were monitored by the extraction method previously described, using infrared spectrometry detection of the silane.

The specific acid- and specific base-catalyzed condensation rates of alkyltrialkoxysilanes have been measured in deuterium oxide at 35°C. The reactions were monitored using a ^{13}C NMR technique. It has been observed that a change in the ^{13}C chemical shift of the methylene carbon adjacent to silicon accompanied the hydrolysis and condensation of alkyltrialkoxysilanes. ^{29}Si NMR spectroscopy was used to assign the ^{13}C chemical shifts to the various species resulting from hydrolysis and condensation at silicon.

Using these techniques, it was found that hydrolysis of the alkyltrialkoxysilanes was dependent upon the pD of the solution. The hydrolysis reaction proceeded in a stepwise manner to yield alkylsilanetriols. These silanetriols were meta-stable and condensed through a disiloxane species to higher molecular weight oligomers. The mechanisms of hydrolysis and condensation, and the pD dependence of these reactions are discussed. Structure-specific results are presented for different organofunctional silanes.

INTRODUCTION

Alkyltrialkoxysilanes undergo several hydrolysis and condensation reactions in aqueous solution [1,2,3]. Recently, several studies have investigated the hydrolysis reactions of aryl-[4] and alkyltrialkoxysilanes [5] in this medium. The results from these studies suggest that hydrolysis proceeds in a step-wise manner. The first step of hydrolysis to form aryl- or alkyldialkoxysilanol is usually the slowest. Subsequent hydrolysis of these silanols to form silanediols and then silanetriols proceed more rapidly. In dilute aqueous solution, the hydrolysis reactions follow pseudo-first order kinetics and are general base- and specific acid-catalyzed. It was argued that hydroxide anion catalyzed hydrolysis involves an S_N2^{**}-Si (rate determining formation of a pentacoordinate intermediate) or S_N2^*-Si (rate determining breakdown of the pentacoordinate intermediate) mechanism [4,5]. Hydronium ion catalyzed hydrolysis appears to proceed through a mechanism with more S_N2-type character [5].

The condensation reactions of silanetriols in aqueous solution are not as well understood [1,3]. Previously, studies on the mechanism of condensation of silanols to form siloxanes examined the reactions of trialkylsilanols [6], dialkylsilanediols [7-9], aryl-[10] and alkylsilanetriols [11] in aqueous-organic solution. The lack of quantitative rate data on the condensation of alkylsilanetriols in aqueous solution arises because of the difficulty in monitoring the individual condensation reactions. Fourier Transform Infrared and laser Raman spectroscopies have been used to obtain information about the structures of aryl- and alkyltrialkoxysilanes in aqueous solution [12,13]. These techniques, however, have not been used to obtain quantitative rate data, which would assist the understanding of the mechanisms by which silanetriols form siloxane oligomers.

In an effort to better understand the chemistry of alkyltrialkoxysilanes in dilute aqueous solution, a study was carried out to examine the influence of pH on the hydrolysis and the condensation of γ-glycidoxypropyltrimethoxysilane. Of particular interest was the development of ^{13}C and ^{29}Si NMR spectroscopic methods capable of measuring individual condensation reaction rates. The quantitative rate data thus obtained are discussed in terms of hydrolysis and condensation mechanisms.

EXPERIMENTAL

Materials

The alkyltrialkoxysilanes employed in this study were

obtained commercially from Union Carbide Corporation under the trade names A-187, A-1100 and A-174 and used without further purification. The other reagents were obtained commercially from Fisher Scientific Company and Norell. Glass distilled water was employed throughout.

<u>Kinetic Measurements</u>

The rates of hydrolysis of the γ-glycidoxypropyltrialkoxysilane were followed by an extraction method [5]. An appropriately buffered solution was prepared by adding a weighed sample of the buffer to a 100 ml volumetric flask. Enough potassium chloride was added to bring the ionic strength to 1.0. The pH was adjusted to the desired value and the solution was diluted to mark with distilled water. After thermal equilibrium at 25.0°C, the reaction was initiated by mixing a weighed sample of the γ-glycidoxypropyltrimethoxysilane with the buffered solution to yield a final concentration of 0.03 M silane. The time to mix the silane is generally less than 30 sec. There is no indication that rate of silane addition influences the rate of hydrolysis. At recorded time intervals, 10 ml aliquots were removed from the reaction mixture, extracted with 25 ml of n-hexane, separated and dried over anhydrous Na_2SO_4. After filtering, the organic layer was concentrated, quantitatively transferred to a 5 ml volumetric flask and diluted to mark with n-hexane. The extraction procedure generally takes 2 min. Control experiments on samples held for 16 h or more suggest that once the silane has been extracted into the n-hexane, further hydrolysis does not occur. The concentration of the extracted material was determined by measuring the difference in absorbance between the baseline at 1700 cm^{-1} and, for example, the Si-O-C stretching frequency at 1100 cm^{-1}. A Beckman Infrared 4240 spectrophotometer and potassium bromide solution cells were used. The pH of the solution was measured at the end of the reaction.

The values of k_{obsd} were determined from the least-squares slope of $-\ln(A_t - A_\infty)$ versus time with correlation coefficients generally ≥ 0.99. The spontaneous rate constants for hydrolysis at a particular pH were determined by extrapolation of the observed rate constants in buffered solution to zero buffer concentrations. These data are given in Table I.

The rates of condensation of alkylsilanetriols were followed by ^{13}C NMR spectroscopy. An appropriately buffered solution was prepared by adding a weighed sample of sodium acetate to a 5 ml volumetric flask, adjusting the pD to the desired value, and then diluting to mark with deuterium oxide. After thermal equilibration at 35.0°C, the reaction was initiated by mixing a weighed sample of the alkyltrialkoxysilane with the buffer solution to yield a final

Table I. The spontaneous rates of hydrolysis of γ-glycidoxypropyl-
trimethoxysilane to γ-glycidoxypropyldimethoxysilanol at
25°C in aqueous solution as a function of pH.

Concn, Substrate[1]	Buffer	pH	k_{spon}
M			(s^{-1})
0.030	acetate	5.00	2.3×10^{-3}
0.030	acetate	6.13	2.2×10^{-4}
0.0210	phosphate	7.16	5.2×10^{-5}
0.030	Tris	8.39	7.3×10^{-5}
0.030	Tris	8.78	1.0×10^{-3}

[1]Substrate = γ-glycidoxypropyltrimethoxysilane

Table II. The observed rates of condensation of alkylsilanetriol
to dialkyltetrahydroxydisiloxane at 35°C in deuterium
oxide as a function of pD.

Substrate	$[RSi(OD)_3]$initial, M	pD	k_{obsd} $(M^{-2}s^{-1})$
$CH_2CHCH_2OCH_2CH_2CH_2Si(OD)_3$ (epoxide)	0.258	2.15	1.2×10^{-3}
	0.243	3.72	8.3×10^{-5}
	0.491	4.05	5.3×10^{-5}
	0.250	4.07	5.0×10^{-5}
	0.245	4.46	3.6×10^{-5}
	0.246	5.27	1.0×10^{-4}
	0.241	6.64	$\sim 5 \times 10^{-4}$
$CH_2C(CH_3)CO_2CH_2CH_2CH_2Si(OD)_3$	0.210	3.97	5.0×10^{-5}
	0.210	4.10	3.4×10^{-5}
$D_3\overset{+}{N}CH_2CH_2CH_2Si(OD)_3$	0.334	5.26	3.9×10^{-4}

concentration of between 0.21 M and 0.49 M silane. At recorded time intervals, spectra of the solutions were taken on a Varian CFT-20 NMR spectrometer equipped with a broad-band decoupler, and a 16 K memory computer. Chemical shifts were measured from hexamethyldisiloxane (external). Typically, a pulse width of 8 μsec, no pulse delay, sweep width of 4000 Hz, and pulse acquisition time of 1.024 sec were used.

The values of k_{obsd} for the disappearance of alkylsilanetriol were determined from the least-squares slope of $1/[RSi(OD)_3]$ − $1/[RSi(OD)_3]_0$ versus the midpoint of the recorded time interval with correlation coefficients generally ≥ 0.98. The concentrations of the alkylsilanetriol were determined by monitoring the relative intensity of the peak due to the carbon adjacent to the silicon, as shown in Figure 2. These data are given in Table II.

^{29}Si NMR spectra assisted in the ^{13}C NMR chemical shift assignments of alkylsilanetriols and dialkyltetrahydroxydisiloxanes. ^{29}Si NMR spectra were taken on a JEOL FX90 NMR spectrometer equipped with a broad-band decoupler, and 64 K memory capacity computer. Chemical shifts were measured from TMS (external). Typically, a pulse width of 10 μsec, pulse delay of 150 μsec and sweep width of 5000 Hz were used. $Gd(NO_3)_3$ was used as a relaxation agent.

RESULTS AND DISCUSSION

The rates of hydrolysis of γ-glycidoxypropyltrimethoxysilane were determined by an extraction method described in the Experimental Section. As previously discussed [5], this extraction method monitors loss of starting material and, therefore, provides information about the first step of hydrolysis. The spontaneous hydrolysis rate constants thus obtained are presented as a pH-rate profile in Figure 1. The hydrolysis appears to be both hydroxide anion and hydronium ion catalyzed since the slopes of the plot at pH values above and below the rate minimum around pH 7.0 are drawn to be +1 and −1, respectively.

^{13}C NMR spectra of 0.241 M solution of γ-glycidoxypropyltrimethoxysilane at pD 6.64 and 35°C that were taken at various reaction times, show no significant concentration of γ-glycidoxypropyldimethoxysilanol nor γ-glycidoxypropylmethoxysilanediol. Instead, it was observed that a peak with chemical shift 5.53 ppm, which corresponds to the alpha-carbon of the starting material, gradually decreases in relative intensity, while new peaks at 9.48 ppm and then 9.81 ppm slowly form. ^{29}Si NMR spectrum of this solution indicated that the chemical species represented by the peaks at 9.48 ppm and 9.81 ppm are γ-glycidoxypropylsilanetriol and bis-(γ-glycidoxypropyl)tetrahydroxydisiloxane, respectively. The pseudo-first

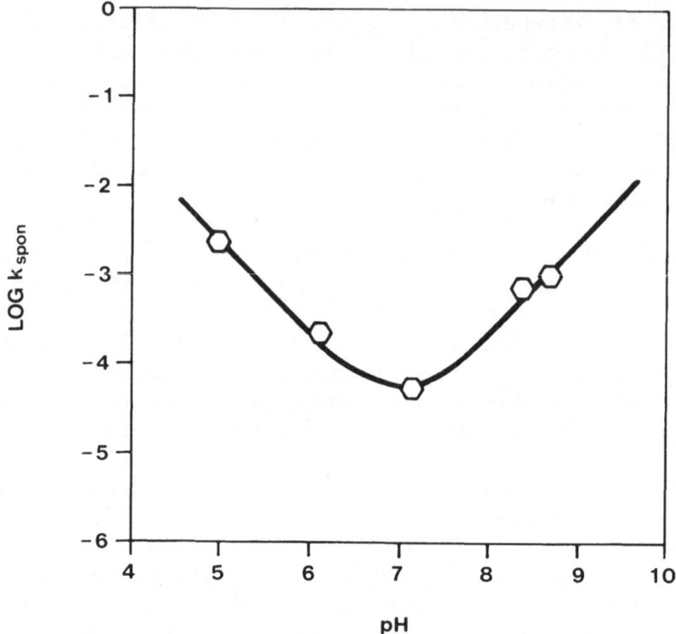

Fig. 1.   pH-Rate profile for the first hydrolysis step of γ-glycidoxypropyltrimethoxysilane under conditions described in
the text. The line is calculated from the second-order
rate constants given in the text.

order rate constant for the disappearance of the peak at 5.53 ppm
is 1.9×10^{-4} sec^{-1}. The spontaneous rate constant for hydrolysis
at pH 6.64, taken from the pH-rate profile (Figure 1) is 6.4×10^{-5}
sec^{-1}. These data suggest that, indeed, the extraction method monitors the slow, first step of hydrolysis. The factor of 3 difference in the k_{spon} between the two techniques is consistent with the
different conditions under which the measurements were made.

The rates of disappearance of the γ-glycidoxypropyltrimethoxysilane are found to follow the equation:

$$- \frac{d[silane]}{dt} = k^{HO}[HO^-][silane] + k^{H}[H^+][silane] + k_b[B][silane]$$

where [B] is the concentration of the conjugate base of the buffer
calculated from pH, pK and total buffer concentration. Since the
hydrolysis reactions are general base-catalyzed, the spontaneous
rate constants for hydrolysis at a particular pH have been determined by extrapolation of the observed rate constants in buffered

solutions to zero buffer concentrations. The spontaneous hydrolysis rate constants, reported in Table I, were then used to calculate the values of k^{HO} and k^{H}. The second order rate constants for
the hydroxide anion and hydronium ion catalyzed hydrolysis are 2.5
x 10^2 sec^{-1} M^{-1} and 2.3 x 10^2 sec^{-1} M^{-1}, respectively. Similar
results have been observed for phenyl-tris-(2-methoxyethoxy)silane
and a series of alkyl-tris-(2-methoxyethoxy)silanes [4,5].

It has been argued that hydroxide anion catalyzed hydrolysis
of phenyl- and alkyltrialkoxysilane proceed by a bimolecular displacement reaction (S_N2^{**}-Si or S_N2^*-Si) with a pentacoordinate intermediate, as shown in Scheme I [4,5]. In accord with these mechanisms [5], substituents which stabilize developing negative
change on silicon in the transition state and decrease steric
crowding should increase the rate of hydroxide anion catalyzed hydrolysis of silanes. The hydroxide anion catalyzed rate of hydrolysis of γ-glycidoxypropyltrimethoxysilane is approximately a
factor of 10 larger than n-propyltris-(2-methoxyethoxy)silane. The
rate difference arises because of the different leaving groups; the
alkyl groups, γ-glycidoxylpropyl and propyl, have very similar
steric and polar effects and therefore, should have only minor
effects on the hydrolysis rate. The rate difference arises in
spite of the fact that the steric and polar effects of the leaving
groups act in opposite directions. The methoxy leaving group is
smaller and less electron withdrawing than the 2-methoxyethoxy
leaving group. If the hydrolysis reaction proceeds by the proposed
S_N2^{**}-Si or S_N2^*-Si mechanism, the steric effects of the leaving
group must be significant.

It has been proposed that the hydronium ion catalyzed hydrolysis of alkyltrialkoxysilanes proceed by a bimolecular displacement reaction with significant S_N2-type character, as shown in
Scheme II. (The reaction mechanism may involve a pentacoordinate
intermediate). In accord with this mechanism, substituents which
decrease steric crowding around silicon should increase the hydronium ion catalyzed hydrolysis rate. Substituents which stabilize
developing negative change in the transition state, also increase
hydrolysis rate, but to a significantly smaller degree than the hydroxide anion catalyzed reaction [5]. It is therefore reasonable
that changing the leaving group from 2-methoxyethoxy to methoxy
should increase the rate of hydrolysis. For example, hydronium ion
catalyzed rate constant for the hydrolysis of γ-glycidoxypropyltrimethoxysilane is approximately a factor of 10 larger than for n-
-propyl-tris-(2-methoxyethoxy)silane.

The condensation reaction of the alkylsilanetriols was followed by ^{13}C NMR spectroscopic technique described in the experimental section. The ^{13}C NMR spectra of the carbon adjacent to silicon of 0.250 M γ-glycidoxypropyltrimethoxysilane in 0.25 M acetate
buffer at pD 4.07 at various reaction times are shown in Figure 2.

SCHEME I

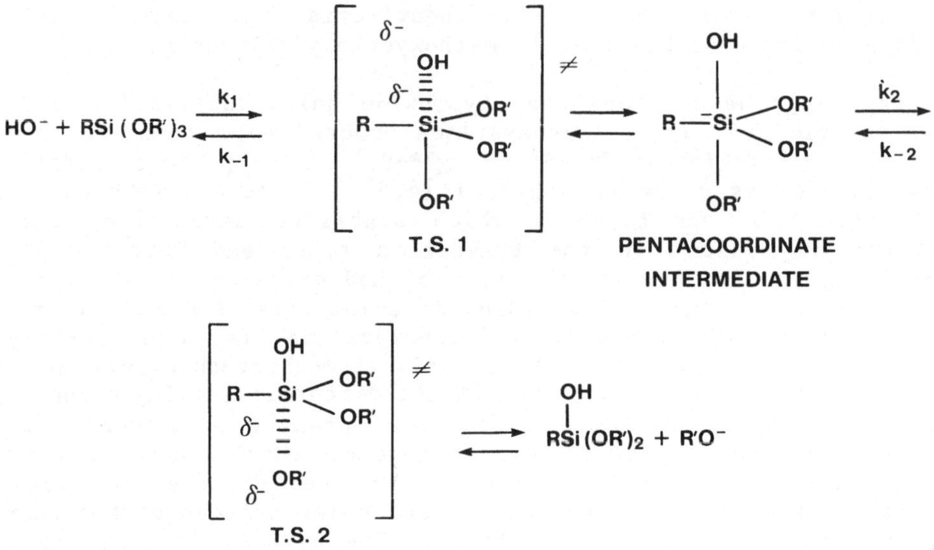
HO⁻ + RSi(OR′)3
k1
k-1
δ-
OH
δ-
R—Si
OR′
OR′
OR′
T.S. 1
≠
OH
R—Si
OR′
OR′
OR′
PENTACOORDINATE
INTERMEDIATE
k2
k-2
OH
R—Si
OR′
OR′
δ-
OR′
δ-
T.S. 2
≠
OH
RSi(OR′)2 + R′O⁻

SCHEME II

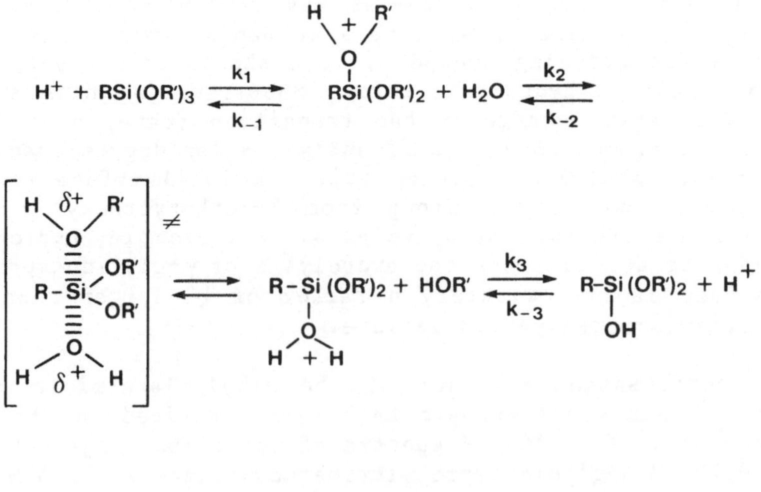
H⁺ + RSi(OR′)3
k1
k-1
H + R′
O
RSi(OR′)2 + H2O
k2
k-2
H δ+ R′
O
OR′
R-Si
OR′
O
H δ+ H
≠
R-Si(OR′)2 + HOR′
O
H + H
k3
k-3
R-Si(OR′)2 + H⁺
OH

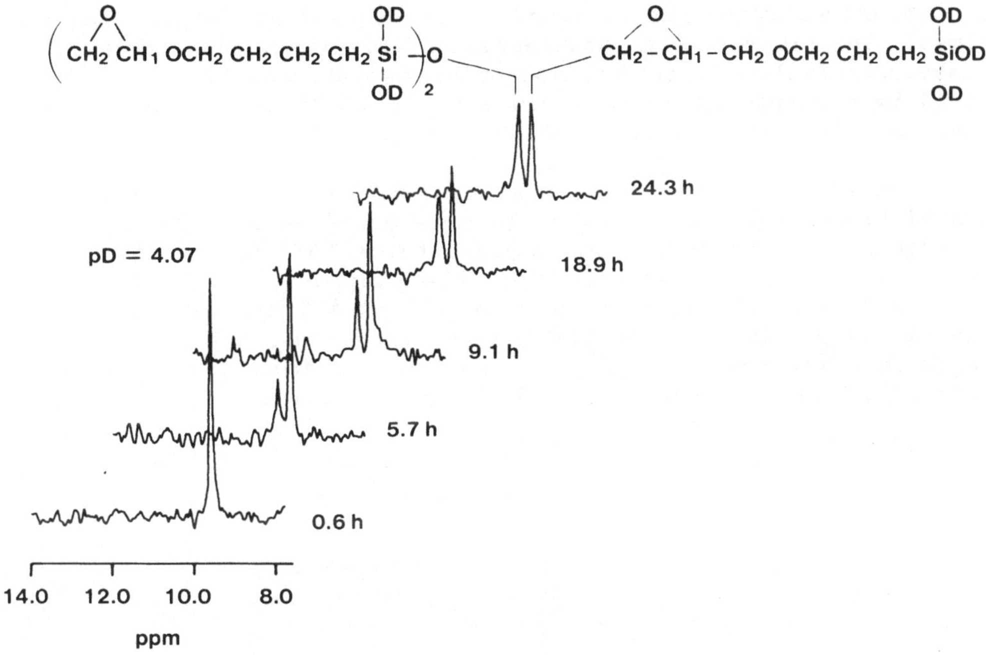

Fig. 2. ^{13}C NMR spectra of the carbon adjacent to silicon are monitoring condensation of γ-glycidoxypropylsilanetriol to bis-
(γ-glycidoxypropyl)tetrahydroxydisiloxane in 0.25M acetate
buffer at pD 4.07 at 35°C. The peaks as 9.48 ppm and 9.81 ppm
correspond to γ-glycidoxypropylsilanetriol and bis-(γ-glycidoxypropyl)tetrahydroxydisiloxane, respectively.

After 0.6 h, the spectrum of the alpha-carbon was a single, sharp
peak at 9.48 ppm. These data indicate that the only detectable
species is γ-glycidoxypropylsilanetriol. Data from the pH-rate
profile given in Figure 1 indicate that the half-life of hydrolysis
at pH 4.07 in aqueous solution and 25°C is 3.5 sec. These two
facts imply that the hydrolysis reactions have reached completion
and that γ-glycidoxypropylsilanetriol can be generated _in situ_. At
later reaction times, the peak at 9.48 ppm decreases in relative
intensity, while a new peak forms at 9.81 ppm. ^{29}Si NMR spectra
indicates that the peak at 9.81 ppm is bis-(γ-glycidoxypropyl)te-
trahydroxydisiloxane, as is shown in Figure 3. The second order
rate constants for the disappearance of 0.25 M and 0.49 M γ-glycidoxypropylsilanetriol were 5.0 x 10^{-5} sec^{-1}M^{-1} and 5.3 x
10^{-5}sec^{-1}M^{-1}, respectively, showing that the condensation of γ-glycidoxypropylsilanetriol to bis-(γ-glycidoxypropyl)tetrahydroxydisi-
loxane is second order in γ-glycidoxypropylsilanetriol. Further,
condensation of bis-(γ-glycidoxypropyl)-tetrahydroxydisiloxane to
low molecular weight oligomers was not observed under these reaction conditions and reaction times of less than 18.9 h. However,
the condensation of bis-(γ-glycidoxypropyl)tetrahydroxydisiloxane

to low molecular weight oligomers was observed at longer reaction
times. For example, the condensation of bis-(γ-glycidoxypropyl)te-
trahydroxydisiloxane to low molecular weight oligomers was indicated by the appearance of a new peak at 10.06 ppm in the ^{13}C NMR
spectrum after 24.3 h, as is shown in Figure 2.

The observed rate constants for condensation of γ-glycidoxy-
propylsilanetriol thus obtained are presented as a pD-rate profile
in Figure 4. The condensation appears to be both deuteroxide anion
and deuterium ion catalyzed since the slopes of the plot at pDs
above and below the rate minimum around pD 4.5 are found to be +1
and -1, respectively. The third order rate constants for the deuteroxide anion and deuterium ion catalyzed condensation are 5.3 x
$10^4 \text{sec}^{-1}\text{M}^{-1}$ and 5.0 x 10^{-1} $\text{sec}^{-1}\text{M}^{-1}$, respectively.

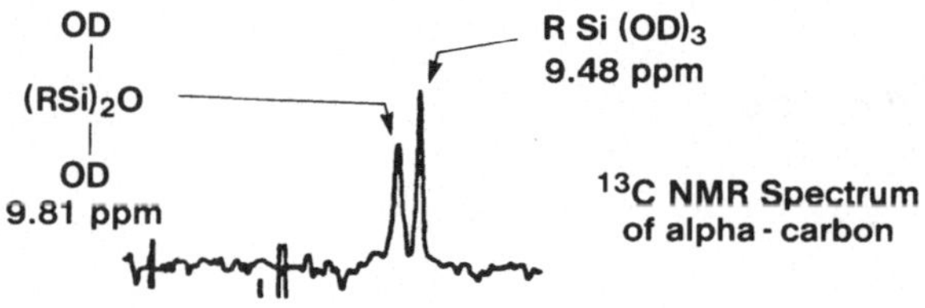

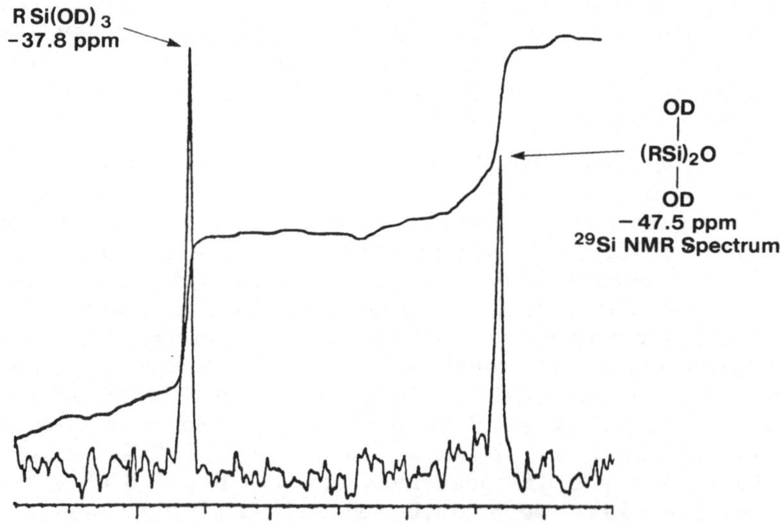

Fig. 3. ^{13}C NMR spectrum of the carbon adjacent to silicon and ^{29}Si
           NMR spectrum showing similar extent of condensation of γ-
           glycidoxypropylsilanetriol to bis-(γ-glycidoxypropyl)tetra-
           hydroxydisiloxane after 19 hours of reaction at pD 4.07,
           35°C and in deuterium oxide.

The oxirane ring was observed in ^{13}C NMR spectrum to undergo ring opening under acidic conditions. For example, the oxirane ring opening was significant when the pD of the reaction solution was 2.15. The effect of the oxirane ring opening on the observed rate of condensation of γ-glycidoxypropylsilanetriol to bis-(γ-glydoxypropyl)tetrahydroxydisiloxane is unknown. The deviation between the calculated line and the observed rate constant for the condensation reaction at pD of 2.15, shown in Figure 4, may have resulted because of the interaction of the diol with the silanetriol group. The influence of the diol on the rate of hydrolysis, however, is probably not significant because the hydrolysis reaction was observed to reach completion before significant oxirane ring opening had occurred.

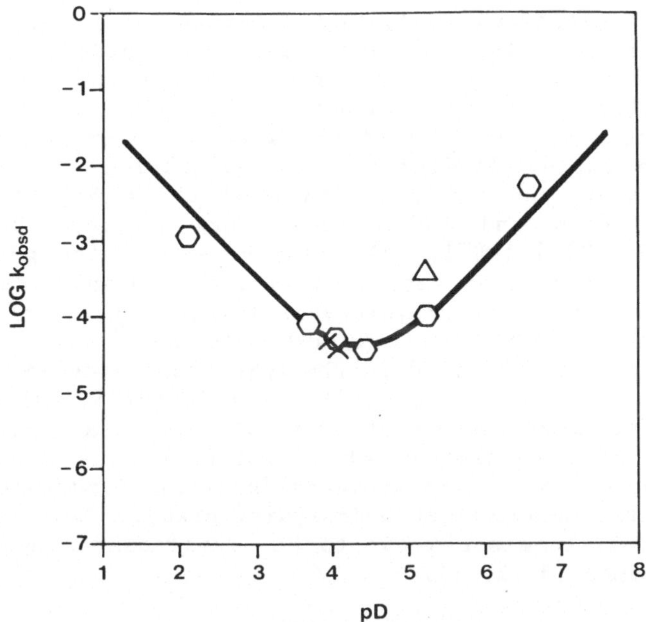

Fig. 4.   pD-Rate profile is shown for the condensation of γ-glycidoxypropylsilanetriol (0) to bis-(γ-glycidoxypropyl)tetrahydroxydisiloxane in 0.25 M in acetate buffer and 35°C. The line is calculated from the third order rate constants given in the text. The k_{Obsd} for the condensation of γ-aminopropylsilanetriol (Δ) and γ-methacryloxypropylsilanetriol (X) at various pD are also included.

The rates of condensation of γ-glycidoxypropylsilanetriol are found to follow the equation:

$$-\frac{d[\text{silanetriol}]}{dt} = k_c^{DO}[DO^-][\text{silanetriol}]^2 + k_c^D[D^+][\text{silanetriol}]^2$$

Since the reactions were carried out under constant acetate buffer concentration, it was not determined whether or not these reactions are general acid- or general base-catalyzed.

The k_{obsd}'s for the condensation of γ-aminopropylsilanetriol at pD 5.26, and γ-methacryloxypropylsilanetriol at pD 3.97 and 4.10 are also included in Figure 4. The k_{obsd} for these compounds are similar to the k_{obsd} for γ-glycidoxypropylsilanetriol at comparable pD. These data imply that the organofunctional group, i.e., glycidoxypropyl, γ-methacryloxypropyl and protonated γ-aminopropyl, do not significantly affect the condensation reactions. These organofunctional groups have similar inductive and steric effects.

These results provide insight into the deuteroxide anion and deuterium ion catalyzed condensation mechanisms. For example, the deuteroxide anion catalyzed rates are second order in silanetriol and first order in deuteroxide anion. These facts are in accord with the mechanism proposed in Scheme III. In this mechanism, the reversible reaction of silanetriol with deuteroxide anion is assumed to be rapid, leading to an equilibrium concentration of silanolate anion, the concentration of which is proportional to the deuteroxide anion and silanetriol concentration, $RSi(OD)_2O^- = (k_1/k_{-1})\,[RSi(OD)_3]\,[OD^-]$. The slower second step gives the observed deuteroxide anion catalyzed rate of condensation of silanetriol to the dialkyltetrahydroxydisiloxane, $-d[\text{silanetriol}]/dt = k_2[RSi(OD)_2O^-][RSi(OD)]_3 = (k_1k_2/k_{-1})[RSi(OD)_3]^2\,[OD^-]$. Under the reaction condition described in the text, the reaction goes to completion. The further condensation of silanetriol with dialkyltetrahydroxydisiloxane was not observed at short reaction times. The slower rate of condensation of silanetriol with dialkyltetrahydroxysiloxane may be due to increase in steric hindrance around the silicon of the dialkyltetrahydroxydisiloxane. Nucleophilic displacement at silicon are known to be significantly susceptible to steric hindrance [3,15,16].

<u>SCHEME III</u>

$$R-Si(OD)_3 + OD^- \overset{k_1}{\underset{k_{-1}}{\rightleftharpoons}} R-Si(OD)_2O^- + D_2O$$

fast

$$R-Si(OD)_2O^- + RSi(OD)_3 \overset{k_2}{\underset{k_{-2}}{\rightleftharpoons}} R-Si(OD)_2-O-Si(OD)_2R + OD^-$$

slow

Grubbs [6] proposed that the condensation of trimethylsilanol in methanol occurred by a mechanism of rearward attack in which an electron donor molecule approaches the backside of the central atom, undergoing displacement of one of its substituents; an S_N2-Si type mechanism. Swain [14] et al, have proposed that silicon forms pentacoordinate intermediates; S_N2^{**}-Si or S_N2^*-Si mechanism. The present results are consistent with both of these hypotheses and do not distinguish between them.

Correspondingly, the mechanism for hydronium ion catalyzed condensation is consistent with the mechanism in Scheme IV. For example, the reactions are second order in silanetriol and first order in deuterium ion. All the protonated species are assumed to be in non-rate determining equilibrium. The present results are consistent with either S_N2^{**}-Si, S_N2^*-Si or S_N2-Si mechanisms.

<u>CONCLUSION</u>

A combination of an extraction method and a ^{13}C and ^{29}Si NMR technique was used to obtain quantitative rate data on reactions of alkyltrialkoxysilanes in aqueous solution. It was found that hydrolysis of γ-glycidoxypropyltrimethoxysilane proceeds through a stepwise process in which the first step is the slow step. These reactions are catalyzed by both deuteroxide anion and deuterium ion, with the rate minimum occurring at approximately pH 7.0. The condensation of γ-glycidoxypropylsilanetriol gives as a first product bis-(γ-glycidoxypropyl)tetrahydroxydisiloxane. The reactions are catalyzed by hydroxide anion and hydronium ion with the rate minimum occurring at approximately pD 4.5. These reactions were found to be second order in silanetriol and first order in deuteroxide anion or deuterium ion.

<u>ACKNOWLEDGEMENTS</u>

We thank R. A. Budnik for the use of the Varian CFT-20 NMR spectrometer and helpful comments, and B. Su for taking the ^{29}Si NMR spectra of the γ-glycidoxypropylsilanetriol.

<u>SCHEME IV</u>

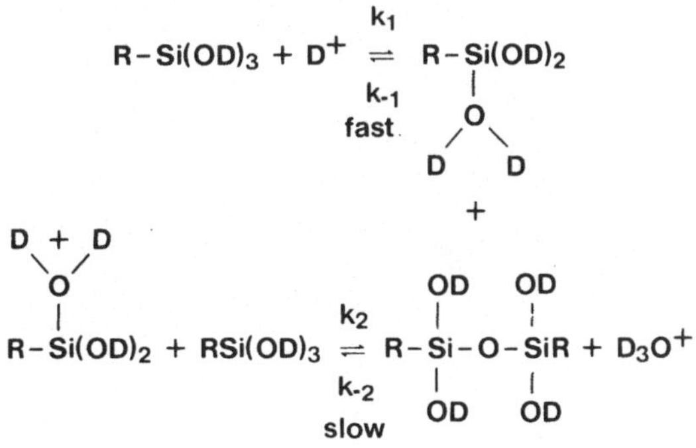

$$R\text{-}Si(OD)_3 + D^+ \underset{\substack{k_{-1} \\ \text{fast}}}{\overset{k_1}{\rightleftharpoons}} R\text{-}Si(OD)_2$$

$$R\text{-}Si(OD)_2 + RSi(OD)_3 \underset{\substack{k_{-2} \\ \text{slow}}}{\overset{k_2}{\rightleftharpoons}} R\text{-}Si\text{-}O\text{-}SiR + D_3O^+$$

REFERENCES

1. E. P. Plueddemann, "Silane Coupling Agents," Plenum, New York, Chapter 3 (1982).
2. L. H. Sommer, "Stereochemistry, Mechanism and Silicon," McGraw-Hill, New York, Chapter 11 (1965).
3. M. G. Voronkov, V. P. Meleshkevick, Y. A. Yuzkelevski, "The Siloxane Bond," Consultants Bureau, Plenum, New York, pp.323-349 and 375-380 (1978).
4. K. J. McNeil, J. A. DiCaprio, O. A. Walsh and R. J. Pratt, J. Am. Chem. Soc., $\underline{102}$, 1859 (1980).
5. E. R. Pohl, Proc. 38th Ann. Tech. Conf., Reinforced Plastics/Composites Inst., Section 4-B (1983).
6. W. T. Grubb, J. Am. Chem. Soc., $\underline{76}$, 3408 (1954).
7. Z. Lasocki, Internat. Symp. Organosilicon Chemistry, Prague, pp. 34-36 (1965).
8. Z. Lasocki, Bull. Acad. Polon. Sci. Ser. Sci. Chim., $\underline{12}$, 227 (1964).
9. Z. Lasocki and S. Chrzczonowicz, J. Poly. Sci., $\underline{59}$, 259 (1962).
10. J. F. Brown, Jr., J. Am. Chem. Soc., $\underline{87}$, 4317 (1965).
11. J. B. Brown, Jr. and L. H. Vogt, Jr., J. Am. Chem. Soc., $\underline{87}$, 4313 (1965).
12. H. Ishida and J. L. Koenig, Appl. Spectrosc., $\underline{32}$, 462 (1978).
13. H. Ishida and J. L. Koenig, Appl. Spectrosc., $\underline{32}$, 469 (1978).
14. C. G. Swain, R. M. Esteve, R. H. Jones, J. Am. Chem. Soc., $\underline{11}$, 965 (1949).
15. E. Akerman, Acta Chem. Scand., $\underline{10}$, 298 (1956).
16. J. Chojnowski, and S. Chrzczonowicz, Internat. Symp. Organosilicon Chemistry Sci. Comm. Prague 4 (1965).

STRUCTURE AND PROPERTIES OF SILANE PRIMERS FOR ADHESIVE BONDING OF
METALS

F. J. Boerio, C. A. Gosselin, J. W. Williams,
R. G. Dillingham and James M. Burkstrand*

Department of Materials Science & Metallurgical
   Engineering
University of Cincinnati
Cincinnati, Ohio 45221

*Perkin-Elmer Corp.
Physical Electronics Div.
6509 Flying Cloud Drive
Eden Prairie, Minnesota 55344

ABSTRACT

The structure and properties of thin films formed by γ-amino-
propyltriethoxysilane (γ-APS) adsorbed onto iron, titanium, and
aluminum from dilute aqueous solutions is described. γ-APS was adsorbed onto iron and titanium mirrors at pH 10.4 and 8.0 to form
siloxane polymers. The structures of the siloxane polymers were
only slightly affected by the substrate and the pH. However, the
properties of the films as primers for improving the wet strength
of iron/epoxy and titanium/epoxy lap joints depended significantly
on the substrate and on the pH at which the films were formed. It
was concluded that the properties of the films depended more on the
orientation of the γ-APS molecules at the oxide surface than on the
overall structure of the films. The structure of films formed by
γ-APS adsorbed onto aluminum was a strong function of pH. At high
pH values, the oxides on aluminum were etched and aluminosiloxane
polymers were obtained. At intermediate pH values little etching
was observed and siloxane polymers were obtained. The siloxane
polymers formed at pH 7.0 inhibited hydration of the native oxide
on aluminum in water at 60 °C.

INTRODUCTION

Organofunctional silanes have been used as "coupling agents" for improving the wet strength of glass fiber reinforced composites for many years. More recently, it has been shown that silane coupling agents are also very effective primers for improving the wet strength of metal-to-metal adhesive joints when applied to the adherends from dilute solutions prior to adhesive bonding [1-6]. Many theories have been proposed to explain the mechanisms by which silane coupling agents function [7]. However, those mechanisms are still not well understood.

We have been interested in using surface analysis techniques such as reflection-absorption infrared (RAIR) spectroscopy, ellipsometry, and x-ray photoelectron spectroscopy (XPS) to determine the molecular structure of thin films formed by organosilanes adsorbed onto metals from dilute aqueous solutions and relating that structure to the properties of such films for improving the hydrothermal stability of metal/epoxy adhesive joints. The purpose of this paper is to compare the results obtained for one important silane, γ-aminopropyltriethoxysilane (γ-APS), and for iron, titanium, and aluminum adherends.

RAIR is a very effective technique for obtaining infrared absorption spectra of thin films on metal mirrors by reflecting parallel polarized infrared radiation from the mirrors at large, almost grazing angles of incidence. Properly designed RAIR experiments are capable of providing a great deal of information about the chemical and physical structures, including orientation, of thin films formed by organic compounds adsorbed onto the oxidized surfaces of metals. RAIR experiments can also provide much information about the oxide films on metals. Ellipsometry is a very sensitive technique that can be used to determine the thickness and refractive index of films formed on reflecting surfaces. XPS is, of course, a very useful technique for determining the chemical structure of the surface regions (outermost few tens of angstroms) of a solid. The principles and applications of RAIR [8], ellipsometry [9], and XPS [10] have been thoroughly described in the literature and no further discussion will be given here.

EXPERIMENTAL

Sample mirrors for surface analysis were prepared by mechanically polishing enameling iron, titanium -6A1, 4V, and 2024 aluminum coupons. γ-APS films were prepared by immersing freshly polished mirrors into 1% aqueous solutions for approximately 15 minutes, withdrawing the mirrors, and blowing off the excess solution using a strong stream of nitrogen. Infrared spectra were obtained using a Perkin-Elmer Model 180 infrared spectrophotometer

and reflection accessories from Harrick Scientific Co. X-ray photoelectron spectra were obtained using a Physical Electronics spectrometer and magnesium K_α radiation. The effects of sample charging were eliminated by referring the C(1s) binding energy for hydrocarbons to 284.6 ev. Ellipsometry was performed using a Rudolph
Research Model 436 ellipsometer.

Metal/epoxy lap joints were prepared according to ASTM standard D1002. Metal adherends were polished, rinsed, and dried.
Pairs of adherends were then bonded together using an adhesive consisting of an epoxy resin (Epon 828, Shell Chemical Co.) and a tertiary amine curing agent (Ancamine K-61B, Pacific Anchor Chemical
Co.). Prior to bonding, some of the adherends were pretreated by
immersion in 1% aqueous solutions of γ-APS for about fifteen
minutes and then blown dry with nitrogen. Other adherends were
left untreated as a control. All of the lap joints were then aged
in a water bath at 60°C. At appropriate intervals, joints were removed from the water bath and tested with an Instron to determine
their residual strength.

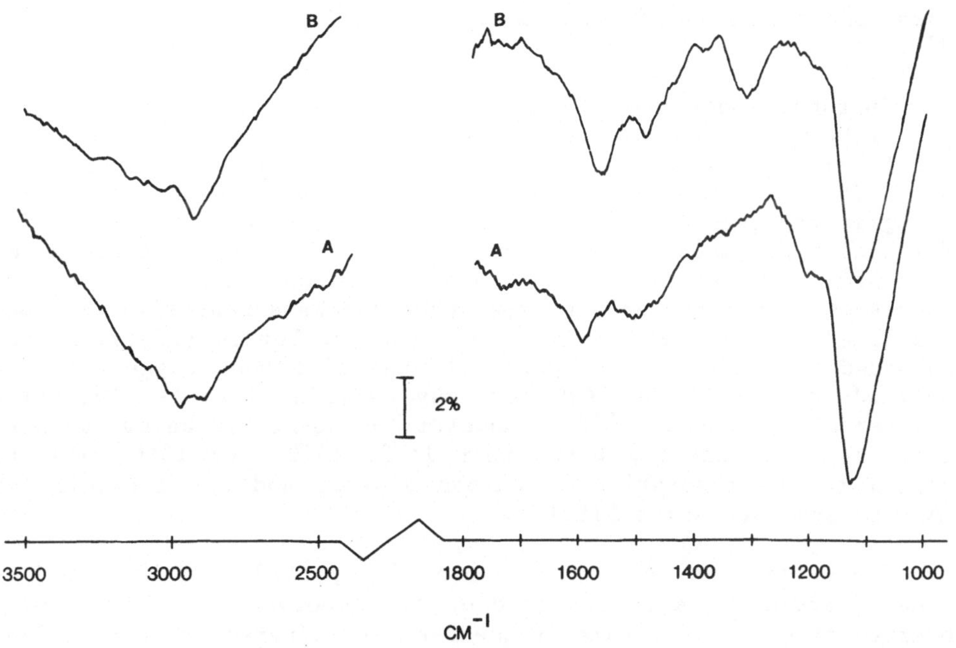

Fig. 1.   Infrared spectra of thin films formed by γ-APS adsorbed
          onto iron mirrors from 1% aqueous solutions at (A)-pH 8.0
          and (B)-pH 10.4.   Adapted by permission from reference 3.

RESULTS AND DISCUSSION

Results obtained from ellipsometry indicated that the mechanically polished metal mirrors were typically covered with natural oxide films approximately 5 nm in thickness. For iron mirrors, the oxide was mostly Fe_2O_3 [11]. The oxides on aluminum and titanium mirrors were mostly amorphous Al_2O_3 [8] and TiO_2 [12], respectively. Ellipsometry also indicated that the γ-APS films deposited on the mirrors as described above were approximately 10 nm in thickness.

As shown in Figure 1A, infrared spectra of thin films formed by γ-APS adsorbed onto iron mirrors from aqueous solutions that were acidified to pH 8.0 by addition of HCl were characterized by a strong band near 1120 cm^{-1} and by weak bands near 1600 and 1500 cm^{-1} [3]. Transmission infrared spectra of γ-APS are dominated by intense bands near 1105, 1080, and 960 cm^{-1} that are characteristic of SiOC stretching vibrations [13]. The absence of these bands from the spectrum shown in Figure 1A and the presence of the band near 1120 cm^{-1}, which was assigned to an SiOSi vibration [3], indicates that the γ-APS was highly hydrolyzed in solution and polymerized on the surface during drying to form a siloxane polymer. The bands near 1600 and 1500 cm^{-1} are characteristic of NH_3^+ deformation modes, indicating the formation of amine hydrochlorides as expected [3].

Infrared spectra of films formed by γ-APS adsorbed onto iron mirrors from aqueous solutions at pH 10.4 [3,13,14] were characterized by strong bands near 1570 and 1110 cm^{-1} and by weaker bands near 1470 and 1300 cm^{-1} (see Figure 1B). The band near 1110 cm^{-1} was again assigned to an SiOSi vibration, indicating the formation of siloxane polymers on the iron surface. The origin of the bands near 1570, 1470, and 1300 cm^{-1} has been the subject of considerable discussion. It has been suggested that the band near 1570 cm^{-1} may be related to amine groups that are hydrogen bonded [13,15] or coordinated to silicon atoms [13]. It has also been suggested that the bands near 1570 and 1470 cm^{-1} may be related to the formation of internal zwitterions [14]. However, as described below, we have recently shown that the bands near 1570, 1470, and 1300 cm^{-1} are related to the absorption of carbon dioxide and the formation of amine bicarbonate salts [16,17].

The assignment of the bands near 1570, 1470, and 1300 cm^{-1} to amine bicarbonate salts is based on two observations. Bands were observed at nearly the same frequencies in infrared spectra of laurylamine that had been exposed to atmospheric carbon dioxide for long times [16]. However, when the laurylamine was recrystallized in a nitrogen atmosphere, these bands disappeared and only bands near 1610 and 1455 cm^{-1}, that are typical of an aliphatic primary amine, were observed [16]. In addition, the bands near 1570, 1470, and 1300 cm^{-1} were not observed in infrared spectra of thin films formed by γ-APS adsorbed onto iron mirrors from aqueous solutions at pH 10.4 in a nitrogen atmosphere (see Figure 2A). However, when films formed in a nitrogen atmosphere were exposed to air, the bands near 1570, 1470, and 1300 cm^{-1} gradually reappeared (see Figure 2B).

Films formed by adsorption of γ-APS onto iron mirrors at pH 10.4 were stable in a dry atmosphere (see Figure 3A) but a transition was noted during exposure to atmospheric moisture (see Figure 3B). The band originally observed near 1110 cm^{-1} gradually increased in frequency to about 1140 cm^{-1} and a shoulder developed near 1040 cm^{-1}, indicating additional polymerization of the adsorbed silane. The band near 1470 cm^{-1} gradually decreased in intensity and the bands near 1570 and 1300 cm^{-1} gradually increased in frequency to about 1590 and 1330 cm^{-1}, respectively, indicating some desorption or rearrangement of the absorbed carbon dioxide.

The stability of the oxide on iron mirrors during immersion in aqueous solutions of γ-APS was also investigated. Freshly polished iron mirrors were oxidized in air at 300°C for 15 min, producing an oxide about 22.5 nm in thickness and probably consisting of Fe_3O_4 and γ-Fe_2O_3 [11]. The infrared spectrum of the oxide was characterized by a band near 650 cm^{-1} and by shoulders near 700 and 600 cm^{-1} (see Figure 4A). The spectrum shown in Figure 4B was obtained after the same mirrors were immersed in a 1% aqueous solution of γ-APS at pH 10.4 for thirty minutes, rinsed in distilled water, and dried in nitrogen. When the spectrum shown in Figure 4B was subtracted from that shown in Figure 4A, it was observed that the shoulder near 600 cm^{-1} was reduced in intensity (see Figure 4C) but the bands near 700 and 650 cm^{-1} were not. These results may indicate that γ-Fe_2O_3 is etched during adsorption of γ-APS onto iron from aqueous solutions at pH 10.4 and that γ-APS films formed by adsorption from such solutions may contain some Fe(III) ions. By comparison, little etching of the oxide was observed during immersion of iron mirrors in aqueous solutions of γ-APS that had been acidified by addition of HCl.

Films formed by γ-APS adsorbed onto titanium mirrors were very similar in structure to those formed on iron mirrors [4]. Polysiloxane films contaminated by absorbed carbon dioxide were obtained from aqueous solutions at pH 10.4. Siloxane polymers having protonated amino groups were formed by adsorption from solutions acidified by addition of HCl. However, little evidence was observed for etching of the air-formed oxide on titanium mirrors during immersion in aqueous solutions of γ-APS regardless of pH [4].

                                      F. J. BOERIO ET AL.

Fig. 2. Infrared spectra of thin films formed by γ-APS adsorbed
         onto iron mirrors from 1% aqueous solutions at pH 10.4 in
         a nitrogen atmosphere and (A)maintained in nitrogen and
         (B)exposed to air.

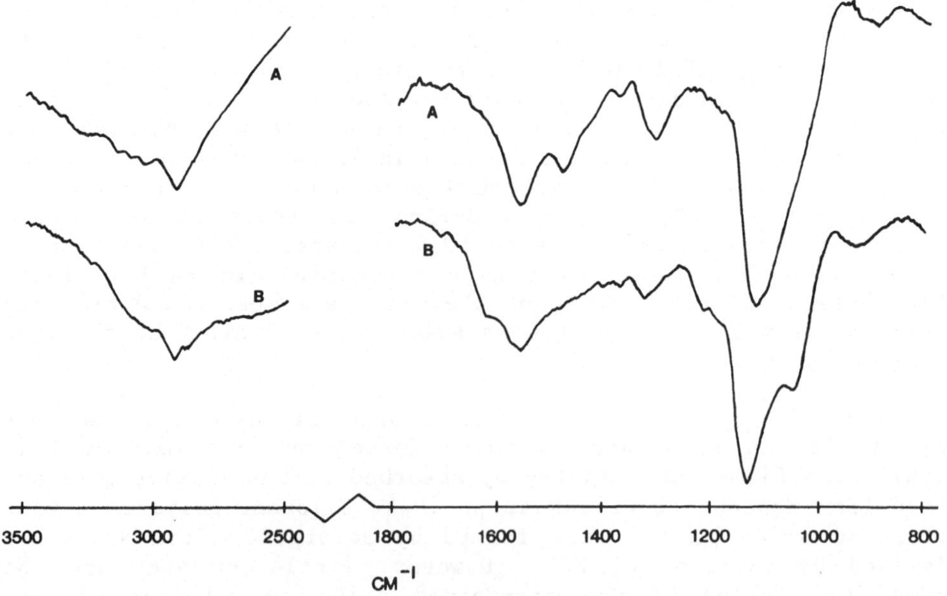

Fig. 3. Infrared spectra of thin films formed by γ-APS adsorbed
         onto iron mirrors from 1% aqueous solutions at pH 10.4:
         (A)as formed and (B)after exposure to lab atmosphere.

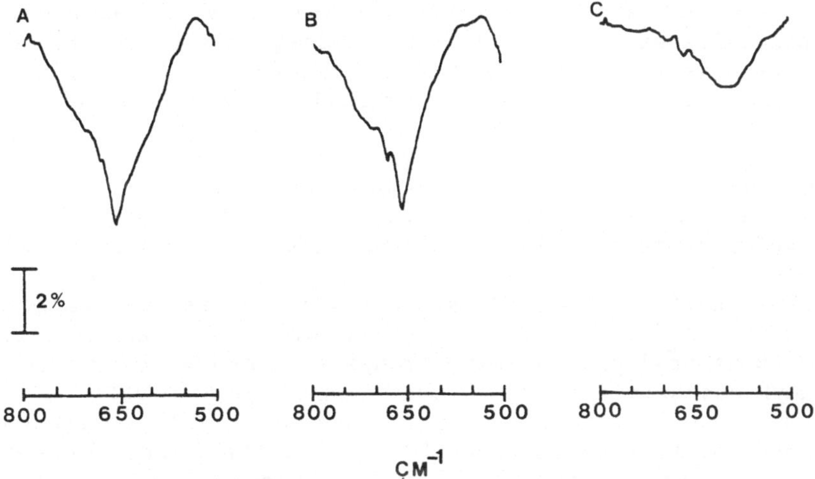

Fig. 4.   Effect of immersion in 1% aqueous solution of γ-APS at pH
          10.4 on infrared spectra of air-formed oxide on iron mir-
          rors: (A)before immersion, (B)after immersion, and (C)
          difference.

γ-APS films were extremely useful primers for improving the hydrothermal stabilities of iron/epoxy [3] and titanium/epoxy [4] lap
joints. Iron/epoxy lap joints prepared without using γ-APS primers
retained only about 25% of their original strength after 60 days in
water at 60°C. Joints prepared from adherends primed with γ-APS at
pH 10.4 retained about 50% of their strength after such exposure.
However, iron/epoxy lap joints prepared from adherends primed with
γ-APS at pH 8.0 were the most durable and retained about 75% of
their strength after 60 days (see Figure 5). γ-APS was also an effective primer for titanium/epoxy lap joints. The breaking
strength for titanium/epoxy joints prepared without γ-APS primers
decreased slowly as a function of immersion time and was about 960
psi after 60 days in water at 60°C. The strength of titanium/epoxy
lap joints prepared from adherends primed with γ-APS at either pH
10.4 or 8.0 hardly changed during similar exposure to water and was
about 1750 psi after 60 days (see Figure 6).

The results described above indicate that the overall structures of films formed by γ-APS adsorbed onto iron and titanium are
similar. However, the properties of the films as primers for improving the hydrothermal stabilities of iron/epoxy and tita-
nium/epoxy adhesive joints are very different. γ-APS primers are
more effective on iron/epoxy lap joints when applied from aqueous
solutions at pH 8.0 than when applied at pH 10.4. Lowering the pH
at which the γ-APS primer is applied to the adherends from 10.4 to

8.0 has no effect on the hydrothermal stability of titanium/epoxy
lap joints. The effectiveness of the primer may, therefore, depend
more on the orientation of the γ-APS molecules at the oxidized surface of the metals than on the overall structure of the primer
films [4].

The isoelectric points of the oxidized iron and titanium surfaces are near 10.0 and 6.0, respectively [18]. The pk_a's of the
silanol and protonated amine groups are near 3.0 [18] and 10.0, respectively. When an iron mirror is immersed in an aqueous solution
of γ-APS at pH 10.4, the oxide surface will consist of nearly equal
numbers of $FeOH_2^+$ and FeO^- groups and will have nearly zero net
charge. About half of the amine groups on γ-APS will be protonated
(NH_3^+) and essentially all of the silanol groups will be dissociated (SiO^-). Therefore, at pH 10.4 some γ-APS molecules may
adsorb onto positive sites through the silanolate ions while others
adsorb onto negative sites through the amino groups. At pH 8.0,
which is below the isoelectric point of most iron oxides, the oxidized surface of the iron mirrors will consist largely of $FeOH_2^+$
groups. In this case the γ-APS molecules may adsorb mostly through
the silanolate ions. γ-APS molecules may orient differently when
adsorbed onto iron from aqueous solutions at pH 10.4 and 8.0 [4].

When titanium mirrors are immersed in aqueous solutions of
γ-APS at either pH 10.4 or 8.0, the surface of the oxide will consist mostly of TiO^- groups. As before, about half the amino groups
on γ-APS will be protonated (NH_3^+) while most of the silanol groups
will be dissociated (SiO^-). At pH 10.4 or 8.0 the γ-APS molecules
may be adsorbed onto negative sites through the NH_3^+ groups. γ-APS
molecules may orient similarly when adsorbed onto titanium from
aqueous solutions at either pH 10.4 or 8.0 [4].

The behavior of γ-APS on the oxidized surface of aluminum is
very different. As noted above, the air-formed oxides on iron and
titanium are relatively stable during immersion in aqueous solutions of γ-APS. However, the oxidized surface of 2024 aluminum
mirrors was rapidly etched in basic solutions of γ-APS but the rate
of etching decreased as the pH of the solutions was decreased
[8,19].

Infrared spectra of films formed by γ-APS adsorbed onto aluminum mirrors from aqueous solutions acidified to pH 6.8 were similar to those of thin films formed on iron and titanium substrates
at comparable pH values (see Figure 7A). Such spectra were characterized by absorption bands near 1600 and 1500 cm^{-1} that were assigned to protonated amine groups and by a strong band near 1120
cm^{-1} that was assigned to the SiO stretching vibrations of a siloxane polymer [8].

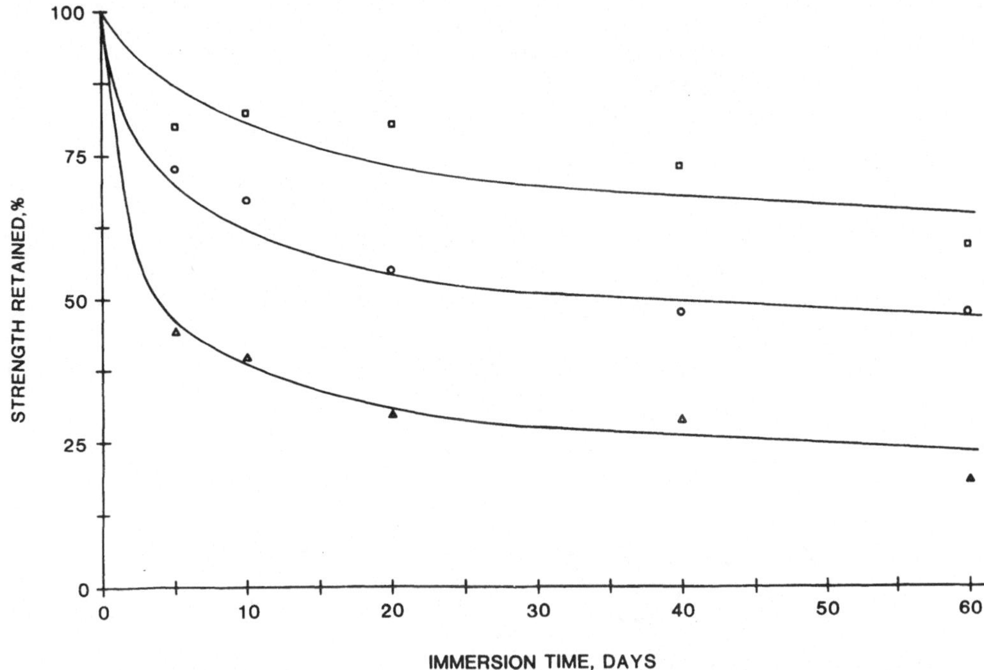

Fig. 5. Breaking strength of iron/epoxy lap joints as a function
of immersion time in water at 60°C: Δ-no silane primer,
O-γ-APS primer applied at pH 10.4, and ⎺-γ-APS primer
applied at pH 8.0. Adapted by permission from ref. 3.

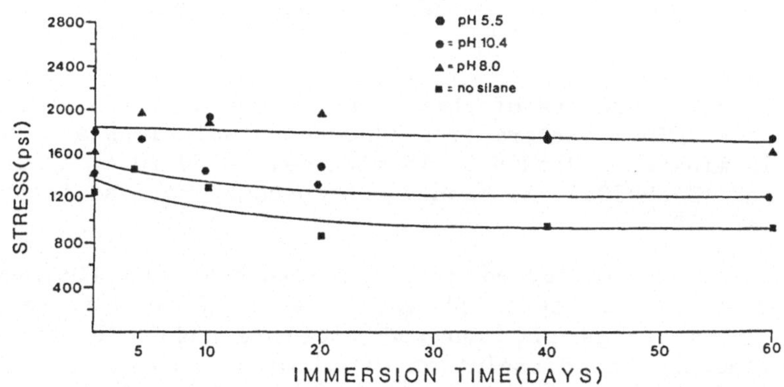

Fig. 6. Breaking strength of titanium/epoxy lap joints as a function of immersion time in water at 60°C. Adapted by permission from ref. 4.

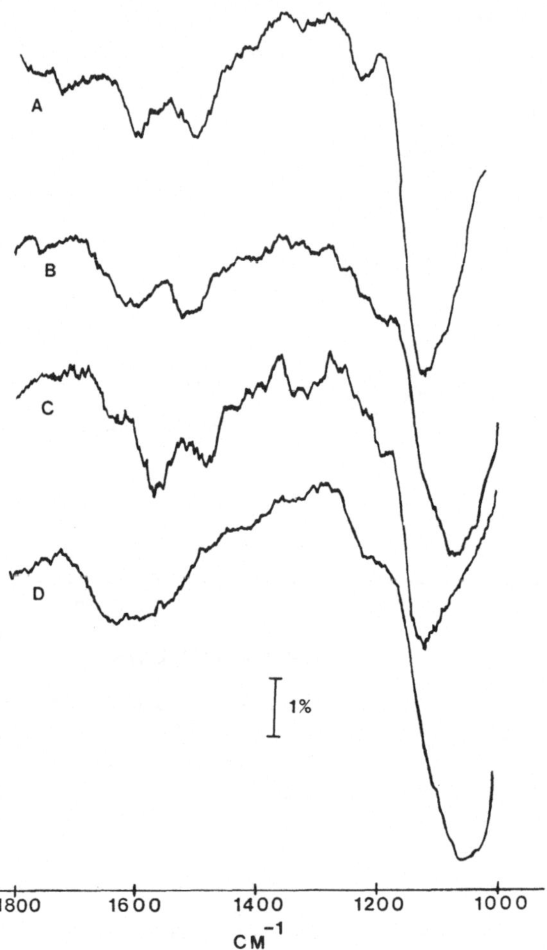

Fig. 7.   Infrared spectra of thin films formed by γ-APS adsorbed on-
         to aluminum mirrors from 1% aqueous solutions at (A)pH 6.8,
         15 minutes, (B)pH 8.5, 15 minutes, (C)pH 10.4, 1 minute,
         and (D)pH 10.4, 15 minutes. Adapted by permission from
         ref. 19.

RAIR spectra of films formed by γ-APS adsorbed onto aluminum from
aqueous solutions at higher pH values were quite different from
those of films formed on iron and titanium under similar conditions, reflecting the dissolution of aluminum oxides (see Figures
7B, 7C, and 7D). Films formed by adsorption of γ-APS onto aluminum
mirrors at pH 8.5 also contained protonated amino groups as indicated by absorption bands near 1600 and 1500 cm^{-1}. However, the
siloxane stretching mode was observed near 1080 cm^{-1}. This mode is
usually near 1120 or 1130 cm^{-1} for γ-APS films on aluminum under
conditions where little dissolution of the oxide occurs. The band

near 1080 cm^{-1} for films formed on aluminum at pH 8.5 may be related to formation of an aluminosiloxane polymer [8,19].

The infrared spectra of films formed on aluminum at pH 10.4 depended strongly on adsorption time. Films formed during a one minute adsorption were similar to those formed on iron and titanium at pH 10.4. However, films formed during 15 minutes adsorption at pH 10.4 were characterized by an infrared band near 1060 cm^{-1}, again indicating dissolution of the oxide and the likely formation of an aluminosiloxane polymer [8,19].

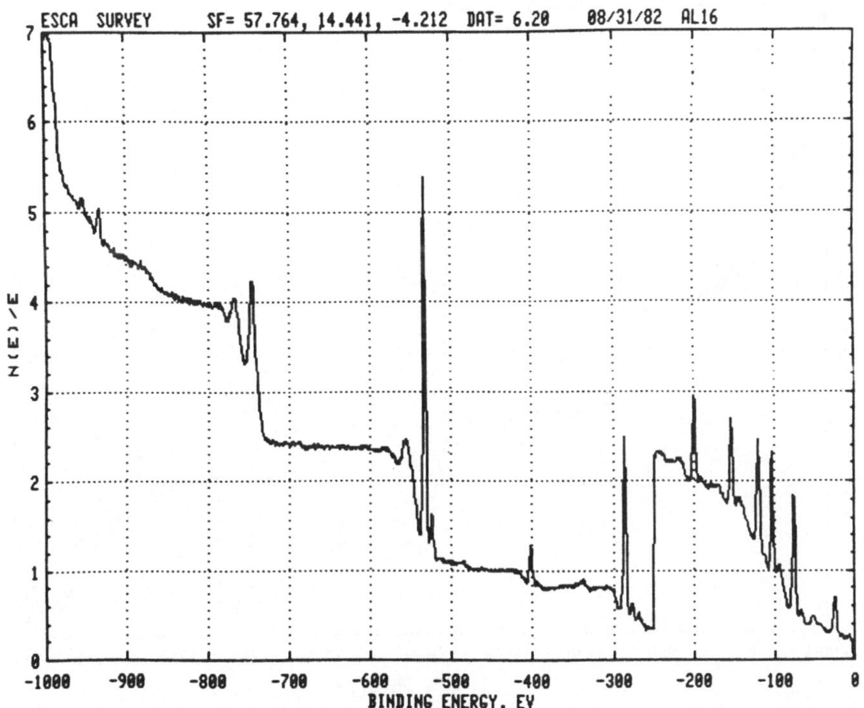

Fig. 8.   X-ray photoelectron spectra of thin films formed by γ-APS adsorbed onto aluminum mirrors from 1% aqueous solution at pH 8.5 for 15 minutes.

XPS spectra of mechanically polished 2024 aluminum mirrors were characterized by strong bands near 531.1 ev and near 74.0 eV that were assigned to the O(1s) and Al(2p) binding energies of Al_2O_3 and by very weak bands near 952.4 and 931.4 eV that were assigned to either metallic copper or Cu(I) ions (2024 alloy contains about 4.5% copper). XPS spectra of thin films formed by γ-APS adsorbed onto such mirrors for 15 minutes at pH 8.5 were generally similar (see Figure 8) to spectra of the "bare" mirrors but additional bands characteristic of N(1s), Cl(2p), and Si(2p) electrons were observed near 400.2, 199.0, and 101.7 ev, respectively. The Si(2p) binding energy was near that expected for a siloxane polymer [20]. The N(1s) peak had components near 399.8 and 400.6 ev, indicating the presence of both free and protonated amine groups [20]. The presence of both chlorine and protonated amine groups indicates formation of an amine hydrochloride. No band shifts that could be attributed to formation of aluminosiloxane polymers were observed.

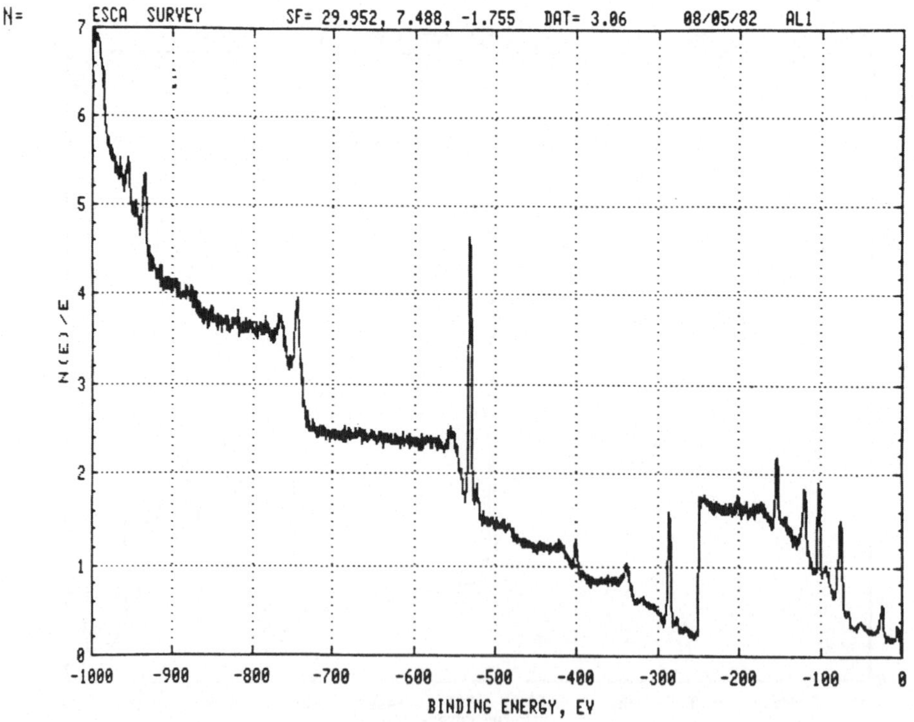

Fig. 9. X-ray photoelectron spectra of thin films formed by γ-APS adsorbed onto aluminum mirrors from 1% aqueous solution at pH 10.4 for 15 minutes.

The XPS spectra of thin films formed by γ-APS adsorbed onto aluminum mirrors for 15 minutes at pH 10.4 were similar to spectra of films formed during 15 minutes at pH 8.5 except for a few features (see Figure 9). The Cl(2p) band was not observed and the N(1s) band had only one component, near 399.8 ev, indicating free amino groups. The intensity of the copper 2p peaks near 952.4 and 931.4 ev increased significantly, indicating etching of the oxide and an accumulation of Cu(I) ions in the γ-APS films. Again, however, no band shifts attributable to the formation of aluminosiloxane polymers were observed.

Some evidence for coordination of Cu(I) ions to amino nitrogen atoms was obtained for films formed by γ-APS adsorbed onto aluminum for 15 minutes at pH 10.4. The x-ray induced copper Auger band was split into components near 339.4 and 336.9 ev. The peak near 336.9 ev is characteristic of Cu(I) ions in Cu_2O [20]. The additional peak near 339.4 ev is indicative of Cu(I) ions in an environment other than the oxide. A similar peak has been observed for γ-APS adsorbed onto copper at pH 10.4 [17] and assigned to Cu(I) ions coordinated to amino nitrogen atoms. Observation of the peak near 339.4 ev for γ-APS adsorbed onto 2024 aluminum at pH 10.4 reflects the ability of γ-APS to coordinate certain metal ions.

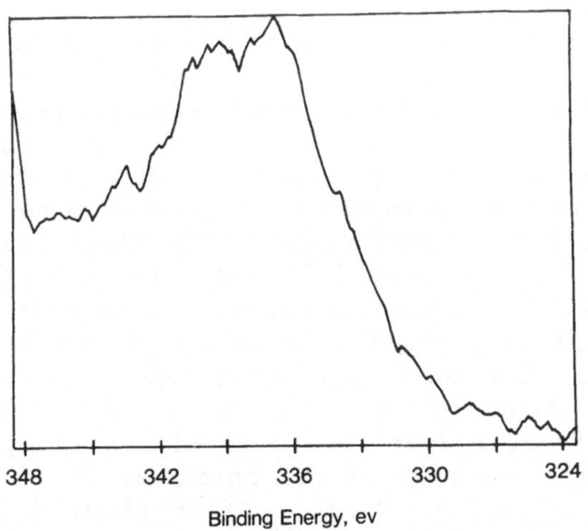

Fig. 10. X-ray induced copper Auger spectra for films formed by γ-APS adsorbed onto 2024 aluminum for 15 minutes at pH 10.4.

The hydrothermal stability of aluminum/epoxy adhesive joints has been extensively studied [1,5,6]. Recently it has been shown that the decrease in strength observed for such joints during exposure to water at elevated temperatures is related to hydration of the oxide (either natural or anodic) to form pseudoboehmite, a poorly crystallized hydroxide having the approximate composition $Al_2O_3 \cdot 2H_2O$ [21]. The pseudoboehmite adheres only poorly to the aluminum substrate and its formation leads to a rapid degradation of joint strength.

It is known that certain ions, including phosphate and silicate, can significantly inhibit the hydration of aluminum oxides at elevated temperatures [22] and it has recently been suggested that the excellent hydrothermal stability of the oxide produced by anodizing aluminum in phosphoric acid is related to the formation of a thin film of $AlPO_4$ on the surface of the oxide that inhibits hydration [21]. Organofunctional silanes such as γ-APS are known to be effective primers for improving the hydrothermal stability of aluminum/epoxy adhesive joints [1,5,6]. We were interested in determining the extent to which these primers inhibited hydration of aluminum oxides. Some preliminary results are described below.

Several sample mirrors were prepared by mechanically polishing 2024 aluminum as described previously. Thin films ($\cong 5$ nm) of γ-APS were then applied to half of the mirrors by immersing the mirrors in dilute aqueous solutions of γ-APS at pH 7.0. After 15 min, the mirrors were withdrawn and the excess solution was blown off using a strong stream of nitrogen. The mirrors were then heated in an oven at $100°$ C for 20 min to thoroughly polymerize the silane film. The remaining (control) mirrors were not treated with γ-APS but they were heated in the same way as the silanated mirrors. All of the mirrors were then immersed in a beaker of distilled, deionized water at $60°C$. At appropriate intervals, pairs of mirrors (silanated and control) were removed from the beaker and dried. The mirrors were then examined with the ellipsometer to determine the ellipsometric parameters Δ and Ψ for the hydroxide layers. A computer program was used to determine the thickness of the hydroxide layers assuming that the refractive indices of the hydroxide and substrate were 1.58 and 1.14 (1-4.06i), respectively [23]. The results are shown in Figure 11. Hydration of the control mirrors proceeded very slowly during the first 7.5 min. However, for times between 7.5 and 10.0 min, hydration was very fast and hydroxide layers about 70 nm in thickness were obtained after 10 min. By comparison, hydration of the silanated mirrors was always very slow and little change in the thickness of the oxide/hydroxide layer was observed after 10 min. After about 2 h immersion, the hydroxide layer on the silanated mirrors was still very thin (less than 10 nm) but that on the control mirrors had increased in thickness to a few hundred nanometers. These results indicate that

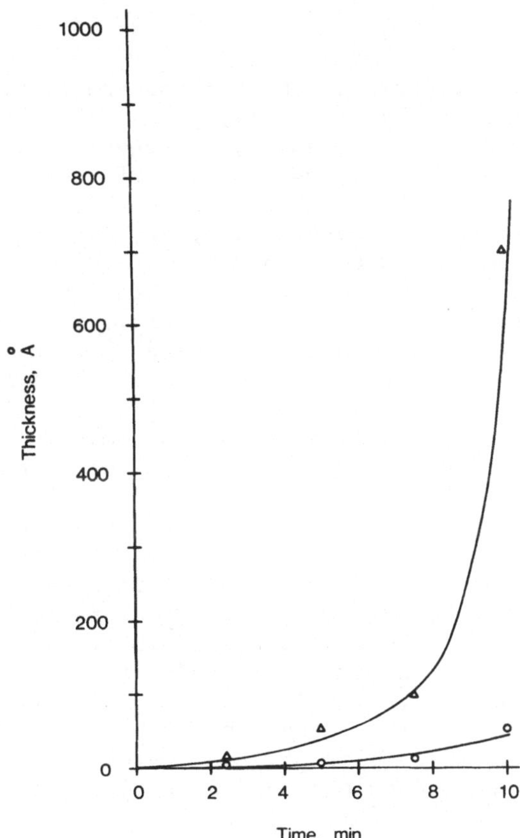

Fig. 11. Hydration of air-formed oxide on aluminum mirrors during
immersion in distilled, deionized water at 60°C: Δ-no
silane primer; O-γ-APS primer applied at pH 7.0.

organosilane primers such as γ-APS may improve the wet strength of
aluminum/epoxy adhesive joints by inhibiting hydration of the
oxide.

ACKNOWLEDGEMENTS

This work was supported in part by a grant from the Office of
Naval Research. The assistance of Armco, Inc., Timet, Inc., and
Union Carbide Corp. by providing iron, titanium, and γ-aminopro-
pyltriethoxysilane samples is also acknowledged.

REFERENCES

1. R. L. Patrick, J. A. Brown, N. M. Cameron and W. G. Gehman, Appl. Polymer Symp. 16, 87 (1981).
2. M. E. Schrader and J. A. Cardamone, J. Adhesion 9, 305 (1978).
3. F. J. Boerio and J. W. Williams, Appl. Surf. Sci. 7, 19 (1981).
4. F. J. Boerio and R. G. Dillingham in Proc. Intl. Symp. Adhesive Joints: Formation, Characteristics, and Testing, K. L. Mittal, ed., to be published, 1983.
5. F. J. Boerio and C. A. Gosselin, Proc. 36th Ann. Tech. Conf., SPI Reinforced Plastics/Composites Inst., Sec. 2G, 1981.
6. A. J. Kinloch, W. A. Dukes and R. A. Gledhill, in Adhesion Science and Technology, L. H. Lee, ed., Plenum Press (1975), p. 597.
7. P. W. Erickson and E. P. Plueddemann, in Composite Materials, Vol. 6, L. J. Broutman and R. H. Krock, eds., Academic Press, New York (1974), Ch. 1.
8. F. J. Boerio and C. A. Gosselin in Instrumental and Physical Characterization of Macromolecules, C. D. Craver, ed., American Chemical Soc., Washington, D.C., in press, 1983.
9. F. L. McCrackin, E. Passaglia, R. R. Stromberg and H. L. Steinberg, J. Res. Natl. Bur. Stds. 67A, 363 (1963).
10. W. M. Riggs and M. J. Parker, in Methods of Surface Analysis, A. W. Czanderna, ed., Elsevier Sci. Pub. Co. (1975), Ch. 4.
11. A. B. Winterbottom, J. Iron Steel Inst. 165, 9 (1950).
12. S. Thibault, Thin Solid Films 35, L33 (1976).
13. F. J. Boerio, L. H. Schoenlein, and J. E. Greivenkamp, J. Appl. Polymer Sci. 22, 203 (1978).
14. F. J. Boerio, L. Armogan, and S. Y. Cheng, J. Colloid Interface Sci. 73, 416 (1980).
15. C. H. Chiang, H. Ishida and J. L. Koenig, J. Colloid Interface Sci. 74, 396 (1980).
16. F. J. Boerio and J. W. Williams, Proc. 36th Ann. Tech. Conf., SPI Reinforced Plastics/Composites Inst., Sec. 2F, 1981.
17. F. J. Boeriod, J. W. Williams and J. M. Burkstrand, J. Colloid Interface Sci. 91, 485 (1983).
18. E. P. Plueddemann, in Composite Materials, L. J. Broutman and R. H. Krock, eds., Vol. 6, Academic Press (New York), 1974, Ch. 6.
19. F. J. Boerio, C. A. Gosselin, R. G. Dillingham and H. W. Liu, J. Adhes. 13, 159 (1981).
20. C. D. Wagner, W. M. Riggs, L. E. Davis, J. F. Moulder and G. E. Muilenberg, "Handbook of X-Ray Photoelectron Spectroscopy," Perkin-Elmer Corp., Eden Prairie, MN, 1979.
21. G. D. Davis, T. S. Sun, J. S. Ahearn and J. D. Venables, J. Materials Sci. 17, 1807 (1982).
22. D. A. Vermilyea and W. Vedder, Trans. Farad. Soc. 66, 2644 (1970).
23. N. McDevitt, Technical Report 73-245, Air Force Materials Laboratory, Wright-Patterson Air Force Base, January, 1974.

THE EFFECT OF γ-APS SUBSTRATE MODIFICATION UPON THE CHEMICAL
ADHESION OF POLY(AMIC ACID-IMIDE) FILMS

D. J. Belton and A. Joshi

Philips Research Laboratories Sunnyvale, Signetics
Corp., Sunnyvale, CA 94086

Lockheed Palo Alto Research Laboratory, Palo Alto,
CA 94304

ABSTRACT

The chemical adhesion of poly(amic acid-imide), PAI, films
upon silicon, silicon nitride, silicon dioxide and aluminum substrates was examined as a function of surface modification by
γ-aminopropyltriethoxysilane, (γ-APS). In these systems it has
been observed that curves describing the kinetics of PAI
dissolution provide a useful means of discriminating differences in
the solvent resistance of the PAI-substrate interphase. A ranking
of the chemical adhesive interaction of PAI films on γ-APS modified
substrates has been shown to go as $Al < Si_xN_y < Si < SiO_2$. This
substrate effect has been shown to act in concert with an effect
arising from the thickness of applied γ-APS layers. These two
effects have been discussed in terms of coherence of the
PAI-γ-APS-substrate interphase, short range γ-APS-substrate
interactions, and the acid-base character of the substrate.

INTRODUCTION

Thin polyimide films have received considerable attention in
recent years for a number of applications within the
microelectronics manufacturing industry [1]. These applications
include insulators in multilevel interconnect systems, passivation
layers, and α-particle barriers. The nature of the polyimide-substrate interphase assumes particular importance from both a litho-
graphic performance, and device reliability standpoint when dielectric applications are considered. A common feature relating the
above stated concerns is the polyimide-substrate chemical

adhesion. It is the intent of this paper to examine phenomenologi-
cally the chemical adhesion of selected polyimide-organo-
silane-substrate systems.

The ability of organosilane materials to function as adhesion
promotors in composite systems is well known. The importance of
the adsorbed organosilane film structure in influencing its ability
to act in this capacity must be considered. Schrader et al. [2]
characterized the structure of γ-aminopropyltriethoxysilane, γ-APS,
films deposited from benzene solutions upon polished pyrex. The
room temperature and boiling water desorption of ^{14}C labeled γ-APS
demonstrated its existance in several distinctly different states
of interaction with the substrate. A physisorbed layer, which
constitutes the bulk of the adsorbed material, is easily removed
during a room temperature water rinse. A second fraction is
removed following several hours exposure to boiling water.
Finally, following long periods of boiling water extraction there
remains an irreversibly adsorbed monolayer. The effect of this
layer structure was demonstrated by Schrader and Block [3] during
the examination of adhesive joint failure in pyrex-resin systems
following boiling water exposure. It was found that the joint life
depended not only upon the amount of adsorbed γ-APS, but on the
structure of the adsorbed film as well. Extraction of a portion of
the adsorbed layers by hot or cold water followed by joint
preparation led the authors to conclude that the physisorbed
material did not contribute significantly to the life of the
joint. This was contrasted by a strong dependence of joint life
upon the amount of chemisorbed material present.

DiBenedetto and Scola [4] investigated γ-APS films on S-glass
fibers with a combination of ion scattering spectroscopy and
secondary ion mass spectrometry. The results indicated that γ-APS
exists in differing states of interaction with itself, as well as
the substrate. Sputter profiling depicted a layer of partially
cured oligomers sandwiched between highly crosslinked polysiloxane
layers.

Ishida and Koenig [5] proposed a model for the structure of
γ-methacryloxypropyltrimethoxysilane, (γ-MPS), adsorbed on E-glass
fibers analogous to that previously described for γ-APS. They
proposed that their model could be generalized for the interphase
structure of many different silane layers. In general, the bulk of
the adsorbed material consists of small oligomers with a few
siloxane linkages rendering them soluble in organic media. Below
this physisorbed layer is a more solvent resistant layer consisting
of oligomers interconnected via more siloxane linkages. In this
layer the lack of organic interconnections renders the materials
extractable by hot water. Finally, near the glass surface is a
hydrolytically stable, chemisorbed layer characterized by an

extensive network structure. Chemisorption occurs via siloxane linkages with substrate SiOH functionalities [5,6,7].

The role of network interpenetration in the adhesive joint strength of sapphire–γ–APS–polyethylene systems was investigated by Sung et al. [8]. This system was chosen in order to negate possible chemical interactions from obscuring the importance of the physical interaction. The peel strength displayed a maxima when plotted against the original γ–APS solution concentration. Interdiffusion was suggested as the mechanism most likely responsible for the observed increase in mechanical strength.

Linde [6] examined the chemical adhesive interactions of γ–APS treated surfaces with a poly(amic acid–imide), PAI, resin. The interaction responsible for substrate adhesion in this system was defined as an incorporation of the alkyl surface bound amine over the aromatic polymer amine in the imide linkage formed upon heating. Polymer scission appears to be a result of this adhesion process. This was corroborated in a study by Greenblatt and coworkers [9] on polyimide adhesion to SiO_2 surfaces treated with γ–APS.

We wish to refer to chemical adhesion as the solvent resistant interaction occuring at or within the PAI–substrate interphase. From the above discussion, it appears that one must be concerned with both network interpenetration, and chemical interactions when considering the application of PAI films on γ–APS–modified substrates. Earlier work alluded to the importance of these phenomenon with respect to the dissolution behavior of PAI films interacting with γ–APS–modified silicon surfaces [10]. In this paper we will extend this work to include a variety of substrate materials that assume importance in the semiconductor manufacturing industry.

EXPERIMENTAL

A. Materials

The polyamic acid solutions used in this work were du Ponts Pyralin materials PI 2545 and PI 2555. Du Ponts VM651, a –APS adhesion promotor, was diluted with a solution of 195:5 methanol: deionized water to the following concentrations:

1) 10 µL VM651/100 ml solvent
2) 25 µL VM651/100 ml solvent
3) 50 µL VM651/100 ml solvent
4) 75 µL VM651/100 ml solvent
5) 100 µL VM651/100 ml solvent

These solutions were allowed to equilibrate overnight prior to use at pH 10.2. The solvent systems employed in this study were Shipleys Microposit 351 and 300. These are sodium hydroxide and potassium hydroxide based aqueous solutions respectively.

Substrate materials were <111> p type test wafers manufactured by Monsanto. Those wafers that were not used as received were coated in the following fashion:

a) SiO_2 films were thermally grown to 600 nm in thickness.
b) Si_xN_y was deposited to a thickness of 100 nm in an AMP 3300 plasma enhanced chemical vapor deposition system.
c) Aluminum was deposited to a thickness of 200 nm in a CHA electron beam evaporator equipped with an Airco-Temescal gun.

B. Measurement of Dissolution Behavior

Unless stated otherwise in context, the substrates of interest were immersed in the appropriate γ-APS solution for times long enough to insure adsorption equilibrium. These times were determined experimentally from adsorption isotherms. The substrates were subsequently dried by flowing clean dry nitrogen over the surface. Poly(amic acid) solutions were then applied by spin coating at 1000 rpm/10 sec, 5000 rpm/30 sec to yield films ranging from 1.9μm (PI2545) to 2.5μm (PI2555). The coated substrates were then placed in convection ovens at the desired temperature for 45 minutes to partially imidize the polymer films. Each wafer was then coated with a positive photoresist material (Hunt HPR 204) and windows 120μm^2 were defined. Sections from each sample were then exposed to the PAI solvent system chosen for incrementally increasing times, and the film thickness was monitored interferometrically with a Nanospec film thickness measuring system (measurement error $\pm$ 6%).

C. Characterization of the APS Film Structure

1. Ellipsometry – The basic equation of ellipsometry expresses the ratio of the complex reflection coefficients, ρ, in terms of the polarizer, Ψ, and analyzer, Δ, angles at extinction.

$$\rho = \tan \Psi \exp i \Delta \qquad\qquad [1]$$

Clean substrates can be characterized by the maximum Δ value obtainable [11]. The presence of a foreign species upon a well-characterized surface will alter the polarization state of the reflected light, and can therefore be detected through a change in Δ. The adsorption/desorption behavior of γ-APS upon the various substrates studied was monitored ellipsometrically, and the results of such are presented in terms of Δ deviations from the clean substrate value, $\delta\Delta$. In all cases, the clean substrate values were obtained immediately following deposition, or in the case of silicon, immediately following precleaning in a buffered HF etchant. All measurements were performed on a Rudolph Research Auto El-II automatic ellipsometer.

2. ESCA - Data was collected upon a PHI-560 electron spectrometer with angular resolved capability. A 300W Mg anode (photon energy 1253.6 eV) was used for photoelectron excitation, and the elemental data were obtained at normal, and grazing (20°) take off angles.

Thickness determinations were made from the integrated areas of selected peaks. The measured electron intensity of an element x in the adsorbate layer, I_x, can be written as:

$$I_x = \kappa S_x n_x \ (1 - e^{-d}/\lambda_{x-A}) \qquad\qquad [2]$$

where n_x is the atom density of the element x in the adsorbate layer and κ is an instrumental factor. S_x is the sensitivity factor of the transition of interest. λ_{x-A} is the escape depth of the transition electron in the adsorbate layer A. d is the adsorbate layer thickness in $\overset{\circ}{A}$.

Likewise, the intensity of the substrate signal, s, is attenuated as follows:

$$I_s = \kappa S_s n_s e^{-d}/\lambda_{s-A} \qquad\qquad [3]$$

By taking the ratio of the adsorbate signal to the substrate signal, the thickness can be calculated.

The escape depths were estimated from the general formula for organic compounds [15]. The atomic density ratio was defined from grazing angle measurements, and stoichiometry considerations. The sensitivity factors were based upon pure elemental values. The estimated values of n and S should therefore lead to an underestimation of the true thickness.

Previous experiments [10] indicated that thickness variations resulting from sample UHV exposures were negligible.

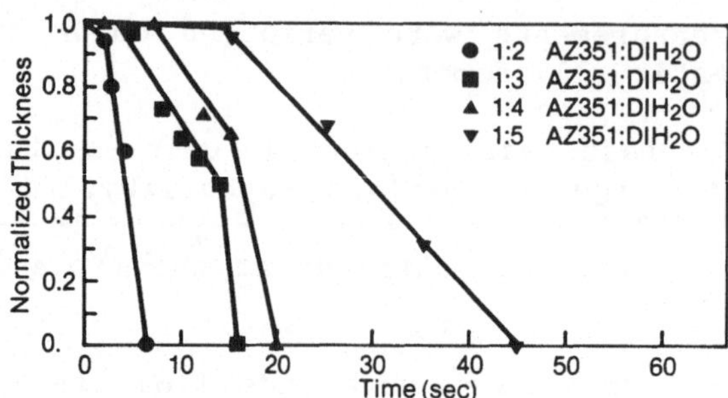

Fig. 1.   Normalized thickness of PAI versus time. The PAI film was
PI-2545 cured at 145°C/45 min. The solvent was Shipley
AZ351 an aqueous NaOH solution diluted with deionized
water.

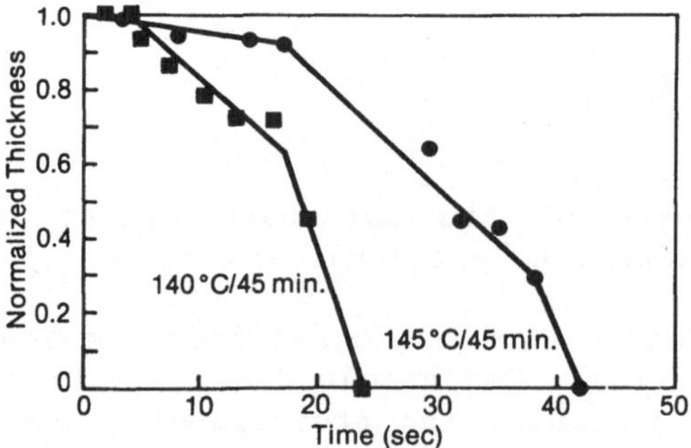

Fig. 2.   Normalized thickness of PAI versus time. The PAI film was
PI-2545 cured for 45 min. The solvent was Shipley AZ351,
an aqueous NaOH solution diluted 1:3 with deionized water.

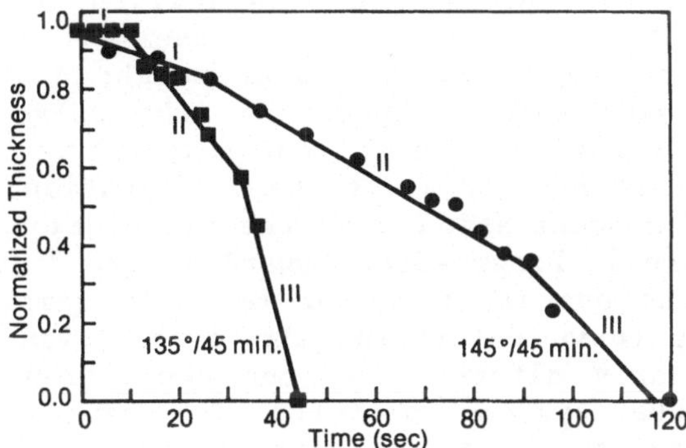

Fig. 3. Normalized thickness of PAI versus time. The PAI film was
PI-2555 cured for 45 min. The solvent was Shipley AZ351,
an aqueous NaOH solution diluted 1:3 with deionized water.

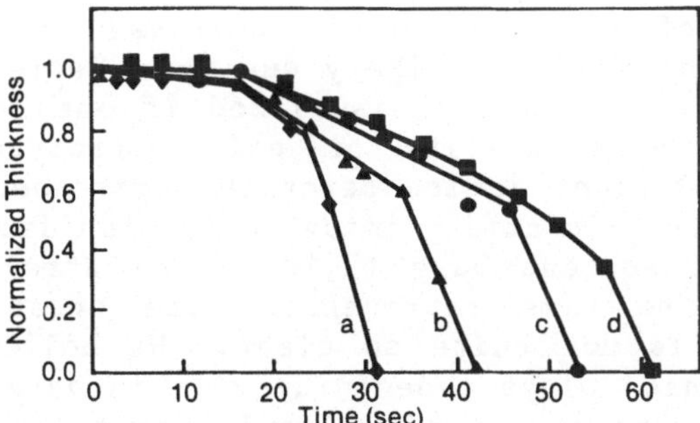

Fig. 4. Normalized thickness of PAI versus time. The PAI film was
PI2545 cured at 150°C/45 min. Surface treatments were:
a) none, b) 5 min immersion in γ-APS followed by a 5000
rpm spin, c) γ-APS solution puddled on substrate followed
by a 5000 rpm spin, d) 5 min immersion in γ-APS of 10_x
concentration followed by a 5000 rpm spin.

RESULTS AND DISCUSSION

A. PAI Dissolution Kinetics

A variety of factors are known to affect the dissolution behavior of PAI films interacting with a given substrate material [12]. Some of these are: the polymer chemical composition, the polymer molecular weight and molecular weight distribtution, the amic acid–imide conversion temperature, the solvent composition, the solvent concentration, the solvent temperature, and the nature of the substrate material or substrate modification. The effect of changing solvent concentration upon the dissolution rate of PI2545 is given in Figure 1. Progression toward a more dilute solvent not only increases the overall time for PAI film removal, but alters the shape of the curve describing the dissolution behavior. The curves are similarly altered when the conversion temperature is increased for the systems pictured in Figures 2 and 3. In practically every case, the exception being the most dilute solution of Figure 1, the PAI films exhibit three regions of differing etch rates. These regions are labeled in Figure 3, and correspond to: I. an inhibition period related to solvent invasion of the polymeric matrix, II. a region of constant rate that is probably due to the effects of gel formation and molecular dissolution, and III. a region characterized by rapid film removal. The appearance of a changing slope (decreasing) in region II, and an increase in time marking the onset of region III are the results of decreasing the ionic strength of the solvent or increasing the amic acid – imide conversion temperature. The major factor contributing to the decreased slope in region II as the ionic strength of the solvent is decreased or the conversion temperature is increased is likely due to a decrease in molecular dissolution. This is easily visualized if one considers, first, the fact that the solubility of polyelectrolytes is known to decrease as solution ionic strength decreases, [13,14] and secondly, that the amic acid – imide conversion decreases polymeric solubility in an aqueous base [7,10]. The effects of the above variables upon the time representing the onset of region III behavior, t_{III}, is not quite so clear. We believe that the PAI substrate interphase plays a deciding role in determining not only t_{III},[10] but also the extent of molecular dissolution occurring in region II (i.e. the initial thickness in region II – the final thickness in region II), as well as the overall time for PAI film removal, t_f.

Figure 4 shows the effects of γ-APS modification of a silicon substrate upon the dissolution behavior of PI2545. The substrate devoid of coupler shows a dramatic loss in film thickness over a rather short time period. Those substrates that were first treated

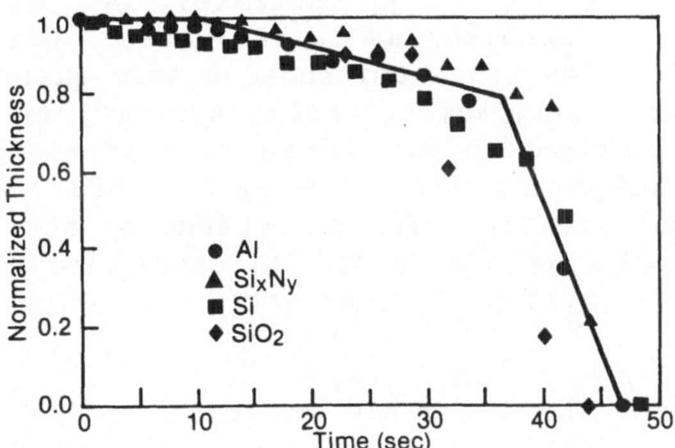

Fig. 5. Normalized thickness of PAI versus time. The PAI film was
PI-2555 cured 140°C/45 min. The solvent was Shipley
Microposit 351, an aqueous NaOH solution diluted 1:2 with
deionized water. Substrate materials were Al, Si_xN_y, Si,
SiO_2. Substrates were without silane modification.

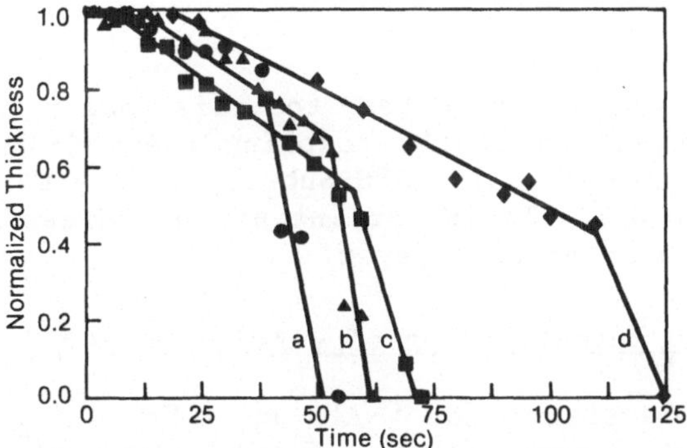

Fig. 6. Normalized thickness of PAI vemsus time. The PAI film was
PI-2555 cured 140°C/45 min. The solvent was Shipley
Microposit 351, an aqueous NaOH solution, diluted 1:2 with
deionized water. Substrates were Al, Si_xN_y, Si, SiO_2.
γ-APS was applied by spin coating from a solution of con-
centration 10 μL/100 mL solvent.

with γ-APS prior to PI2545 application and thermal conversion show both an increase in t_f as t_{III} increases, and an increase in the extent of region II behavior as t_f increases. Curves b and c correspond to γ-APS films whose thicknesses were quite similar in magnitude as measured ellipsometrically, however, the mode of application varied substantially. Curve d corresponds to the same application technique as curve b with a γ-APS thickness approximately four times greater. The importance of both the application technique, as well as the film thickness is readily apparent. These observations will be further treated at a point later in this paper.

If we change the substrate material, and do not perform an organosilane modification the behavior shown in Figure 5 results. In this case the dissolution behavior of PI2555 appears to be substrate independent. Modification of these same substrate materials with γ-APS results in the dramatically altered behavior shown in Figure 6. Once again as t_f increases, t_{III} increases, and the extent of region II behavior increases. The fact that the dissolution behavior is substrate independent for the solvent systems employed, unless an organosilane substrate modification preceeds the polymer film deposition, leads us to conclude that region III behavior can be ascribed to a chemical adhesion failure within the interphase. t_f can, therefore, be used as a measure of the solvent resistant adhesive interaction, and a ranking of the magnitude of this interaction for the substrates studied can be accomplished via Figure 6. This ranking goes as:

$$Al < Si_xN_y < Si < SiO_2$$

The observations presented to this point raise questions regarding the influence of the coupling agent film structure, and the subsequently formed PAI-γ-APS-Substrate interphase upon the PAI dissolution behavior. It is our intent to address these questions during the remainder of this paper.

B. Ellipsometric Investigation of γ-APS Film Structure

Adsorption isotherms for γ-APS upon Si, SiO_2, Si_xN_y, and Al are presented in Figure 7. The figures are plotted as the Δ variation, $\delta\Delta$, versus solution concentration. For each substrate material, as the solution concentration increases the amount of adsorbed organosilane increases. These isotherms are typical of those where physisorption is important. Indeed, as previously discussed, the adsorption of organosilanes on high energy substrates will occur beyond a chemisorbed monolayer. The structure of the adsorbed organosilane at equilibrium is generally multilayered, with a preponderance of physisorbed oligomers [2,5,10]. An indication of the relative quantities of loosely

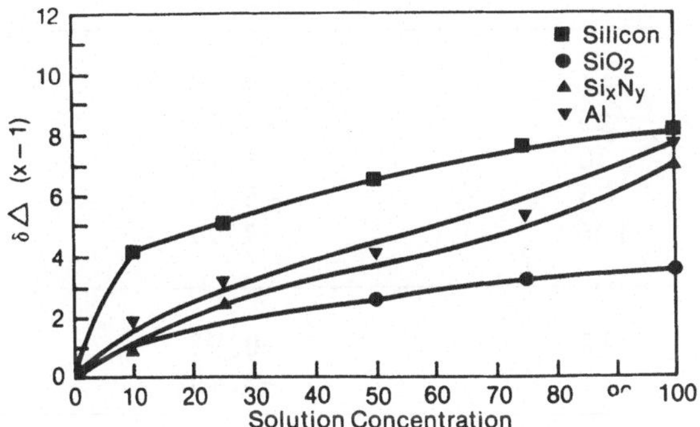

Fig. 7.   The relative amount of γ-APS adsorbed versus original γ-
APS solution concentration. The relative amount adsorbed
is plotted as −δΔ on Al, Si$_x$N$_y$, Si, SiO$_2$.

bound material on a given substrate can be ascertained from the δΔ
data presented in Table 1. Here the δΔ value after the attainment
of adsorption equilibrium is contrasted against the δΔ values
following the γ-APS desorption. (In the case of desorption, a room
temperature water extraction was considered complete upon the
attainment of constant δΔ values). In the above context, loosely
bound refers to the water extractable material, since water
desorption experiments will not truly define only physisorbed
species. Increasing the original γ-APS solution concentration
increases the quantity of room temperature water extractable
material. This is in accord with earlier findings [2,5]. The
effect of this loosely bound layer upon the solvent resistance of a
subsequently formed PAI-γ-APS interphase is demonstrated in Figure
8. Here the time required for complete PAI film removal is plotted
as a function of original γ-APS solution concentration and
substrate material. For each substrate material the time required
for complete PAI film removal increases as the original γ-APS
solution concentration increases. In other words, the chemical
adhesion of the PAI film increases as the thickness of the γ-APS
layer increases for a given substrate.

Table I. $\delta\Delta$ characterization as a function of substrate and subtrate treatments.

Sample	Solution Concentration (L 100mL)	After Deposition	After Desorption
Al	10	−2.48	−1.74
	25	−3.07	−1.42
	50	−4.13	−1.32
	75	−5.14	−1.32
	100	−7.77	−1.35
Si N	10	−0.95	−0.16
	25	−2.49	−0.19
	50	−	−
	75	−3.12	−0.05
	100	−7.00	−0.10
Si	10	−4.18	−3.65
	25	−5.04	−3.11
	50	−6.44	−2.95
	75	−7.50	−2.82
	100	−7.93	−2.70
SiO	10	−1.19	−0.53
	25	−1.74	−0.23
	50	−2.60	−0.36
	75	−3.13	−0.29
	100	−3.41	−0.30

The data offered in Figures 6 and 8 correspond to different organosilane application techniques. Figure 6 corresponds to substrates that were modified by first puddling a 10µL/100mL solution of γ-APS upon the surface, and spinning at 5000 rpm until dry (30 sec). Figure 8 corresponds to PAI films that were applied to substrates which were first allowed to achieve adsorption equilibrium with γ-APS prior to drying in a flowing N_2 stream. It is gratifying to note that the ranking of the chemical adhesion in these separate cases corresponds. It is also interesting that the data in Figure 6 demonstrates a substrate effect, while the data in Figure 8 demonstrates effects from both the substrate, as well as the γ-APS layer thickness. These complementary sets of data allow us to qualitatively separate out the effects of the substrate upon the adsorbed γ-APS structure from the effects of γ-APS thickness upon the subsequent PAI chemical adhesion.

If, as the data suggests, we separate these effects, then we are provided with a convenient means for their discussion. Let us begin with the observed increase in chemical adhesion accompanying increasing γ-APS film thickness. The apparent contradiction of the chemical adhesion data with that of Schrader and Block [3] can be resolved, and a phenomenological description of the PAI dissolution behavior can be offered as follows. Schrader and Block examined the mechanical integrity of bonded systems following a boiling water exposure. In these experiments we are examining the chemical adhesion following a 140°C PAI curing cycle which intuitively should alter the interphase. According to Linde [6] a reaction at this temperature should lead to a quantity of surface bound alkyl imide linkages. This mode of adhesion is diagrammed in Figure 9. It is not unreasonable to assume that the presence of the polymeric amic acid functionalities could, in addition, lead to the incorporation of physisorbed organosilane oligomers into a more complex polymer-silane network structure. Finally, the 140°C curing cycle can lead to further organosilane condensation reactions thereby incorporating additional material into a surface bound network structure. The validity of this latter conjecture is supported by the desorption data presented in Table II. Here, room temperature water extraction data is presented for samples with and without a 140°C/45 min thermal cycle. It appears, therefore, that an increase in the quantity of loosely bound silane enhances the ability of the applied PAI film to both interpenetrate and interact with alkyl amine functionalities. The enhanced mixing and the subsequent enhanced susceptibility to intermolecular chemical linkages will increase the extent of the interfacial region, and could form the basis for the observed, increased stability in the presence of an aggressive solvent.

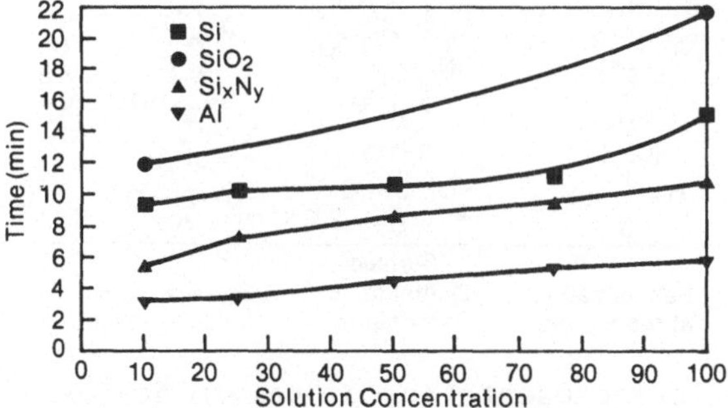

Fig. 8.   Time for the complete removal of PI-2555 cured at 140°C/ 45 min from substrates treated with γ-APS by a 60 min immersion and N$_2$ dry. The substrates were al, Si$_x$N$_y$, Si, SiO$_2$.

Table II.   The effect of curing temperature upon the quantity of room temperature extractable γ-APS.

Original γ-APS Solution Conc. (μL 100 ml)	Curing Temperature	δΔ Adsorption	δΔ Desorption
100	140°C/45 min	-13.26	-10.34
100	R.T.	-13.29	- 1.51
50	140°C/45 min	-7.63	-6.26
50	R.T.	-6.91	-1.30
25	140°C/45 min	-4.12	-3.80
25	R.T.	-4.45	-1.12

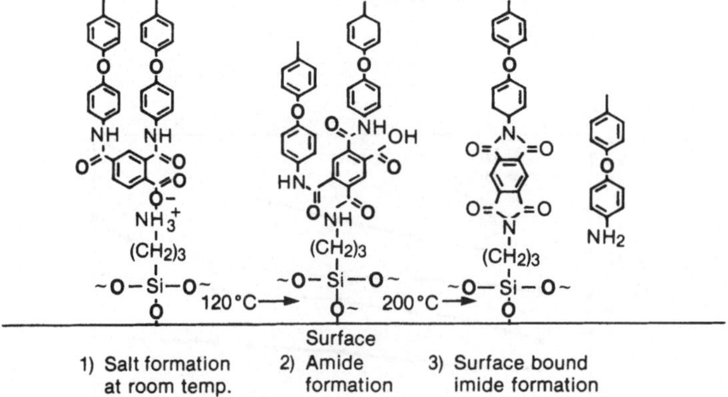

Fig. 9.   Schematic representation of polyimide-γ-APS surface interactions.   (after ref. 6)

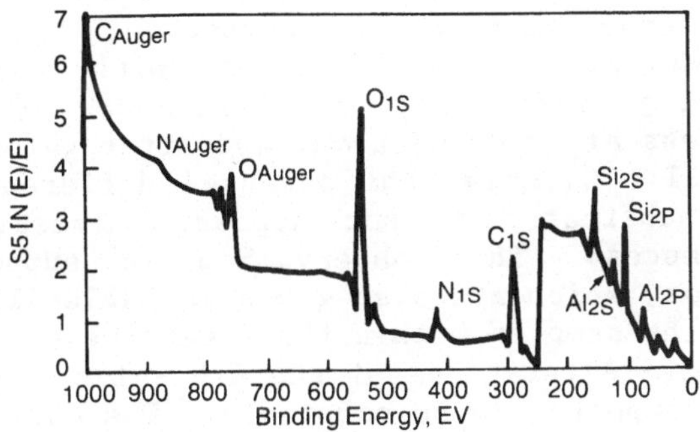

Fig. 10. ESCA spectrum representing the average composition of the
γ-APS modified Al surface.

Table III. γ-APS layer characterization by ESCA. The thickness as
calculated in Å and the carbon to nitrogen ratio of the
adsorbed film are given.

Substrate	Original γ-APS Solution Conc.	After Adsorption THK.(Å)	C/N	After Desorption THK (Å)
Al	10 μL/100 mL	----	---	----
	25 μL/100 mL	14.7	4.2	9.2
	50 μL/100 mL	30.8	2.6	2.7
	75 μL/100 mL	48.3	2.8	6.3
	100 μL/100 mL	54.5	2.7	8.5
Si	10 μL/100 mL	13.5	---	8.0
	25 μL/100 mL	17.0	3.8	7.0
	50 μL/100 mL	26.0	3.0	9.0
	75 μL/100 mL	31.5	2.4	15.5
	100 μL/100 mL	41.0	2.5	9.2

C. ESCA Investigation of Adsorbed γ-APS Layers

ESCA was used to probe organosilane surface films on Si and Al substrates. Spectra were recorded after deposition, and after a 60 minute room temperature water extraction. A representative spectra is shown in Figure 10 along with the appropriate assignments clearly indicating the presence of γ-APS. The calculated thickness after deposition, and after water exposure are given in Table III. Clearly, the observed differences in t_f for these different substrates do not appear to result from γ-APS thickness differences. This observation is supported by the measured C/N ratios, which are also given in Table III (hydrocarbon contamination of the samples within the apparatus was monitored via a reference sample. Alterations of the C/N ratio as a function of contamination was expected to be constant). The observed substrate dependence can certainly be considered in terms of the possible contributions of a number of effects. These include: the effect of the acid-base character of the substrate upon the molecular structure and organization of the γ-APS layer [16], the porosity or homogeniety of the adsorbed film [17], and possible Al-silane complex formations [17,18]. If we consider the similarity in thickness, and quantity of water extractable material present on the Al and Si substrates, and in addition, consider the possible consequences of the 140 °C curing cycle prior to solvent exposure, then one may be able to ignore long range substrate effects upon the reorganized interphase. That is, since the loosely bound silane fractions are similar, the degree of polymer mixing and subsequent inter and intramolecular interactions should also be similar. This would point to the actual γ-APS-substrate integrity in ultimately determining t_f.

The observation that the extent of the PAI-γ-APS interphase, as well as, the magnitude of the γ-APS-substrate interaction determines the overall dissolution time seems an apparent paradox. That is, if the strength of the γ-APS-substrate chemical interaction dominates, then we should observe no thickness effect. The resolution of this dilemma could reside in the ability of increasingly thicker γ-APS layers to decrease porosity or increase the homogeniety of the surface fraction. If so, the thickness dependence could be ascribed to the influence of increasing interphase coherence via inter and intramolecular interactions, as well as, increased substrate-interphase short range effects. The substrate dependence could in turn be rationalized strictly in terms of the strength of the acid-base interactions alone. This latter conjecture is the subject of future investigations within our laboratory.

SUMMARY AND CONCLUSIONS

The kinetics of polyimide film dissolution are characterized by distinct regions of etching behavior. These regions have been ascribed to: I. solvent invasions of the polymeric matrix, II. molecular dissolution, and III. loss of chemical adhesion. The appearance of the kinetic curves can be manipulated by developer concentration, amic acid-imide conversion temperature, interphase structure, and the substrate material. The chemical adhesion can be monitored by the time required for complete PAI film removal, t_f. t_f increases with respect to substrate materials according to:

$$Al < Si_xN_y < Si < SiO_2$$

In addition, t_f increases with increasing thickness as measured ellipsometrically, and with ESCA.

REFERENCES

1. First technical conference on polyimides: Synthesis, Characterization and Applications, K. L. Mittal Program Chairman, Society of Plastics Engineering, Ellenville, N.Y., Nov. 10-12, 1982.

2. M. E. Schrader, I. Lerner, and F. J. D'Oria, Modern Plastic, 45, 195 (1967).

3. M. E. Schrader and A. Block, J. Polym. Sci., Part C, 34, 781 (1971).

4. A. T. DiBenedetto and D. A. Scola, J. Colloid Interface Sci., 64, 480 (1978).

5. H. Ishida and J. L. Koenig, J. Polym. Sci., Polym. Phys. ed., 18, 1931 (1980).

6. H. G. Linde, J. Polym. Sci., Polym. Chem. ed., 20, 1031 (1982).

7. Y. K. Lee and J. D. Craig, Org. Coatings and Plastics Chem. Preprints, vol. 43, 451, (1980).

8. N. H. Sung, A. Kaul, S. Ni, C. S. P. Sung, and I. J. Chin, Proc. 36th Ann. Conf. Reinforced Plastics/Composites Inst., SPI, Session 2-B, (1981).

9. J. Greenblatt, C. J. Araps, and H. R. Anderson Jr., First Tech. Conf. on Polyimides, SPE, extended abstracts, 58, (Nov. 1982).

10. D. J. Belton, P. van Pelt, A. E. Morgan, Silicon Processing, ASTM STP 804, D. C. Gupta ed., American Society for Testing and Materials, 273 (1983).

11. W. E. J. Neal, in "Surface Contamination: Genesis, Detection, and Control," K. L. Mittal ed., Plenum, New York, (1979), p. 749.

12. P. van Pelt, D. Belton, S. DiIorio, Electrochem. Soc., exten-
     ded abstract, 81-2, 593 (1981).
13. D. Belton and S. I. Stupp, Macromolecules, $\underline{16}$, 1143 (1983).
14. F. Oosawa, "Polyelectrolytes," Marcel Dekker, New York, (1971),
     chapter 11.
15. M. P. Seah and W. A. Dench, Surf. Interface Anal., $\underline{1}$, 2 (1979).
16. H. Ishida, Polymer Preprints, $\underline{24}$, 198 (1983).
17. H. Ishida, private communication, March, 1983.
18. F. J. Boerio, Polymer Preprints, $\underline{24}$, 204 (1983).

SILANE COUPLING AGENTS FOR BASALT FIBER REINFORCED POLYMER COMPOSITES

R. V. Subramanian and Kuang-Hua H. Shu

Polymeric Materials Section
Materials Science and Engineering
Washington State University
Pullman, WA   99164-2720

ABSTRACT

Basalt fiber, a new mineral fiber reinforcement comparable in strength and modulus to E-glass fibers, has been produced from naturally occurring basalt rock. The interfacial bond strength in basalt fiber-polymer systems has been investigated using a single fiber pull-out test method.

Basalt fibers, about 90 µm in diameter, were specially drawn for fiber pull-out tests and were treated with a number of ionic and nonionic silane coupling agents under a variety of experimental conditions, changing solution pH and concentration, aging time, and fiber treatment time. The values of pull-out stresses were measured for treated and untreated fibers embedded both in epoxy and polyester matrix resins. The surfaces of treated and untreated fibers and those of pulled-out specimens were examined by scanning electron microscopy. Details of debonding and modes of failure were revealed in scanning electron micrographs.

The measured pull-out stresses are higher than those reported for E-glass fibers. The effect of iron and other metal oxides present in basalt fibers is manifested in the effect of pH of silane treatment on pull-out stress. The existence of different isoelectric points (IEPS) for different sites on the basalt fiber surface is indicated. The controlling effects of silane hydrolysis, condensation, orientation on the fiber surface, and chemical bonding to the fiber and polymer are revealed in the trends of interfacial bond strengths with experimental variables. The contribution of radial compressive stresses caused by thermal mismatch, and resin shrinkage during curing to the measured pull-out stress is also evident.

The results of pull-out stresses show that silane coupling agents are effective in improving interfacial bond strength in basalt fiber systems and that basalt fiber has excellent potential as a reinforcing fiber for polymer composites.

INTRODUCTION

Basalt

The distribution of basalt is worldwide. Basalt is a dense, finer-grained igneous rock, consisting of basic plagioclase (usually labradorite), augite, and magnetite. In North America many thousands of square miles in Washington, Oregon and Idaho are covered with basaltic lava. The Columbia Basalt Plateau, approximately 100,000 square miles in area, is one of the largest flows of flood basalt in the world. The typical composition of basalt is (1):

SiO_2: < 56.00%; Al_2O_3 : 10.50 to 22.00%; TiO_2: < 5.50%; Fe_2O_3: <6.00%; FeO: 2.50 to 15.00%; MnO:< 1.00%; MgO (FeO>10%): 2.00%; MgO (FeO<10%): > 3.00%; CaO: 5.00 to 15.00%; Na_2O: < 5.50%; P_2O_5: <1.50%.

Basalt Fiber

Mineral fibers from basalt, mainly basalt wool, have been made in Eastern European countries and in the USSR [2-6]. The advantages of basalt as a raw material in making staple fibers are its relatively homogeneous chemical nature, its large scale availability, and its ability to form fibers from the molten state. We have been engaged in studying the feasibility of drawing basalt fibers suitable for reinforcement of polymer matrices [7-11].

In our work, basalt fibers were drawn from molten basalt in an electrically heated platinum-rhodium bushing, as described in detail elsewhere [7]. After improvements by process modifications, the measured virgin fiber strengths are in the neighborhood of 3.5 GPa (500,000 psi), close to that of E-glass fibers [9]. The Young's modulus of the fibers from different basalt rocks, measured by thin-line ultrasonics, vary between 78-90 GPa, and are higher than that for E-glass fibers. The values are comparable to 90 GPa reported for fibers made from Berestovets basalt in the USSR, but the strengths are far higher than 0.75 GPa which was obtained by the Soviet scientists [4].

Basalt Fiber Reinforcement

The results of our investigation of epoxy polymer composites have shown that basalt fibers are comparable to E-glass as reinforcement for thermosetting polymers [10]. Being a mineral fiber, basalt should also be susceptible to surface modification by silane coupling agents. Composite tensile strengths confirmed this expectation. Indications were also obtained that interlaminar shear, and impact strengths were improved by application of an aminosilane coupling agent. Since the effects of silane treatment in improving polymer adhesion to reinforcing fibers is more directly seen in fiber pull-out tests [12-15] this technique was adopted for detailed study [16]. The results of this study are reported in this paper.

EXPERIMENTAL METHODS

Resin

The polymer resins used were CIBA Araldite 6004 epoxy resin, and ROHM & HAAS Paraplex P-43 unsaturated polyester resin. In the case of the epoxy resin, 11 phr diethylenetriamine (DETA) was used as curing agent. This resin was either left to cure at room temperature for five days or in a 70°C oven for 24 h. For unsaturated polyester, 1 wt% Lupersol Delta X (methyl ethyl ketone peroxide) was used as initiator, and 0.5 wt% of 6 wt% cobalt naphthenate as promoter. Cobalt naphthenate was well mixed with the unsaturated polyester before Lupersol Delta X was added to the mixture. Curing was completed in an oven at 70°C for 24 h, and the specimen was slowly cooled in the oven. For both resins, vacuum (water suction) was applied for four minutes to remove air bubbles trapped in the resins while mixing the curing agents.

Coupling Agents

The coupling agents used are listed in Table 1. These silanes were diluted into aqueous solutions. The conditions of silane treatment and curing of the resin are summarized in Table 2. To adjust the pH of the silane solution to 2, hydrochloric acid was added; to obtain pH 4, acetic acid, and to obtain pH 11, ammonium hydroxide were added.

Fiber

The fiber used in these experiments was basalt fiber drawn at 1380 °C from X-6 basalt rock of the following composition: SiO_2: 49.10%; Al_2O_3: 13.80%; TiO_2: 3.16%; Fe_2O_3: 4.00%; FeO: 10.00%; MnO: 0.21%; CaO: 9.43%; MgO: 5.25%; K_2O: 1.26%; Na_2O: 3.09%; P_2O_5: 0.68%. A slow speed of drawing (9.4 m/min) was employed to obtain a

Table 1. Silane Coupling Agents

Trade Name	Chemical Structure
A1100[a]	$H_2N-(CH_2)_3-Si-(OCH_2CH_3)_3$
A-187[a]	$CH_2-CH-CH_2-O-(CH_2)_3-Si-(OCH_3)_3$ (epoxide)
A-174[a]	$CH_2=C-C-O-(CH_2)_3-Si-(OCH_3)_3$ (with CH_3 and O)
A-189[a]	$HS-(CH_2)_3-Si-(OCH_3)_3$
Z-6032[b]	$CH_2=CH-C_6H_4-CH_2-NH_2^{+}-(CH_2)_2-NH-(CH_2)_3-Si-(OCH_3)_3 \cdot Cl^-$
A-6031[b]	$CH_2=C-C-O-(CH_2)_2-N(CH_3)_2^{+}-(CH_2)_3-Si-(OCH_3)_3 \cdot Cl^-$ (with CH_3 and O)
Z-6050[b]	Polyamino functional silane
Z-6075[b]	$CH_2=CH-Si-(OOCCH_3)_3$

a: Union Carbide
b: Dow Corning

fiber diameter about 90 µm which facilitated handling of the fiber in pull-out tests.

The fiber was cut into 6 cm lengths, and washed with acetone and then FREON TF (CCl_2FCF_2Cl) briefly. It was air-dried and treated by dipping into the silane solution for different periods of time as indicated in Table 2. During fiber treatments, the silane solution was stirred slowly, after which the treated fibers were briefly washed with distilled water and then left in a 105°C oven for 4 h to promote the condensation reactions and to remove excess water. The treated fibers were embedded in resin immediately after drying.

Fiber Pull-Out Specimens

The experimental method was to cast a basalt fiber in the center of a resin disc, and then to pull the fiber out of the resin. From the pull-out force and the resin-fiber contact area, the pull-out stress was calculated. The basic features of the specimen are shown in Figure 1a.

The actual specimen is shown in more detail in Figure 1b, which includes a 1.5 cm long, 0.7 cm inner diameter PVC tube for holding resin and mercury during casting and curing, a U-shaped card for holding the basalt fiber in place at the center of the tube, and a piece of scotch tape for sealing the bottom of the PVC tube. The specimen was prepared as described below.

The inner wall of the PVC tube was sprayed with FREKOTE 33 release agent and the tube was baked in a 100°C oven for 15 min to increase the durability of the release agent. A single basalt fiber was placed along the axis of the PVC tube and threaded through the hole at the center of the scotch tape closing the bottom of the tube. The upper part of the fiber was glued to the U-shaped card with DEVCON 5-minute Epoxy glue. The center hole of the scotch tape with the lower end of the fiber sticking out of it was sealed with Dow Corning High Vacuum Grease.

About 0.2 ml mercury was poured into the PVC tube and then 0.055 ml of resin was added on top of the mercury. (Mercury was in contact with the resin during curing both at room temperature and at 70°C.) When the resin was fully cured, mercury was released by peeling off the scotch tape.

Beneath the resin, the mercury forms a convex surface and does not wet the inner wall of the PVC column. But the resin wets the wall of the column. Therefore the resin disc is much thinner at the center than at the edges as shown in Figure 1c. Before the pull-out test, the fiber end which sticks out of the bottom of the resin disc was cut off.

The fiber was pulled out from the resin disc by using a Floor Model Instron Tensile Testing Instrument, Type TT-CM-L and a load cell B (2000 g capacity). In order to hold the PVC tube on the tensile tester without squeezing the column directly, a wooden jig, as shown in Figure 2, was constructed and used. The crosshead speed during fiber pull-out was selected after careful experimentation. Too high a speed caused fiber failure before pull-out, and a very low speed required too much time. After many trials, a crosshead speed of 0.02 cm/min was selected as the optimum for this experiment.

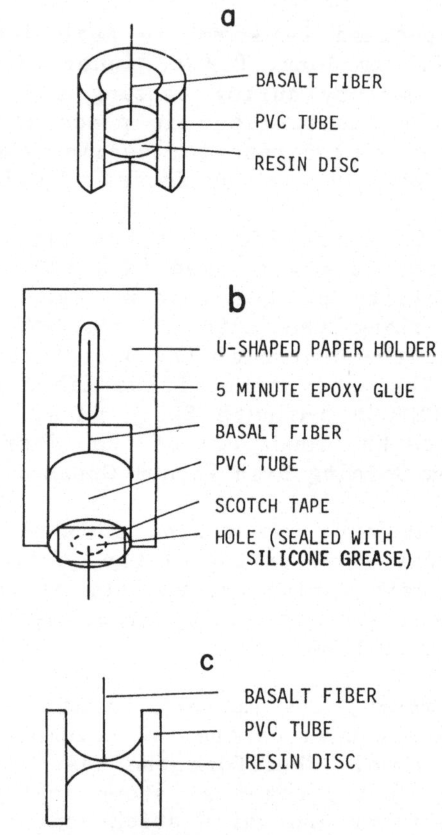

Fig. 1.   Single fiber pull-out specimen
          (a)   Basic Features
          (b)   Specimen Preparation
          (c)   Fiber Embedded in Resin Disc

Table 2. Silane Coupling Agent Treatment[1] of Basalt Fiber for
          Single Fiber Pull-Out Test

Resin	Silane	pH	Age of Solution (h)	Treatment Time (min)
Epoxy	none	--	----	------
	none[2]	--	----	------
	A-1100	10.6	2	0.5, 1, 10
	A-174	4	2	0.5, 1, 10
	A-187	4.5	2	0.5, 1, 10
	A-187	4.5	0.25, 0.5, 1	10
	A-187/amine	10.7	0.25, 0.5, 1	10
	Z6050	2	0.5, 1, 2	10
		4.4	0.5, 1, 2	10
		6.9	0.5, 1, 2	10
		9.9	0.5, 1, 2	10
	Z6032[3]	2	2	1, 10, 60
	Z6032[4]	2	2	1, 10, 60
		4.4	2	1, 10, 60
		7.6	2	1, 10, 60
		10	2	1, 10, 60
		11	2	1, 10, 60
Polyester	none	--	----	------
	A-174	4	2	0.5, 1, 10
	A-189	4.2	2	0.5, 1, 10
	Z6075	3.1	2	0.5, 1, 10
	Z6031	2	2	0.5, 1, 10
		4.5	2	0.5, 1, 10
		7.7	2	0.5, 1, 10
		9	2	0.5, 1, 10
		10	2	0.5, 1, 10

[1]All silanes were in aqueous solution 1wt%; treated fibers
were dried at 105°C for 4 h; the curing condition for resin was
70°C, 24 h. Exceptions are stated in other footnotes.

[2]25°C, 5-days curing.

[3]1.0wt% solution; 60°C 1 h drying; 25°C/5 days curing.

[4]0.1wt% solution; 60°C 1 h drying; 25°C/5 days curing.

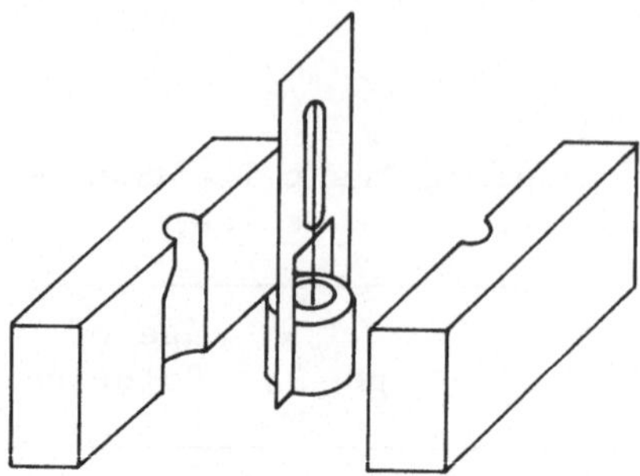

Fig. 2.   Jig for Holding Fiber Pull-Out Specimen.

Pull-Out Stress

The pull-out stress is given by $\sigma_A = P/\pi dl$, where P is the pull-out force, d the diameter of the fiber, and l the length of the fiber actually embedded in the resin disc. The diameter of the basalt fiber was measured by an American Optical Company type 1-60 microscope and a Vickers Image Splitting Eyepiece. The length of the fiber which was embedded in the resin disc was measured by an Unitron TMS-1911 measuring microscope and a Wide Field Filar Micrometer, MODEL FWE. Both the fiber diameter and the embedded fiber length were measured after the pull-out test. The fiber surfaces were examined by scanning electron microscopy to investigate failure modes.

At least twelve specimens were prepared for each condition of fiber pull-out tests. Fiber fracture rather than pull-out occurred in some of these specimens. However, fiber pull-out was observed usually in eight to ten specimens, and the average pull-out stresses were calculated only from the values measured for fiber pull-out. The ratio of standard error to the mean is presented in the tables under the column coefficient of variation and is expressed in percentage.

RESULTS AND DISCUSSION

Single Fiber Pull-Out Stresses

Pull-out stresses were measured for bare fibers and for silane treated fibers. A variety of silanes with reactive and unreactive funtional groups were employed with epoxy and polyester matrix resin. The conditions of silane treatment were varied. The treatment time, the age of the solution, and the pH of the solution were mainly the variables employed with each silane.

The results obtained using silanes A-1100, A-174, and A-187 in an epoxy matrix are presented in Table 3, and those obtained using A-174, A-189 and Z-6075 with a polyester matrix are in Table 4. The effect of amine catalyst in silane treatment is summarized in Table 5 for A-187/epoxy systems. Results obtained using Z-6050, a polyamine functional silane are given in Table 6 for pH varied between 2 and 10. Table 7 contains the results of pull-out stresses for Z-6032 silane used with epoxy matrix, and Table 8 for Z-6031 with a polyester matrix, both at different pH conditions. These results are discussed in succeeding sections.

<u>Fiber Pull-Out Test Method</u>

The techniques of preparation and testing of fiber pull-out specimens described here were developed with a number of significant modifications and improvements of other reported techniques in order to overcome some of the basic limitations of the method. The critical length for stress transfer, l_c, is related to the interfacial shear stress, τ, as $l_c/d = \sigma_f/2\tau$, where d and σ_f are respectively the diameter and ultimate fracture stress of the fiber. Hence, for fiber pull-out to occur, the interfacial bond should be weak so that a large embedded length can be used or, for small fibers and strong bonding, the resin disc should be thin enough to reduce the embedded length to less than the critical stress transfer length. When the thickness of the resin disc exceeds l_c, the ultimate fracture stress of the fiber will be reached and fiber fracture, and not fiber pull-out, will be observed. Flaws in the fiber will also lead to failure in tension.

In other investigations using fiber pull-out tests with glass-resin systems, it was found difficult to obtain thicknesses of the disc less than 1 mm [12]. Hence glass rods of large diameter were used in preparing the specimens. In the present investigation, using a mercury support to form a resin disc, resin thickness of 0.25 mm was consistently achieved by an adaptation of the method reported by Favre and Perrin [13]. The technique is useful thus even for thin carbon fibers [14], though the basalt fibers used in the present study were specially drawn at slow speeds to be about 90 μm in diameter. In this manner, a large fraction of the prepared specimens failed by fiber pull-out rather than fiber fracture.

Even when fiber pull-out did occur, it was not a clean pull-out, as can be seen in the SEM of pulled-out fibers (Figure 3). A resin core is seen to be attached to the fiber which had been originally formed by wetting of the fiber by the resin, and curving up of the resin surface. Failure is invariably initiated somewhere below the top of the resin core, some distance away from the fiber, and the crack subsequently reaches the interface and follows the

Table 3. Single Fiber Pull-Out Stress of Silane[1] Treated and Untreated Basalt Fiber Embedded in an Epoxy Resin[2]

Silane	pH	Fiber Treatment Time (min)	No. of Specimens	$\overline{\sigma}_A$ (MPa)	Coeff. of Variation %	% Improvement[5]
none[3]	--	--------	8	17.3	32	0
none[4]	--	--------	13	19.1	30	0
A-1100[3]	10.6	0.5	4	47.3	26	173
		1	5	39.3	10	127
		10	4	35.3	9	104
A-174[4]	7.6	0.5	11	17.6	23	2
		1	11	18.3	32	6
		10	9	20.2	30	17
A-187[4]	4.5	0.5	9	23.9	26	25
		1	9	24.2	21	27
		10	8	20.6	22	8

[1]Aqueous solution (1wt%) aged for 2 h.

[2]Araldite 6004 was cured with 11 phr DETA.

[3]Treated fiber was dried at 25°C for 1 h, and the resin was cured at 25°C for five days.

[4]Treated fiber was dried at 105°C for 4 h, and the resin was cured at 70°C for 24 h.

[5]Based on untreated fiber.

Table 4. Single Fiber Pull-Out Stress of Silane-Treated[1] and Untreated Basalt Fiber Embedded in a Polyester Resin[2]

Silane	pH	Fiber Treatment Time (min)	No. of Specimens	$\overline{\sigma}_A$ (MPa)	Coeff. of Variation %	% Improvement[3]
none	--	--------	10	24.6	27	---
A-174	4	0.5	6	43.1	10	75
		1	3	52.3	38	113
		10	3	52.9	11	115
A-189	4.2	0.5	9	27.0	13	10
		1	9	26.7	17	9
		10	11	23.7	12	4
Z-6075	3.1	0.5	10	29.2	17	19
		1	7	28.4	17	15
		10	7	21.9	27	11

[1] Aqueous solution (1wt%) aged for 2 h. Treated fiber was dried at 105°C for 4 h.

[2] ROHM & HAAS P-43 was cured at 70°C for 24 h with 1wt% MEKP as initiator and 0.5wt% of 6wt% cobalt naphthenate as promotor.

[3] Based on untreated fiber.

Table 5. Single Fiber Pull-Out Stress of A-187-Treated[1] Basalt Fiber Embedded in an Epoxy Resin[2]

Silane	Age of Solution (h)	No. of Specimens	$\overline{\sigma}_A$ (MPa)	Coeff. of Variation %	% Improvement[3]
A-187	0.25	4	31.0	10	62
	0.50	6	35.5	18	86
	1	5	49.0	18	157
A-187	0.25	7	37.2	19	94
+	0.5	4	31.5	32	65
n-butyl amine	1	5	38.8	11	103

[1]The silane coupling agent was diluted into 1wt% aqueous solution. Fiber treatment time was 10 min. 1:1 molar ratio of n-butyl-amine was used to catalyze A-187. The treated fiber was dried at 105°C for 4 h.

[2]Araldite 6004 was cured with 11 phr DETA at 70°C for 24 h.

[3]Based on untreated fiber.

Table 6. Single Fiber Pull-Out Stress of Z-6050-Treated[1] Basalt Fiber Embedded in an Epoxy Resin[2]

pH	Age of Solution (h)	No. of Specimens	$\bar{\sigma}_A$ (MPa)	Coeff. of Variation %	% Improvement[3]
2	0.5	11	18.5	24	-3
	1	8	21.5	20	13
	2	7	31.1	9	63
4.4	0.5	8	23.2	17	21
	1	11	22.4	19	17
	2	8	26.1	9	37
6.9	0.5	9	23.1	29	21
	1	9	26.0	29	36
	2	7	30.9	30	62
9.9	0.5	8	27.5	11	44
	1	7	26.3	34	38
	2	8	25.2	23	32

[1] Fiber treatment time was 10 min. The treated fiber was dried at 105°C for 4 h.

[2] Araldite 6004 was cured with 11 phr DETA at 70°C for 24 h.

[3] Based on untreated fiber.

Table 7.   Single Fiber Pull-Out Stress of Z-6032-Treated[1] Basalt Fiber Embedded in an Epoxy Resin[2]

pH	Fiber Treatment Time (min)	No. of Specimens	$\overline{\sigma}_A$ (MPa)	Coeff. of Variation %	% Improvement
2[3]	1	2	21.5	33	24
	10	6	25.8	15	49
	60	7	25.5	24	47
2	1	5	21.6	10	25
	10	2	24.6	7	42
	60	4	23.0	17	33
4.4	1	7	27.9	18	61
	10	2	31.5	2	82
	60[4]	12	15.0	20	-13
7.4	1	4	24.2	12	40
	10	5	26.8	38	55
	60	1	19.0	--	10
9.9	1	18	28.8	17	66
	10	3	29.8	14	72
	60	2	25.2	6	46
11.1	1	5	24.2	13	40
	10	7	23.6	33	36
	60	11	18.2	22	5

[1]Aqueous solution (0.1wt%) aged for 2 h.   [2]Araldite 6004 was cured with 11 phr DETA at 25°C for 5 days.   [3]Aqueous solution (1wt%) aged for 2 h.   [4]The coupling agent solution aged for 24 h. [5]Based on untreated fiber.

Table 8. Single Fiber Pull-Out Stress of Z-6031-Treated[1] Basalt Fiber Embedded in a Polyester Resin[2]

pH	Fiber Treatment Time (min)	No. of Specimens	$\overline{\sigma}_A$ (MPa)	Coeff. of Variation %	% Improvement[3]
2	0.5	6	35.3	7	36
	1	10	34.0	17	38
	10	11	37.6	13	53
4.5	0.5	10	29.7	14	21
	1	10	32.0	12	30
	10	10	28.3	19	15
7.7	0.5	10	24.7	22	41
	1	9	33.9	17	38
	10	11	39.0	28	59
9	0.5	10	37.5	17	52
	1	12	35.9	24	46
	10	9	38.9	25	58
10	0.5	8	28.7	17	17
	1	10	30.6	12	24
	10	10	32.6	14	33

[1] Aqueous solution (1wt%) aged for 2 h. Treated fiber dried at 105°C for 4 h.

[2] ROHM & HAAS P-43 was cured at 70°C for 24 h with 1wt% MEKP as initiator and 0.5wt% of 6wt% cobalt naphthenate as promotor.

[3] Based on untreated fiber.

fiber direction until debonding is completed. This phenomenon has
been observed also in other studies using the fiber pull-out tests
[12-14]. Subramanian et al. [14] have shown that the energy to
initiate the crack in the resin therefore forms a significant fraction of the measured pull-out stress for small diameter (5-10 μm)
carbon fibers (small surface area), though it is negligible for the
larger diameter (>50 μm) fibers having a larger surface area. In
the case of basalt fibers of 100 μm diameter used in the present
study, the pull-out stress was therefore calculated by dividing the
pull-out load by the area of contact of the embedded fiber with the
resin. Since the interfacial area was calculated from the length
of the fiber measured from the resin core to the end of the fiber,
the irregular nature of the surface profile near the junction of
the resin core with the fiber introduced some uncertainty in the
measurements of the debonded length of the fiber.

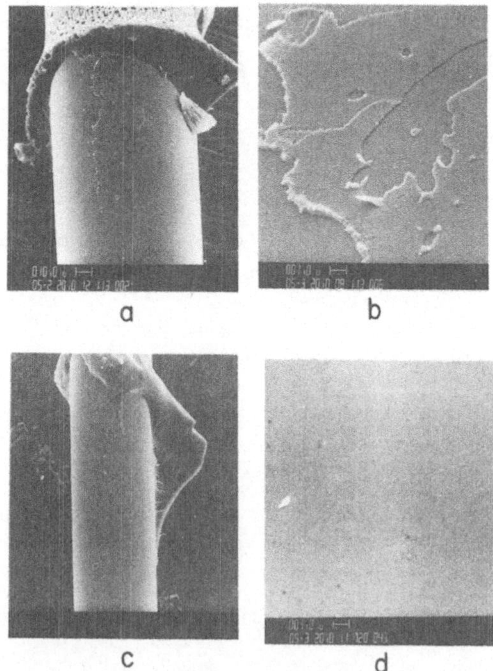

Fig. 3.   SEM of basalt fibers after pull-out: (a,b) APS treated
          fiber, epoxy resin disc (500X and 5000X), respectively;
          (c) MPS treated fiber, polyester resin disc (300X) and
          (d) bare fiber surface for comparison.

Other possible causes for experimental error and the observed coefficient of variation of the results can be enumerated. Entrapped voids, as seen in Figure 3 in the resin or at the interface due to imperfect wetting, could give rise to difference in stress concentration. Difference in residual stresses in the cured specimens, and changes in mode of failure from adhesive (interfacial) to cohesive (resin) failure caused by differences in surface treatments also contribute significantly to the variable errors in measurement of fiber pull-out stress.

The failure mode in the fiber pull-out test is thus quite complex, and it has been found that the test gives only a lower bound of the interfacial shear strength. For example, the value obtained (23 MNm^{-2}) in the pull-out test for glass-polyester systems was less than half of the bond strength (50–60 MNm^{-2}) measured by the short beam shear test or by debonding of single filaments completely embedded in the matrix [12]. Many reasons for this observation may be considered. The characteristic profile of the resin disc curving up to meet the fiber surface can be seen in scanning electron microgaphs, as in Figure 3. It has been suggested [12] that this may be considered as the lower half of an imaginary elliptical crack in contact with the glass rod and that the stress concentration facilitates failure when the disc is under tensile loading. Furthermore, the residual stresses produced in the pull--out specimen due to curing of the resin are different from those in the other types of specimens. A further explanation for the low values obtained in pull-out tests is the assumption made in calculating the interfacial shear strength that a uniform shear exists along the length of the embedded fiber. It has been pointed out that calculated values of the interfacial shear strengths are likely to be low [17,18].

In view of the above observations it is significant that the interfacial shear strength exceeding 50 MPa (Table 4), determined for basalt fiber by the pull-out test in our study, is much higher than the reported value of 23 MPa for E-glass [12] obtained by the same method. In both cases, similar polyester matrix resins were used and the same silane coupling agent γ—methacryloxypropyltrimethoxysilane (A-174) was applied on the fiber, though the treatment solution used by Shortall and Yip contained tetradecylpyridinium bromide, a boundary lubricant, and polyvinyl acetate, a film forming, adhesion promoting polymer as added constituents. In fact, the higher values of the pull-out stress in Tables 3 to 8 are of the order of short beam shear strength reported for E-glass polyester composites in references [12] and [19]. It should be relevant therefore to measure the short beam shear strengths of basalt fiber polyester composites and compare them with the results of pull-out tests.

Surface Treatment of Basalt Fiber

It is evident from the results of Tables 3 to 8 that the specific effects of silane treatment depend on the conditions of application of the silane, and are related to their mechanism of interaction with the basalt fiber surface and the polymer resin. The first step in the application of the silane, $R-Si-(OR)_3$, is the hydrolysis of the alkoxy groups $(-OR)_3$ of the silane to yield silane triols [20]. The silanol groups so formed can then condense among themselves slowly to form oligomers united by $-Si-O-Si-$ covalent bonds. The condensation of silanols can also involve the surface silanol groups on basalt fiber, in which case the silane becomes covalently bonded to the fiber.

Both the rate and degree of hydrolysis, as well as of condensation, depend on the conditions of application involving concentration, treatment, and aging times, and pH of the silane solution. The effects of these variables on the interfacial shear strength are therefore to be traced to their effect on silane hydrolysis and condensation. A fast hydrolysis reaction readily makes available the silanol groups needed for bond formation with the fiber surface. However, if the silanols formed are allowed to undergo self-condensation, it will lead to the deposition of a thick layer of oligomeric silanols which will form a distinct interphase region between fiber and matrix, and the properties of this interphase will determine the results of the fiber pull-out test.

The effect of the organofunctional group R on the silane is dependent upon its reactivity with the polymer matrix [20]. The specificity of silanes in improving the interfacial shear strength is thus derived from their ability to coreact with the polymer matrix and form covalent bonds. When covalent bonds are formed by the silane with both the fiber surface and the polymer matrix, a molecular bridge is formed between fiber and polymer which could enhance the interfacial shear strength considerably. The individual results of the pull-out test are discussed in detail in the following section on the basis of these general considerations.

Table 3 compares the results of the average fiber pull-out stress from an epoxy resin for three silane treatments with that obtained using no treatment for the fibers. Of these three silanes, A-1100, the aminopropyltriethoxysilane (APS), has an amine group, and A-187, the glycidoxypropyltrimethoxysilane (GPS), has an epoxy functional group (Table 1), both of which can coreact with the epoxy resin. On the other hand, A-174, the methacryloxypropyltrimethoxysilane (MPS) carries the methacryloxy group which cannot coreact with the epoxy polymer. It is understandable therefore that APS and GPS are effective coupling agents for the epoxy matrix

and increase fiber pull-out stresses, while MPS does not. Similarly, MPS, which can coreact with polyester matrix, yields, in polyester systems, fiber pull-out stresses (> 50 MPa) exceeding those
obtained using APS with the epoxy matrix (Table 7), thus confirming
the specificity and effectiveness of covalent bond formation of silanes with the matrix polymer in improving the interfacial shear
strength.

The differences observed in Table 3 between GPS and APS,
which both react with the epoxy resin, merit further explanation.
The silane solutions used in this experiment were aged for 2 h to
allow hydrolysis to proceed before bringing them into contact with
the fiber surface. APS increases the fiber pull-out stress after
only 30 sec of reaction with the basalt surface, indicating the
high reactivity of the silanol groups with the hydroxyls attached
to the silicon atoms on the basalt surface. It is also seen from
Table 3 that longer treatment times, which can cause larger numbers
of silane molecules to be layered on the surface, are clearly not
optimum for efficient increase in interfacial shear strength. The
influence of the $-NH_2$ group of APS is highly evident in the reactivity observed above. The amine group on the third carbon atom from
silicon is thought to stabilize the low molecular weights of silanols by internal chelation to form a stable six membered ring
[21,22] and a multiply hydrogen-bonded structure is proposed by
Ishida and co-workers [22] for the intramolecular interaction:

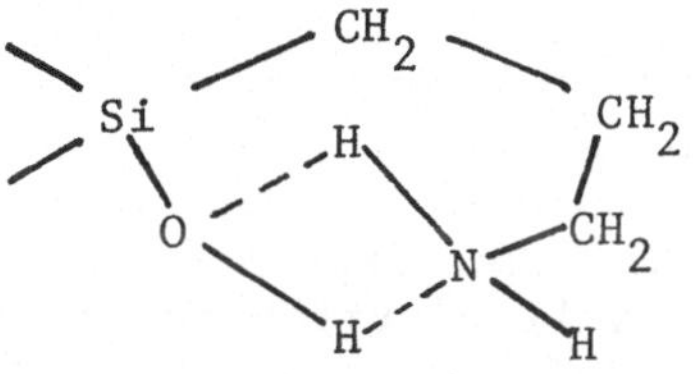

In aqueous solutions of 1 wt% or less, the aminosilane contains
large amounts of silanetriols [22]; above 0.15% by weight, oligomers are formed by the self (amine)-catalyzed condensation of silanols. Schrader has shown that the dimers and oligomers of γ-amino-
propylsilane can be chemisorbed on glass [23]; condensation of silanols of the hydrolyzed aminosilane with those on the glass surface would also be expected to be catalyzed by amine group on the
silane. Only short treatment times are required in this case for
silane pickup on the glass surface. Longer treatment time leads to
a thicker layer which is clearly not as efficient in increasing
fiber pull-out stress (Table 3). The internal hydrogen bonding in
the aminosilane reduces the functionality available for formation
of polysiloxanols. The stability of the APS solution without the
formation of insoluble gels accounts for the retention of its efficiency even after 2 h of aging.

Since GPS prepared under similar conditions shows poor efficiency and not much change in reactivity with time of treatment
(Table 3), it was surmised that with this silane, self-condensation
might have proceeded to a great extent beyond the optimum desirable
levels. The fiber pull-out stress was therefore measured after
10-minute treatment with solutions which had been aged for shorter
times, from 10 to 60 min (Fig. 4). A progressive increase in
stress with aging time of GPS solutions up to 60 min indicated slow
hydrolysis of the silane producing the necessary silanol groups for
reaction with the fiber surface. The high values of stress produced by solutions which had aged for 60 minutes confirmed the earlier surmise that aging for 2 h was not the optimum condition for
applying GPS.

Since the hydrolysis is known to be catalyzed by amines [24],
it was attempted to increase the rate of hydrolysis of GPS by
adding n-butyl amine. Moderately high fiber pull-out stresses were
then obtained after only 10 min of aging time in the presence of
amine. Also, the pull-out stress did not change significantly with
aging up to 60 min (Fig. 5); and the lower stress compared to the
highest in Figure 4, obtained in the absence of amine, showed that
the amine had, as described above, not only promoted hydrolysis and
effectively reduced the optimum aging time, but had also promoted
self-condensation of silanes, resulting in reduction of the measured stress. This was confirmed when the silane solution in the
presence of amine precipitated out after 60 min.

The scanning electron micrographs, shown in Fig. 6 to 8, of
fibers taken after various silane treatments, provide visual evidence for the above conclusions. In the absence of amine catalyst,
the silane coating on the fiber from a 15-minute old solution (6a)
is barely visible while well distributed patches appear after immersion in a solution aged for 60 min (6b). The pictures in Figure
6c and 6d show that in the presence of n-butyl amine, the silane
coating formed from a 15-minute-old solution is clearly discernible
in many patches, and from a 60-minute-old solution, it is very
thick indeed, covering the whole surface in many lumps or islands.
The corresponding SEM of pulled-out fibers subjected to GPS treatment are shown in Figures 7 and 8. The stronger adhesion of fibers,
when 60-minutes aging was used in the absence of amine, is seen in
the residue of resin sticking to the pulled-out fiber, and in the
increased pull-out stress, 53.8 MPa for the 60-minute aging, compared to 33.6 MPa for the 15-minute aging of the silane solution
(Fig. 7a,b). On the other hand, the appearance of pulled-out
fibers treated with GPS aged for different times in the presence of
butyl amine shows very little difference, and the measured stresses
were also similar, 34.9 MPa (15-minutes aging) compared to 39.2 MPa
(60-minutes aging) (Fig. 8a,b). Figure 7c, d and Figure 8c, d show
these comparisons of the same pulled-out fibers at higher (5000X)
magnification.

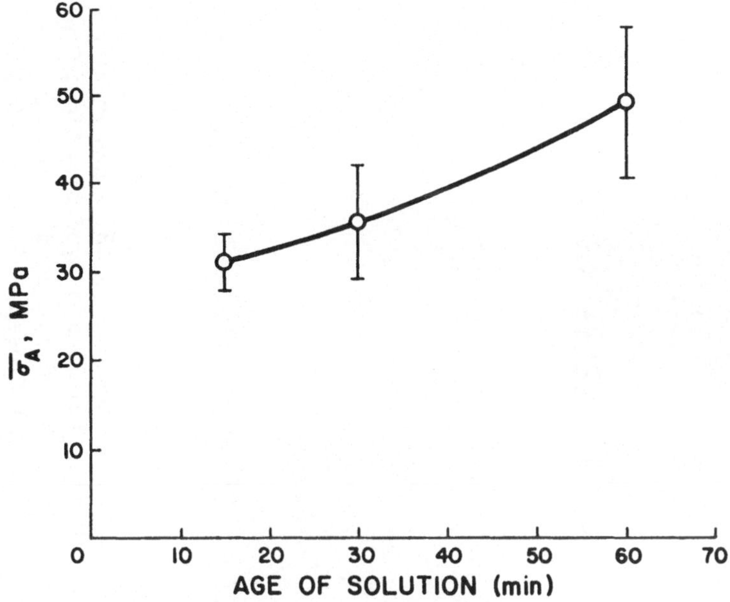

Fig. 4. Pull-out stress of basalt fiber treated (10 min) with A-187 silane solutions after varying times of aging.

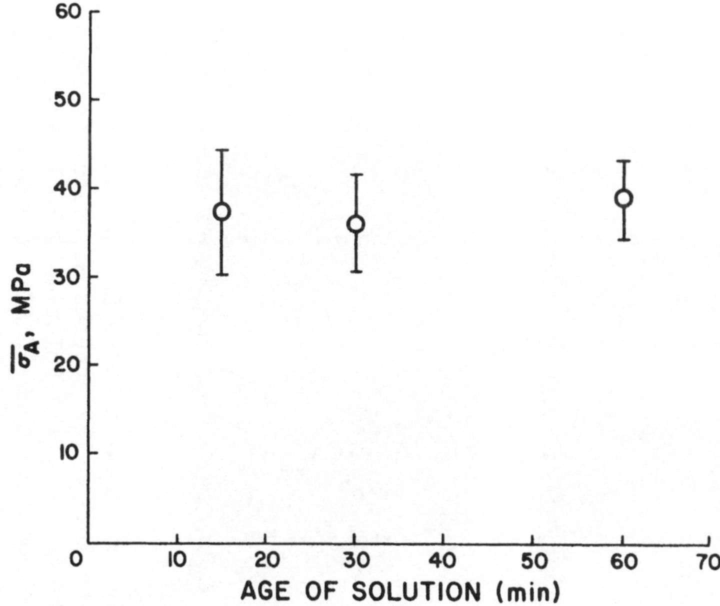

Fig. 5. Pull-out stress of basalt fiber treated (10 min) with A-187 silane/n-butyl amine solutions after varying times of aging.

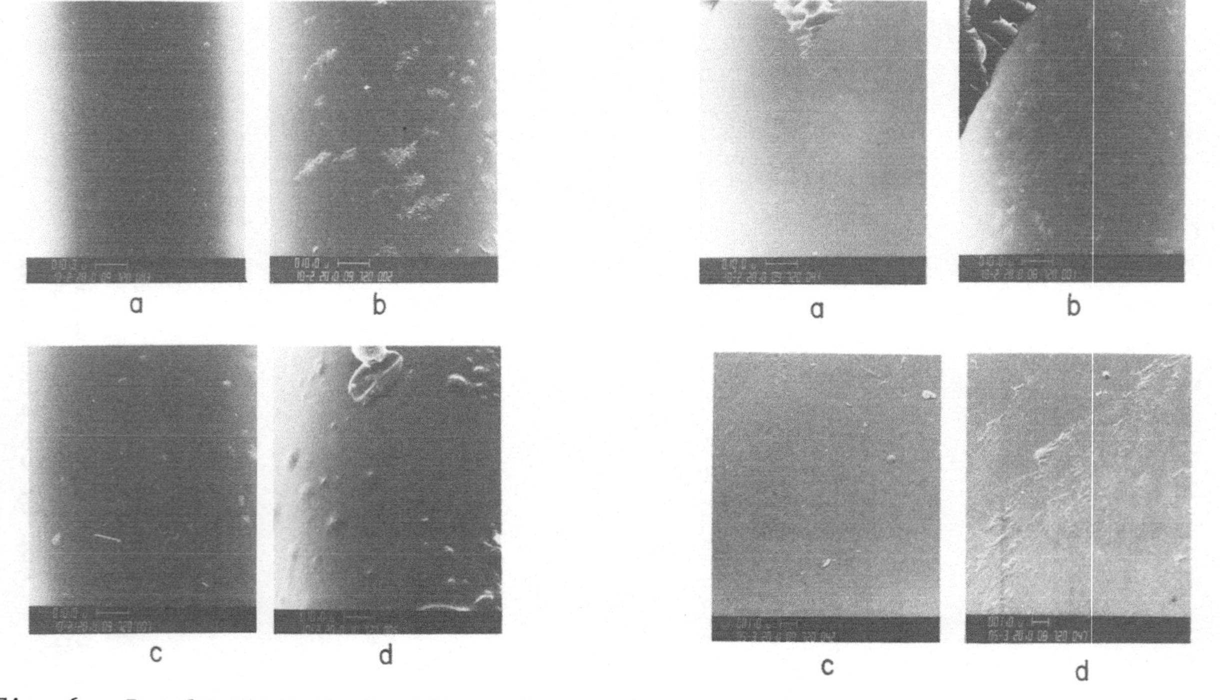

Fig. 6. Basalt Fiber Surface Treated with lwt% A-187 Aqueous Solution for 10 min (1000X). (a) Solution aged 15 min. (b) Solution aged 60 min. (c) n-Butyl amine catalyzed solution aged 15 min. (d) n-Butyl amine catalyzed solution aged 60 min.

Fig. 7. Pulled-out Specimen Surface of Basalt Fiber Treated with lwt% A-187 Aqueous Solution for 10 min. (a) 15 min aged solution; σ_A=33.6 MPa (1000X) (b) 60 min aged solution; σ_A=53.8 MPa (1000X) (c) Closeup of (a) (5000X) (d) Closeup of (b) (5000X).

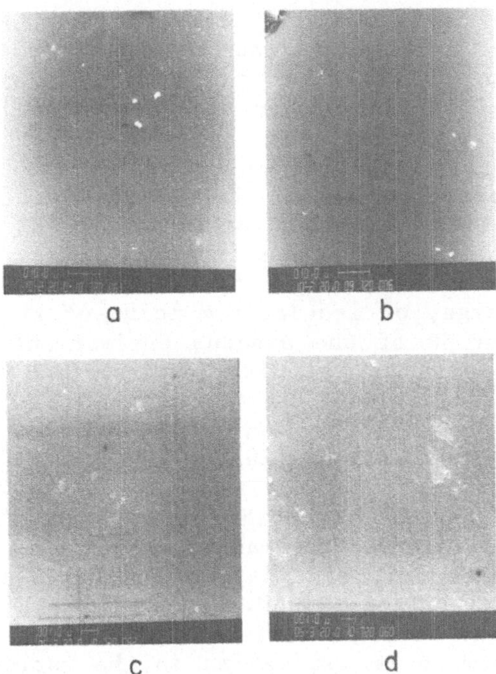

Fig. 8.   Pulled-out Specimen Surface of Basalt Fiber Treated with
          1wt% A-187 + n-Butyl amine Aqueous Solution for 10 min.
          (a) 15 min aged solution; σ_A=34.9 MPa (1000X); (b) 60 min
          aged solution; σ_A=39.2 MPa (1000X); (c) closeup of (a),
          (5000X; (d) closeup of (b), (5000X).

In addition to the aging and treatment times, another experimental
variable that is important as a controlling parameter on the effectiveness of silanes is the pH of the solution. This was studied in
the case of two cationic silane coupling agents Z-6031 (Table 8)
and Z-6032 (Table 7) and a polyaminofunctional silane Z-6050 (Table
6). Of these, Z-6032 has shown outstanding promise as a universal
coupling agent for polyesters, epoxies, and other thermosetting
resins because of its unique combination of structural features
[25].

     First of all, as seen in the structure shown in Table 1,
Z-6032 possess a vinyl functional group capable of coreacting with
a polyester, as well as an amine group that can react with an epoxy
resin. Secondly, the amine group on the third carbon atom from silicon can chelate internally with silanol, imparting stability to
aqueous solutions as in the case of APS discussed earlier. Furthermore, a bulky organic group on the nitrogen enhances solution
compatibility with organic polymers. Z-6031 is similar, but is
present as the quaternary ammonium salt rather than as the hydrochloride, and carries the methacryloxy functional group.

The combined presence of cationic and anionic functional groups must have a strong influence on the deposition and orientation of the silane on the fiber surface. This orientation of the coupling agent will be influenced by electrokinetic effects at the interface. Diffusion of cations from the surface into the aqueous phase, (because the latter's high permitivity lowers attractive forces between ionic species on the surface) will leave behind an anionic surface which can orient cations. The H^+ and OH^- ion concentrations in the aqueous phase impose an electrokinetic surface potential (zeta potential) on the hydrated mineral surface, and the point of zero charge, or isoelectric point of the surface (IEPS), is related to the pH of the aqueous environment. In an aqueous medium of higher pH than the IEPS, the surface will be anionic, while at a lower pH, it will be cationic. Thus, ionic functional silanes can be expected to be much more sensitive to the pH of the solution than nonionic silanes [20].

The results for the variation of pull-out stress with pH of silane treatment illustrate the marked effect of such electrostatic interactions (Tables 6,7, and 8), for Z-6031, Z-6032 and for the polyaminofunctional silane Z-6050. It can be seen readily that the effect of treatment time is different at different pH conditions and that there seem to be two maxima in the variation of pull-out stress with pH, one in the acidic and another in the alkaline range. The effect is seen most clearly for Z-6031 and Z-6032 as shown in Figures 9 and 10. The latter effect is considerably different from observations made with cationic silanes on silica [26,20]. The flexural strength of silica reinforced polyester castings was found to decrease drastically above a pH of 2. Since the IEPS of SiO_2 is 2, the surface will be anionic above this pH and it could be expected that the cationic silane will be attracted to the surface and deposited on it "upside down" with the vinyl functional group, which is close to the positive charge on the molecule, anchored near the surface. But from more acidic solutions, the silanol groups, (IEP 3), will be oriented to the silica surface, and the vinyl funtional groups will be properly oriented for optimal interaction with the polyester matrix. The decline of flexural strength with increasing pH of application of the cationic silane on SiO_2 was therefore understandable [27] and consistent with other observations. For example, that the orientation of the silane finish may extend into the resin phase was also suggested by the work of Kahn [28] on orientation of liquid crystals on mineral surfaces treated with silane coupling agents. Not only could the orientation of liquid crystals be controlled by selecting the appropriate silane, but the effectiveness of a given silane was determined by the pH of application as related to the isoelectric point of the surface [29].

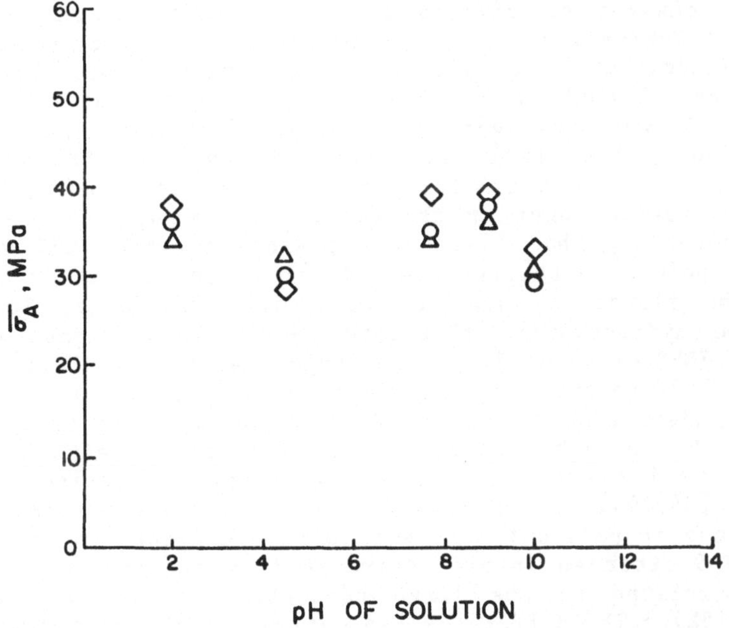

Fig. 9. Pull-out Stress of Z-6031-Treated Basalt Fiber Embedded in an Unsaturated Polyester Resin Versus Solution pH at different Treatment Times, O = 0.5 min, Δ = 1 min, $\Diamond$ = 10 min.

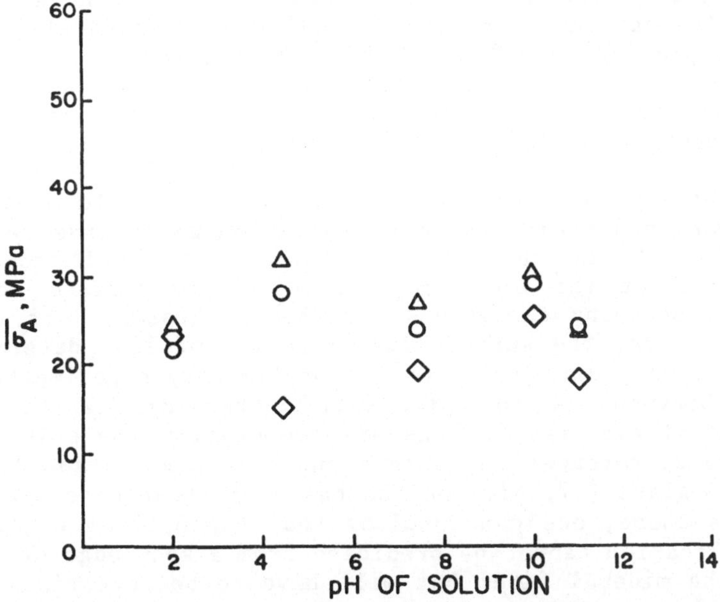

Fig. 10. Pull-out Stress of Z-6032-Treated Basalt Fiber Embedded in an Epoxy Resin Versus Solution pH at different Treatment times: O = 1 min, Δ = 10 min, $\Diamond$ = 60 min.

The behavior of silanes on basalt surface, as opposed to silica and E-glass, must take into account the inclusion of many oxides in addition to SiO_2 in its composition, especially that of iron oxides. Though it has been suggested that the IEPS of mixed oxides is the weighted average of the IEPS´ of the components [30], Michaels and Bolger [31] have observed separate IEPS values for Kaolin surfaces, about two for the silica surfaces and seven for a-luminum hydroxide edges in the crystal platelets. It is concei-vable, therefore, that basalt can possess varying IEPS values at different points on the surface. In this context, it is useful to recall the flotation data for ores using cationic, long chain alkylamine hydrochloride flotation aids. In the case of Fe_2O_3 having an IEPS of about 7, it was found that the alkylamine surfac-tant gave 100% recovery of the ore at pH between a 8-12, indicating efficient adsorption of the surfactant on Fe_2O_3 in this pH range [32,33]. On the other hand, anionic flotation agents like alkyl sulfonates and alkyl carboxylates were most effective in acidic solutions [34,20]. It is therefore an intriguing possibility that the increase of pull-out stress in our experiments to reach maxima around pH 9 after an initial decrease from the more acidic condi-tions is related to the large concentration of iron oxides in basalt (~15%). The differences in optimum pH observed with differ-ent silanes is evidently related to their IEPS.

Similar observations have been made with polyester reinforced by basalt powder filler [35]. The best wet strength retention of basalt-filled castings employing Z-6031 was observed at pH 10 and of those employing Z-6032 at pH 6 to 8. Furthermore, in adhesive joints made with metals, the effect of pH of aminosilane treatment on the strength data has been related to the IEPS of the metal oxide substrate [35].

Undoubtedly, the oxidation state of iron, the presence of other oxides, and their degree of hydration also serve to control the behavior of the surface. Furthermore, the bulk composition does not reflect the true composition of the surface which can depend on processing conditions and thermal history. In the case of basalt fibers, the surface concentration of iron determined by Auger electron spectroscopy [16] in preliminary experiments showed variation between 2.6 to 3.3%, with increasing diameter of the fiber (thermal history) and was also lower than the bulk composi-tion. Such differences in surface and bulk composition have been noted for E-glass [37,38]. On the basis of these observations and discussions above, one can conclude that the optimum condition of silane application cannot be predicted from a knowledge of composi-tions of the mineral fiber but will have to be determined experi-mentally under conditions of actual use.

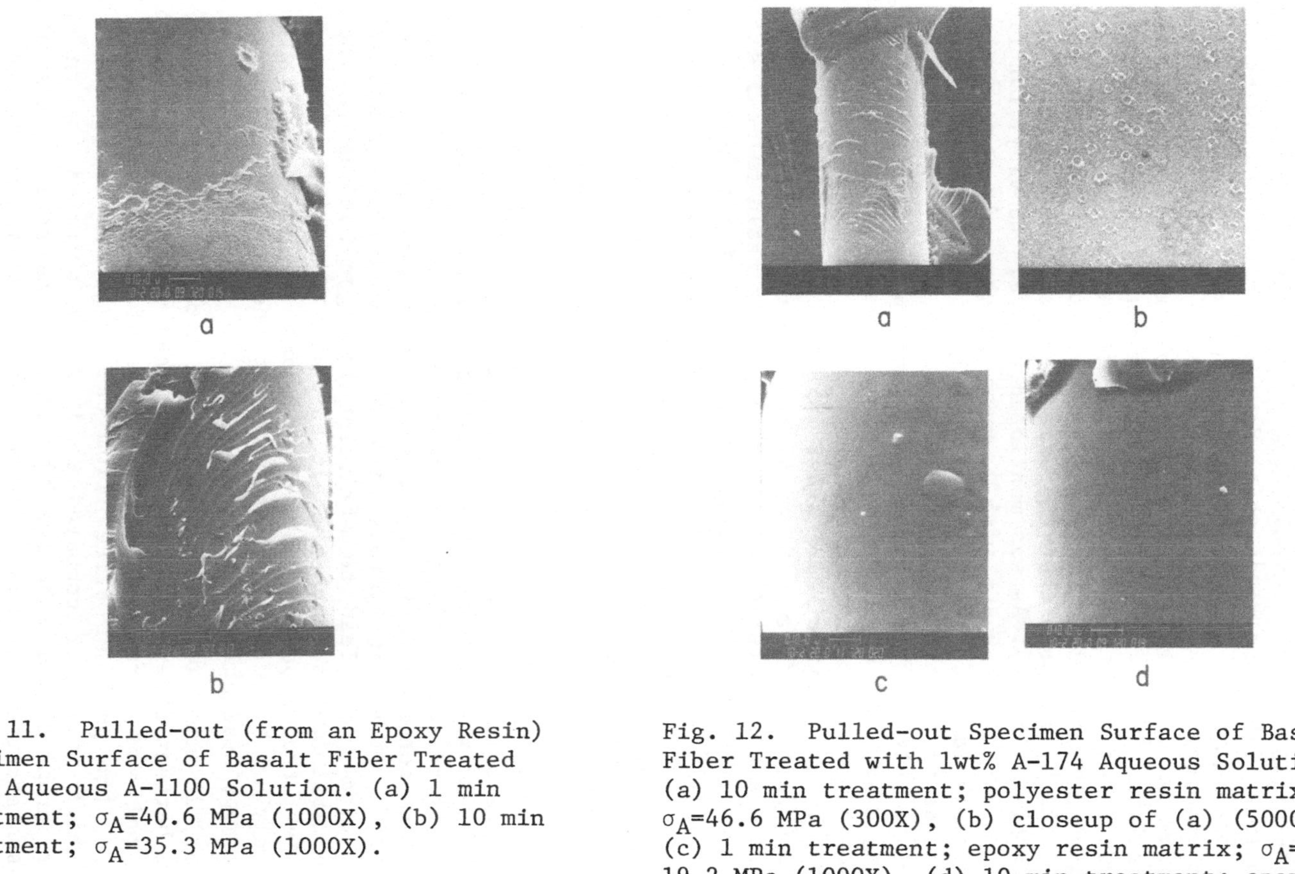

Fig. 11. Pulled-out (from an Epoxy Resin) Specimen Surface of Basalt Fiber Treated with Aqueous A-1100 Solution. (a) 1 min treatment; σ_A=40.6 MPa (1000X), (b) 10 min treatment; σ_A=35.3 MPa (1000X).

Fig. 12. Pulled-out Specimen Surface of Basalt Fiber Treated with 1wt% A-174 Aqueous Solution. (a) 10 min treatment; polyester resin matrix; σ_A=46.6 MPa (300X), (b) closeup of (a) (5000X), (c) 1 min treatment; epoxy resin matrix; σ_A= 19.2 MPa (1000X), (d) 10 min treatment; epoxy resin matrix; σ_A=21.3 MPa (1000X).

<u>Locus of Failure</u>

A comparison of scanning electron micrographs of pulled-out surfaces (Fig. 11-13) reveals some interesting features of debonding between fiber and resin which depend upon the reactivity of the functional group. Thus, the pulled-out surface of APS treated fiber embedded in an epoxy matrix (Fig. 11a,b) shows the residual epoxy resin is left on the surface, confirming the coreaction of the silane coating with the epoxy resin leading to cohesive failure. Similarly, the SEM of MPS-treated fiber embedded in a polyester matrix shows residual polymer on the pullout surface (Fig. 12a,b) but none when the same fiber was embedded in an epoxy matrix (Fig. 12c,d). As was mentioned before, the vinyl functional group of MPS can coreact with styrene in the polyester resin, as found by Ishida and Koenig [39], but not with the epoxy resin. The appearance of the pulled-out surface (Fig. 12c,d) in fact suggests a failure between silane coating and epoxy polymer, with the silane coating remaining tightly bonded to the fiber surface. In this connection it is interesting to note that the vinyl group of MPS could have polymerized under the conditions of drying (105°C/4h/air) to form a tight silane layer into which the interdiffusion and penetration of the bulky epoxy resin molecules could have been minimal. Such homopolymerization of MPS has been proposed by Ishida and Koenig [39] from spectral observations of disappearance of vinyl groups in air, but not in vacuum, during heating of MPS.

Similar differences are noticed in the failure of specimens prepared with untreated fibers also. In the case of polyester resin, the embedded basalt fiber comes out with a "clean" surface suggestive of interfacial failure (Fig. 13a,b). But it is clearly not the case when an epoxy resin is used (Fig. 13c,d), where the coating of residual polymer on the fiber reveals superior adhesion leading to cohesive failure. This is not surprising since the presence of a number of hydroxyl and amino functional groups in the epoxy polymer can lead to efficient hydrogen bonding to the hydrated metal oxides on the basalt surface. Similar bonding of the polyester, though possible through carboxyl or hydroxyl end groups, is likely to be much weaker because of the very small number of such groups available.

However, it would be misleading to make a direct and simple correlation between observed adhesive and cohesive failures and the measured pull-out stresses. In the case of untreated fibers, for example, the pull-out stress from the polyester matrix was 24.6 MPa, which is higher compared to 19.1 MPa obtained for an epoxy matrix which showed cohesive failure in the resin. The higher stresses measured for the polyester can be attributed to the residual radial compressive stress on the fiber created by the shrinking of the polyester resin during curing. The polyester shrinks by

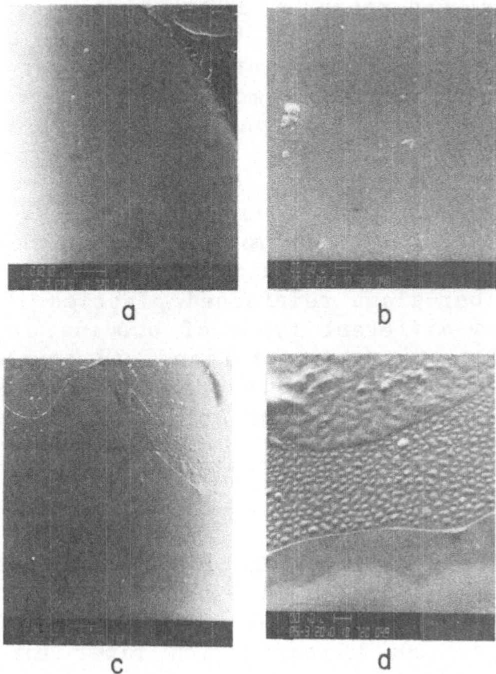

Fig. 13.   Pulled-out Specimen Surface of Untreated Basalt Fiber.
           (a) polyester resin matrix; σ_A=27.2 MPa (1000X), (b)
           closeup of (a) (5000X), (c) epoxy resin matrix; σ_A = 18.8
           MPa (1000X), (d) closeup of (b) (5000X).

about 8% during curing, while the epoxy resin shrinks only 2%. The
radial compressive stress in the later case will therefore be
smaller and contribute less to the frictional resistance to pull-
-out.

     The shrinkage during curing will also be dependent upon the
temperature of curing, and the mismatch between the thermal constants of the fiber and the resin will make additional contribution
to the radial compressive stresses.   Such radial compressive
stresses alone cannot account for the measured interfacial bond
strengths.   For example, for the case of hot curing polyester

system which had undergone a 100°C temperature change during
curing, the compressive stress was found to be only 12.8 MPa
[19,40]. The balance of the interfacial shear strength between
fiber and matrix polymer must come from the various other types of
bonding discussed already, both physical and chemical.

The mechanism of reaction of silanes at the basalt fiber-
-organic polymer interface is thus quite complex, and involves much
more than a simple adhesion promotion between the two phases. As
documented by Plueddemann [20] and discussed extensively by Ishida
and Koenig for fiber-glass reinforced plastics [41], the coupling
mechanism involves different types of bonding, including physical
adsorption, chemisorption, orientation, and covalent bonding discussed above; it also involves the creation of an interphase region
whose structure and properties are of critical importance in controlling composite behavior.

Conclusions

The results of this investigation lead to the conclusion that
the interfacial bond strength in basalt fiber-polymer systems can
be improved by silane coupling agents, and that basalt fiber has
excellent attributes as a reinforcing fiber for polymer resins.
The response of basalt fiber to silane treatment, though similar to
that of glass fiber, is modified significantly by the presence of
iron oxides in basalt. Residual stresses arising from curing and
shrinking of polymer systems contribute a minor fraction of the
pull-out stress; the radial compressive stresses are larger in the
case of polyester than for epoxy resin. The fiber pull-out test is
a useful method for investigating changes in interfacial bond
strengths by different surface treatments applied to basalt
fibers.

ACKNOWLEDGMENTS

This research was supported by the award of a WSU
grant-in-aid. The help of the WSU Geology Department and Dr. P.
Hooper in providing basalt samples and analysis is gratefully
acknowledged.

REFERENCES

1. H. H. Hess and A. Poldervaart, "Basalts," Interscience, New
        York, 1967, p. 221.
2. Chemical and Engineering News, 1973, June 4, p. 49; 1974,
        April 29, p. 18.
3. Kaswant in Kastellaun, Sprechsaal fur Keramik, Glas Email 91,
        577, (1968).
4. E. D. Andreevskaya and T. A. Plisko, Steklo i Keramika 8, 15
        (1963).

5. V. A. Dubrovskii, V. A. Rychko, T. M. Bachilo and A. G. Lysyuk, Steklo i Keramika 12, 18, (1968).

6. V. A. Darenskii, Yu N. Dem'yanenko, P. P. Kozlovskii, K. V. Manzhurnet, A. I. Kukarkin, R. T. Ozhugaryan and K. S. Badolyan, Steklo i Keramika 12, 38, (1968).

7. R. V. Subramanian, H. F. Austin, and R. A. V. Raff, Final Report, Pacific Northwest Regional Commission, Contract No. NR-3001, (1975).

8. R. V. Subramanian, H. F. Austin, and T. J. Y. Wang, Final Report, Pacific Northwest Regional Commission, Contract No. NR-3017, (1976).

9. R. V. Subramanian and H. F. Austin, U. S. Patent 4,149,866 (1979).

10. R. V. Subramanian, H. F. Austin and T. J. Y. Wang, SAMPE Quarterly 8 (11), 1 (1977).

11. R. V. Subramanian and H. F. Austin, "Basalt Fibers" in Handbook of Fillers and Reinforcements for Plastics, Ed. Harry S. Katz and John V. Milewski, Van Nostrand Reinhold, New York (1978), p. 504.

12. J. B. Shortall and H. W. C. Yip, J. Adhesion, 7, 311 (1976).

13. J. P. Favre and J. Perrin, J. Mater. Sci., 7, 1113 (1972).

14. R. V. Subramanian, J. Jakubowski, and F. D. Williams, J. Adhesion 9, 185 (1978).

15. J. P. Favre and M. C. Merienne, Int. J. Adhesion and Adhesives, 1, 311 (1981).

16. K. H. Shu, "Interfacial Bonding in Basalt Fiber-Polymer Composite Systems," Thesis, Washington State University (1978).

17. R. C. DeVekey and F. J. Majumdar, Magazine Concr. Res. 20, 229 (1968).

18. P. Lawrence, J. Mater. Sci., 7, 1 (1972).

19. H. W. C. Yip and J. B. Shortall, J. Adhesion 8, 155 (1976).

20. E. P. Plueddemann, Interfaces in Polymer Matrix Composites, Academic Press, (1974), p. 174.

21. E. P. Plueddemann, SPI Conf. Reinf. Plastics/Composites Div., 19-A (1969).

22. H. Ishida, S. Naviroj, S. K. Tripathy, J. J. Fitzgerald, and J. L. Koenig, J. Polym. Sci. Polym. Phys. Ed., 20, 701 (1982).

23. M. E. Schrader, in Interfaces in Composites, E. P. Plueddemann Ed., Academic Press, (1974) p. 110.

24. R. L. Kaas and J. L. Kardos, Polymer Eng. and Sci., 11, No. 1, 11 (1971).

25. E. P. Plueddemann, "Cationic Silane Coupling Agents for Thermoplastics," in Polymer Plastics Technology and Engineering, Vol. 2, Ed. Louis Naturman, Marcel Dekker, New York (1973).

26. E. P. Plueddemann and G. L. Stark, Mod. Plast., March, 74 (1974).

27. E. P. Plueddemann, Proc. SPI Conf. Reinf. Plastics/Composites Div., 21-E (1973).

28. F. J. Kahn, Appl. Phys. Letters, $\underline{22}$, 386 (1973).
29. F. J. Kahn, and G. N. Taylor, cited in E. P. Plueddemann, Mod. Plast., 76 (March 1974).
30. G. A. Parks, Advances Chem. Series, $\underline{67}$, 121 (1967).
31. A. S. Michales and J. C. Bolger, I & EC Fundamentals, $\underline{3}$, 14 (1964).
32. I. Iwasaki, S. T. B. Cooke and A. F. Columbo, U. S. Bur. Mines Rept. Invest., 5593 (1960).
33. J. C. Bolger and A. S. Michales, "Interface Conversion for Polymer Coatings," Philip Weiss, G. Dale Cheever, Eds., Elsevier, New York (1968).
34. H. J. Modi and D. W. Fuerstenau, Trans, AIME, $\underline{217}$, 381 (1960).
35. R. V. Subramanian and H. F. Austin, Int. J. Adhesion and Adhesives, $\underline{1}$, 50 (1980).
36. F. J. Boerio and F. J. Dillingham, in Proc. Intl. Conf. Adhesive Joints: Their formation, Characteristics and Testing, K. L. Mittal, Ed., Plenum Press (1983), to be published.
37. R. Wong, J. Adhesion, $\underline{4}$, 1971 (1972).
38. J. Patrick and A. K. Rastogi, Amer. Cer. Soc. Bull., $\underline{53}$, (9), 631 (1973).
39. H. Ishida, J. L. Koenig, J. Polym. Sci., Polym. Phys. Ed., $\underline{17}$, 615 (1979).
40. J. H. Hill, Ph.D. Thesis, Cornell University (1967).
41. H. Ishida and J. L. Koenig, Polym. Eng. Sci. $\underline{18}$, 128 (1978).

THEORETICAL ESTIMATION OF THE POSSIBILITY TO REGULATE THE
VISCOELASTIC PROPERTIES AND TENSILE STRENGTH OF FILLED POLYMERS BY
CHANGING THE SIZE OF SUPERMOLECULAR DOMAINS IN THE INTERPHASE LAYERS
OF POLYMER BINDERS

Yu. S. Lipatov, N. I. Korzhuk and V. F. Babich

Institute of Macromolecular Chemistry
Ukrainian Academy of Sciences
252160, Kiev, USSR

It is known that the presence of filler influences markedly
the super-molecular structure of polymer surface layers [1,2]. Mechanical properties of polymers are connected with the supermolecular structure [3,4]. Therefore it is of great importance to
clarify the possibility of changing mechanical properties of filled
polymers by variation of the surface layer structure.

Here, we have made an attempt to discuss the influence of supermolecular structure size on the mechanical properties of the
surface layer of binder in filled polymers. Our other task was to
evaluate the change in mechanical properties of filled polymer when
there are changing mechanical properties of a surface layer of finite thickness on the filler surface.

For our calculations we have chosen very simple models of
filled polymer. The following assumptions have been made: 1) polymer is considered to have a heterogeneous structure [5] and consists of some grains (these grains may be globulae, crystallites,
domains, clusters, etc.), 2) between grains there is an intergrain
layer playing the role of binder [6], and 3) the viscoelastic
properties of grains and intergrain layers are essentially
different. In such a model the volume fraction of intergrain phase
in total polymer is proportional to the grain specific surface.
This model may be considered as a polymeric cube with edge "a" pre-
senting a grain, covered uniformly by the polymer surface layer of
thickness "d" with different viscoelastic properties. From the
assumptions made, it is clear that the intergrain layer thickness
"d" is not dependent on the grain size "a".

The viscoelastic properties of such polymeric materials may be calculated using Takayanagi model [7]. From pure geometrical consideration there should exist the following correlations between parameters of Takayanagi model ϕ and λ and grain size "a" and intergrain thickness "d":

$$\phi = a/(a + 2d); \quad \lambda = \phi^2$$

For the following calculation we accept that polymeric grain has complex elasticity modulus $E^*_{gr} = f(T)$ and mechanical loss tangent $\tan \delta_{gr} = f(T)$, where T is temperature. From our earlier experimental data [1] it follows that the intergrain material has complex modulus $E^*_{igr} = f(T + \Delta T)$ and $\tan \delta_{igr} = f(T + \Delta T)$. It means that the curves to temperature dependences E^*_{igr} and $\tan \delta_{igr}$ are identical to curves for E^*_{gr} and $\tan \delta_{gr}$, being shifted along the temperature axis to lower or higher temperatures by arbitrary chosen value ΔT.

Having taken various "a" and "d", one can model the changes in the grain size and intergrain layer thickness. Changing ΔT enables one to vary the correlation between mechanical characteristics of grains and intergrain material.

In this case the mechanical properties of a two-component system E^*, E', E'' and $\tan \delta$ may be expressed using the corresponding values for each component:

$$E^* = \sqrt{(E')^2 + (E'')^2}; \quad \tan\delta = E''/E'; \quad E' = (1-\lambda)E'_{igr} + \frac{\lambda x}{x^2 + y^2};$$

$$E'' = (1-\lambda)E''_{igr} + \frac{\lambda y}{x^2 + y^2}; \quad x = \frac{\phi E'_{gr}}{(E'_{gr})^2 + (E''_{gr})^2} + \frac{(1-\phi)E'_{igr}}{(E'_{igr})^2 + (E''_{igr})^2}$$

$$y = \frac{\phi E''_{gr}}{(E'_{gr})^2 + (E''_{gr})^2} + \frac{(1-\phi)E''_{igr}}{(E'_{igr})^2 + (E''_{igr})^2}$$

Here, E^* is complex modulus of the bulk material, E' and E'' are real and imaginary parts of complex modulus E^*, E^*_{gr}, E'_{gr} and E''_{gr} are the corresponding characteristics of intergrain material and E^*_{igr}, E'_{igr} and E''_{igr} —the same for grain material. Figure 1 shows typical calculated dependences of E^* and $\tan \delta$ for the model of heterogeneous material of grain structure. These curves have been obtained for the case when grain material has the properties of cured epoxy resin [8] whereas intergrain material has a glass transition temperature 10 lower than the grains. Figure 1 shows that diminishing "a" decreases E^* and regularly shifts curves $E^* = f(T)$ and $\tan \delta = f(T)$ to lower temperatures. If we suppose the

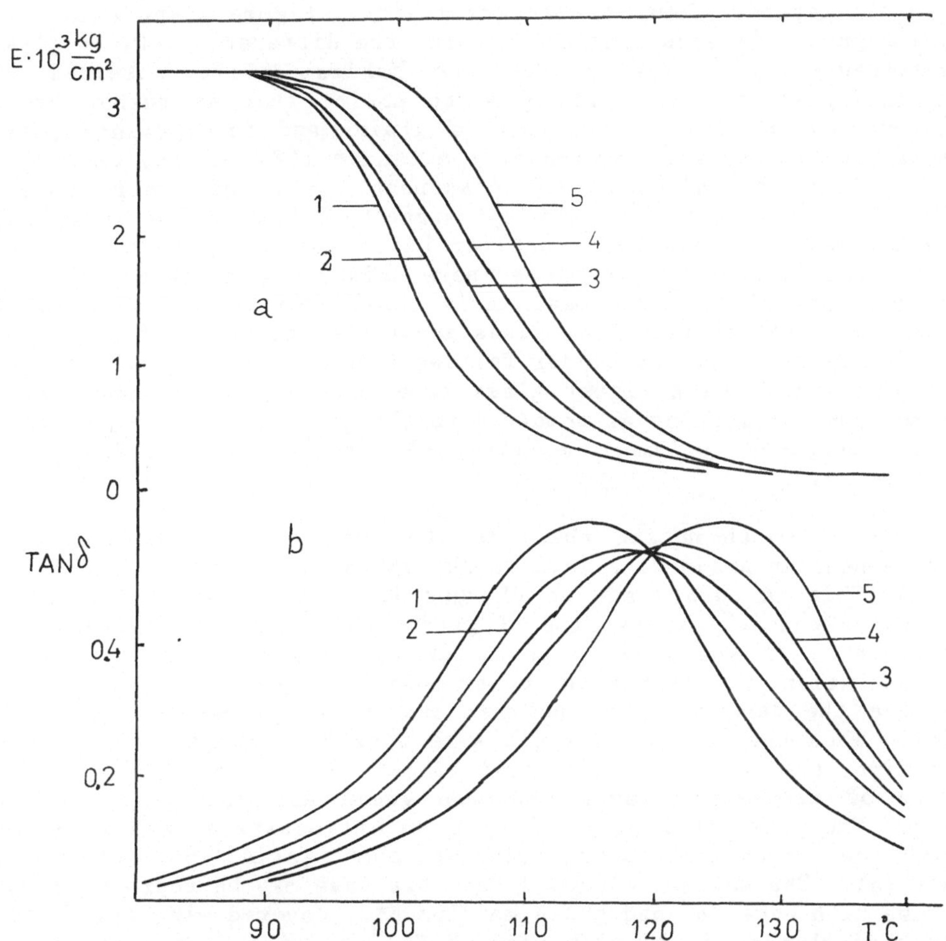

Fig. 1. Typical calculated temperature dependences of complex modulus (a), and mechanical loss tangent (b) for the model of heterogeneous material at various model parameters: "a": 1-1$_o$, 2-2, 3-1, 4-0.5, 5-0.1. T_{gr}=115°C, T_{igr}=125°C.

glass transition temperature of the grain to be lower than the intergrain layer, the curves will be shifted to higher temperatures when "a" decreases.

The analogous calculations have been made for various correlation between the glass transition temperatures of grains and intergrain layers. Some results are given in Figure 2 and Figure 3. From Figure 2 is seen that the greater the difference between glass temperatures, the greater is the maximum shift obtained by decreasing grain size. It is worth noting that at rather great differences in glass temperature of grains and intergrain layers, there appears two $\tan\delta$ maxima in the $\tan\delta = f(T)$ curves, their position being dependent on the component ratio and their properties. The conditions of the appearance of two maxima in binary systems were analyzed by us earlier [9]. Therefore, here we shall consider only those cases where there exists one maximum. Usually the presence of only one maximum is considered to be the sign of homogeneous structure. Some ideas about the influence of the grain size on E* of polymeric binder follows from Figure 3. If the intergrain material has a higher glass temperature and correspondingly higher complex modulus as compared to the grains, diminishing their size increases polymer modulus E* (see curves 1,2,3) and vice-versa.

It is worth noting that, in the model under consideration, the changes in grain size lead to the changes in their volume fraction. This in turn leads to changes in E* and $\tan\delta$ of the model because mechanical properties of grains and intergrain layers are different. If the grain size is changed without changing their concentration, the simple two-phase model of Takayanagi type cannot explain the dependence of polymer mechanical properties on grain size. When the change in grain size does not change their volume fraction, the variation in properties may be the result of the presence of transition layers between grain and intergrain layers [10]. The mechanical properties of such transition layers differ from the mechanical properties of both grain and intergrain material. The mechanical model for this case may be represented as a cube with edge "a" and grain modulus E^*_{gr}, covered with transition layer of thickness "d" with modulus E^*_t and with intergrain layer of thickness "c" and modulus E^*_{igr}. Here, $E^*_{gr} > E^*_t > E^*_{igr}$. From geometrical consideration, it follows that the concentration of grain phase ϕ_{gr}, transition layer ϕ_t and intergrain layers ϕ_{igr} can be expressed as follows:

$$\phi_{gr} = a^3/(a + 2d + 2c)^3$$

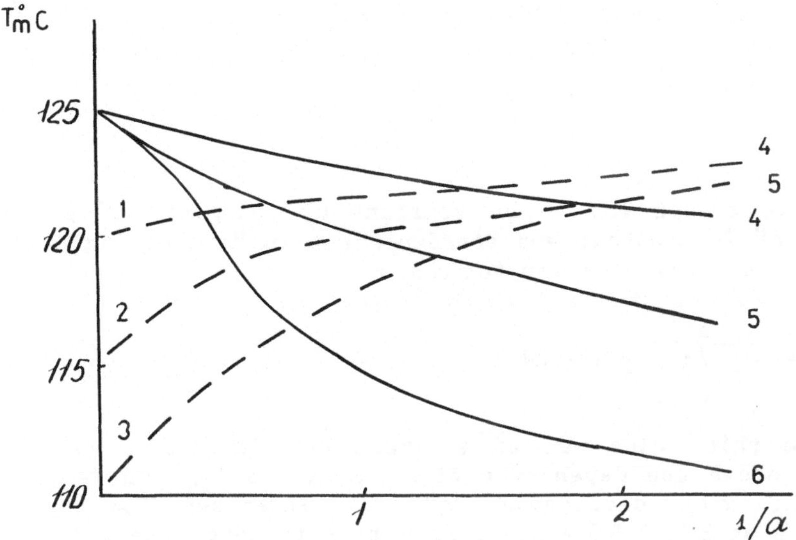

Fig. 2. Dependence of temperature T of maximum tan δ for the model on 1/a at various T_{glass} of components: 1-T_{gr}=120°C, T_{igr}=125°C, 2-115 and 125°C, 3-110 and 125°C, 4-125 and 120°C, 5-125 and 115°C, 6-125 and 110°C.

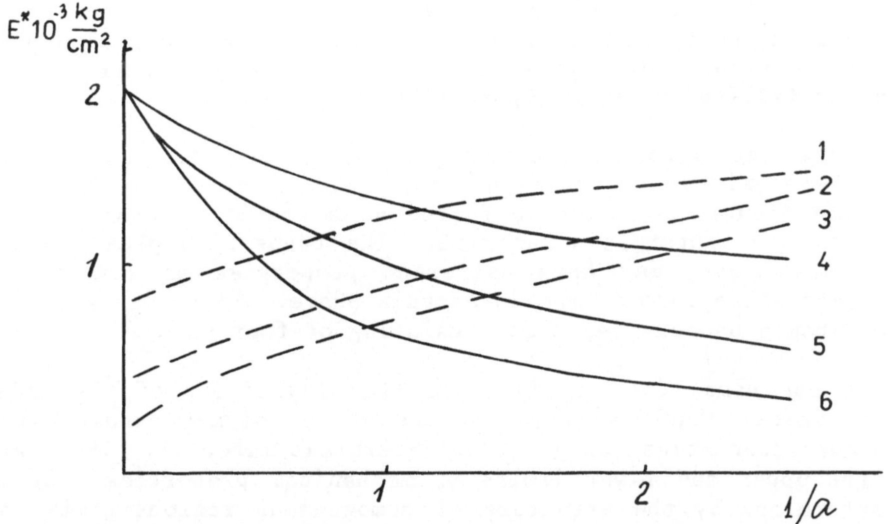

Fig. 3. Dependence E on 1/a at T=110° for the same model (indications see Fig. 2.)

$$\phi_t = [(a + 2d)^3 - a^3]/(a + 2d + 2c)^3$$

$$\phi_{igr} = 1 - \phi_{gr} - \phi_t$$

If we accept ϕ_{gr} to be constant by changing "a" and if we suppose "d" is constant and non-dependent on "a", we can calculate the value "c" from equation for ϕ_{gr} :

$$2c = \frac{a}{\phi_{igr}^{1/3}} - a - 2d$$

From this value we can evaluate the dependence $\phi_t = f(a)$. Figure 4 shows the dependence of ϕ_{gr}, ϕ_t and ϕ_{igr} on "a". It is seen that with decreasing "a", ϕ_t increases, whereas ϕ_{igr} diminishes at $E^*_{gr} > E^*_t > E^*_{igr}$. From this follows, the value E^* for the model will be increased with diminishing "a" due to increasing fraction of more rigid interphase layer and decreasing of fraction of low-modulus intergrain layer. As in the case of absence of transition layers, the character of changing E^* will be dependent on phase properties. At the same time there are essential differences between these cases. The increase in elastic modulus with decreasing "a" takes place only to the point where all the intergrain phase transits into the state of transition layer. In a two-phase system, decreasing "a" at $E^*_{gr} > E^*_{igr}$ diminishes E^* (see Figure 3). All these considerations are in good accordance with general concepts of structure of polymeric systems and filled polymers, where transition layers (especially for polymers, filled with polymeric fillers) play an important role [1].

There can also be proposed another model for the systems where there exists a transition layer. For example, we can assume that this transition layer is formed as a result of transition in both grain and intergrain material. The transition phase, formed from grain phase, may have different properties as compared to transition phase formed from intergrain phase. In these cases, the system should be considered as consisting of four phases.

At the same time, it is clear that regardless of the model, the mechanical characteristics of the whole polymer cannot exceed these characteristics for grain or intergrain material. That means that the upper and lower limits of mechanical properties changing are determined by the structure of homogeneous regions-grains and intergrain layers. The same is valid for glass transition temperatures. That means that if the whole polymer in the filled system is in the state of interphase layer, the binder properties will be

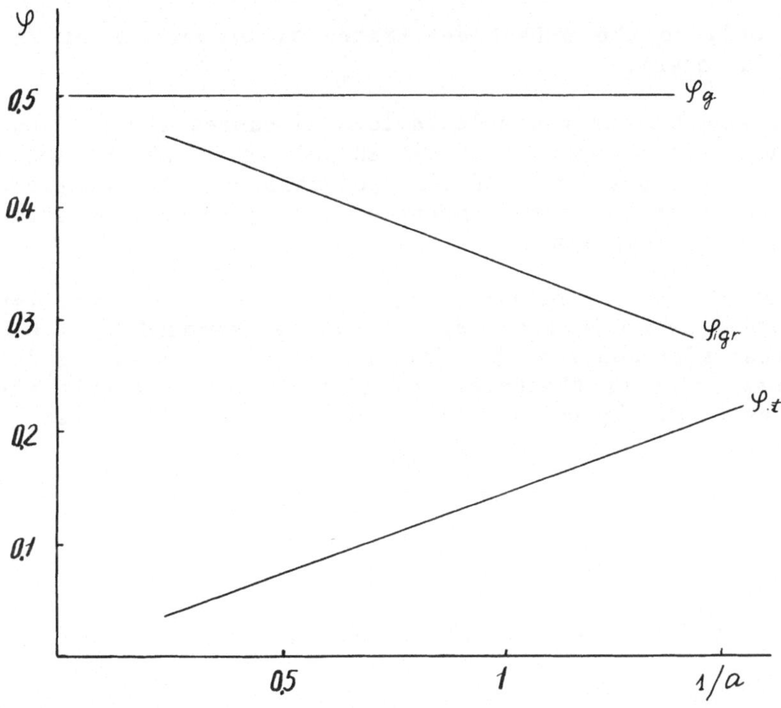

Fig. 4. Dependence of grain phase concentration ϕ_{gr}, intergrain phase ϕ_{igr} and transition phase ϕ_t for three-component model on parameter "a".

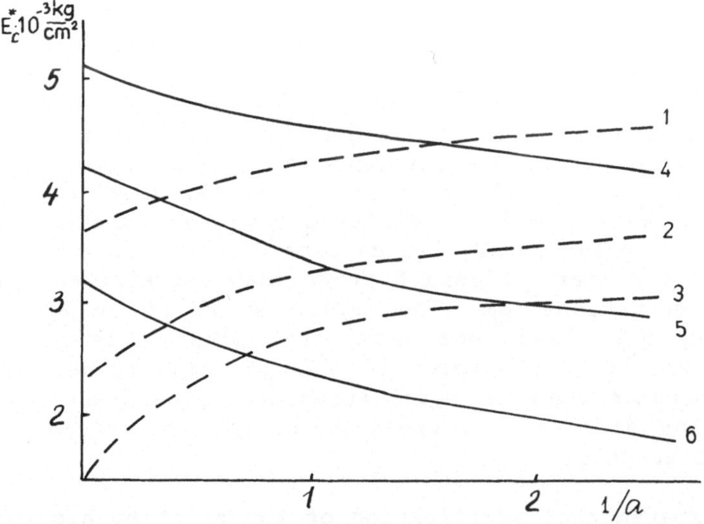

Fig. 5. Dependence cf complex modulus for the model of filled polymer on I/a at T=110°C in presence of transition layers of various structure and properties. Indications see Fig. 2.

changed only to the extent determined by properties of grain and intergrain layers.

The assumptions and calculations discussed above allow us now to consider the properties of filled polymers. Let us analyze the viscoelastic properties of filled polymers and the contribution of interphase layer in these properties taking in consideration their dependence on grain size.

A model for the filled polymers with an interphase layer has been studied by us earlier [9]. It is represented by a cube made of material with modulus E_f^* (filler particle with edge a_f) covered by polymer layer of thickness "d_1" with modulus E_1. This system is covered once more by polymer with modulus E^* and thickness "c".

$$E_c^* = (1-\lambda)E^* + \lambda\left(\frac{1-\phi}{E^*} + \frac{\phi}{E_1^*}\right)^{-1}$$

$$\phi = \frac{a_f + 2d_e}{a_f + 2d_e + 2c}; \quad \lambda = \phi^2$$

$$E_1^* = (1-\lambda)E_e^* + \lambda_1\left(\frac{1-\phi_1}{E_e^*} + \frac{\phi}{E_f^*}\right)^{-1}$$

$$\phi_1 = \frac{a_f}{a_f + 2d_e}; \quad \lambda_1 = \phi_1^2$$

The results of calculations are given in Figure 5. It is seen that modulus E_c^* decreases with growing "a" if $T_{ig} > T_{gr}$ and increases if $T_{ig} < T_{gr}$. The size of grains has more influence with greater ΔT. As the modulus difference between grains and intergrain layers depends on ΔT, let us consider the correlation between E_c^* and E^* of binder. Figure 6 shows such correlation for various filler concentrations ϕ_f. The influence of E^* on E_c^* increases with growing ϕ_t. As in our case, the value E^* is restricted by values E_{gr}^* and E_{igr}^*, the value E_c^* changes also in definite limits which are dependent on the model structure, interphase layer thickness, modulus difference between grain and intergrain phases and filler concentration.

The experimental verification of the relationship derived was made for epoxy binder ED-20 with wide intervals of E^* values. Quartz powder was used as a filler (particle size 1-3 μm). Experimental results are given in Figure 6. It is seen that there exists

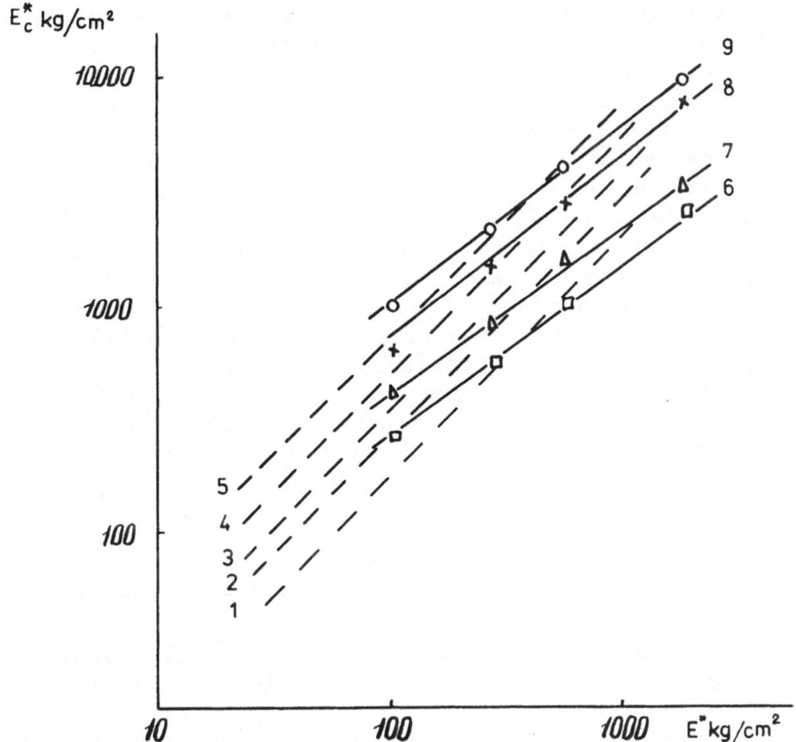

Fig. 6. Calculated dependences of complex modulus of filled polymer on binder complex modulus at various filler concentrations: 1-ϕ_f=0.2, 2-0.4, 3-0.5, 4-0.6, 5-0.7. Curves 6-9-experimental data at ϕ_f=6-0.04, 7-0.11, 8-0.31, 9-0.44.

linear dependence $\log E_c^* = f(\log E^*)$ for a wide interval of changes in E^* and ϕ_f. Here, value $k = \mathrm{tg}\angle$ (slope of straight lines) is close to 1, $k = 0.9$, which verifies the theoretical calculations. In such a way the increase in the elastic modulus of binder by changing the size or shape of supermolecular structure can really increase the modulus of filled polymers according to the following equation:

$$\log E_c^* = k\log E^* + A \quad \text{or} \quad E_c^* = E^{*k} \cdot 10^A$$

The quantitative analysis of tensile strength of filled polymer using Takayanagi model is not realistic, as the ultimate characteristics are connected with local properties, not with averaged characteristics. However, some ideas about influence of "a" value on tensile strength may be gained if we accept that the material was broken after reaching some critical deformation ε_{crit}. In this case, as a characteristic of tensile strength, the value of average stress σ_{crit} can serve, this value corresponding to ε_{crit}. It is clear that dependence of "tensile strength", σ_{crit} on "a" in this case, will be the same as for the dependence E_c^* on "a". More detailed analysis needs to take into account changes in the stress distribution near the surface of filler particles due to changing the properties of the surface layer of binder on the filler particles. Thus, the theoretical consideration shows the real possibility of regulating the properties of filled polymers by changing the supermolecular structure of binder near the filler surface. The physical principles of such changes in properties have been developed earlier [1].

REFERENCES

1. Yu. S. Lipatov, Physical Chemistry of Filled Polymers, British
       Library, 1979.
2. Yu. M. Malinsky, Progress in Chemistry, (Russ.), 1970, 39, 1511.
3. G. L. Slonimsky, V. I. Pavlov, Vysokomolek. soed., 1966, 7,
       1279.
4. L. Z. Rogovina, G. L. Slonimsky, Vysokomolek. soed., 1966, 8,
       2046.
5. Yu. S. Lipatov, Pure and Appl. Chem., 1975, 43, 273.
6. Yu. S. Lipatov, A. N. Kuksin, L. M. Sergeyeva, J. Adhes., 1974,
       6, 259.
7. S. Uemura, M. J. Takayanagi, J. Appl. Polymer Sci., 1966, 10,
       113.
8. Yu. S. Lipatov et al, J. Appl. Polymer Sci., 1980, 25, 1029.
9. V. F. Babich, Yu. S. Lipatov, J. Appl. Polymer Sci., 1982, 27,
       53.
10. Yu. S. Lipatov in "Adhesion and Adsorption of Polymers," part
       B, L.-H. Lee, ed., Plenum, 1980, 601.

POLYMER COMPOSITES OF POLY(p-PHENYLENE TEREPHTHALAMIDE) AND NITRILE

BUTADIENE RUBBER: (I) PREPARATION AND PROPERTIES

Motowo Takayanagi and Kohei Goto*

Dept. of Applied Chemistry, Faculty of
Engineering, Kyushu University
Hakozaki, Higashi-ku
Fukuoka, 812 JAPAN

*Japan Synthetic Rubber Co., Ltd.
Higashi-yurigaoka, Asao-ku
Kawasaki, 214 JAPAN

ABSTRACT

Microfibrillar poly(p-phenylene terephthalamide) (PPTA) was
uniformly dispersed in a matrix of nitrile butadiene rubber (NBR)
by coagulating a homogeneous solution of N-sodium PPTA and NBR in
common solvent of dimethylsulfoxide and N,N-dimethylformamide (DMF)
with acidic water, being accompanied by regeneration of metalated
PPTA. The observation with polarization optical microscope showed
that the image of the polymer composite was uniformly birefringent
under crossed nicols. The electron-microscopic observation on the
residue remaining after extraction of uncured NBR with DMF from the
polymer composite revealed a network of microfibrills with a
diameter of 10 ~ 30 nm. In a cured sheet of the PPTA/NBR composite, PPTA microfibrils tend to align along the milling direction,
giving rise to the anisotropies of modulus and swelling ratio in
organic solvent. The reinforcing effects of microfibrillar PPTA in
NBR are superior in modulus and strength to those of carbon black
at the same reinforcement content. The improvement in tear
strength was noticeable in microfibrillar PPTA-reinforced NBR. The
fractography supported the view that microfibrils of PPTA impede
the crack growth by bifurcation or deflection of crack.

INTRODUCTION

Recently advanced composites reinforced by high performance fibers such as carbon fiber or Kevlar aramid fiber have been developed in the field of engineering materials. Hard and strong fibers are dispersed in a ductile matrix, providing the composite with excellent mechanical properties and heat endurance. The applied stress is uniformly distributed fiber to fiber through the ductile matrix with adhesion at the interface of fiber and matrix. The aspect ratio of the fiber should be larger than the critical aspect ratio [1].

In the case of rubber, the modulus of fiber-reinforced rubber is higher than that of particulate-reinforced (e.g. carbon black) rubber. The tensile strength of short-fiber reinforced rubber is lower than that of particulate-reinforced rubber above Tg. Below Tg, the relation in strength is reversed. The molecular motion of matrix rubber seems to make the situations more complicated in deformation of rubber composites.

In this paper, a new field of polymer composites is explored; rigid rod-like molecules such as poly(p-phenylene terephthalamide) (PPTA) are finely and uniformly dispersed in flexible rubber molecules such as nitrile butadiene rubber (NBR) to extend the basic principle of macroscopic reinforcement mentioned above to the molecular level. It should be noticed that the molecule of PPTA still keeps its rigidity at the molecular level and is expected to reinforce the matrix if the intermolecular adhesion between PPTA and NBR is strong enough to avoid the phase separation as far as possible. Such a polymer composite behaves like a new raw material, being different from the macroscopic fiber-reinforced materials. PPTA molecules have a strong tendency to form liquid crystalline phase in a solvent of sulfuric acid. Avoiding the formation of liquid crystals is preferable to producing a homogeneous texture of polymer composite. A method of preparation of the polymer composite is to prepare an isotropic solution of rigid and flexible molecules at a polymer concentration lower than the critical concentration for liquid crystal formation and to precipitate the composite rapidly from the solution to avoid any liquid crystal formation. By applying this method to the systems of aramid and nylon 6 or 66, we obtained polymer composites successfully [2]. They showed an increased yield stress of about 300% over the matrix nylon by use of 5% aramid. Sulfuric acid was used in the previous paper. For the system of aramid and NBR, sulfuric acid cannot be employed owing to the decomposition of NBR. In such circumstances, it was necessary to find a new solvent for PPTA.

In a previous paper, we found that sodium hydride and di-methylsufoxide (DMSO) were reacted to form sodium methyl sufinyl carbanion in DMSO, in which PPTA was dissolved with formation of N-sodium PPTA [3]. In this paper, metalated PPTA in DMSO and NBR in N,N-dimethylformamide (DMF) were mixed and a homogeneous isotropic solution was prepared. This enabled the preparation of a polymer composite of PPTA and NBR by applying the conventional method of pouring the solution into a large amount of coagulating bath.

EXPERIMENTAL

Materials

NBR samples used in this study were JSR N220S, N230S and N240S (Japan Synthetic Rubber Co., Ltd.), which contain 40 wt%, 35 wt% and 26 wt% bound acrylonitrile (AN) units, respectively.

The PPTA samples were prepared by low temperature polyconden-sation of p-phenylenediamine and terephthaloyl chloride in a mixed solvent of hexamethylphosphoramide and N-methylpyrrolidone [4]. The molecular weights of PPTA were determined by measuring the intrinsic viscosity in 97% sulfuric acid at 303°K and calculated from the viscosity equation of Arpin et al. [5]. They were 4900, 21900 and 25100.

The PPTA fiber, Kevlar 49 (Du Pont de Numours and Co.), was chopped to the length of 0.5 mm. This was used for comparison with the molecular composite. The carbon fiber, Torayca (Toray Co.), was also chopped to the length of 6 mm for the same purpose.

The carbon black was of the type ISAF (Diablack, Mitsubishi Kasei Industry Co., Ltd).

Preparation of Polymer Composite

Polymer composite of PPTA and NBR were prepared by blending N-sodium PPTA in DMSO and NBR in DMF. The mixed solvents formed a common solvent for both components. The following procedure is a typical example for preparation of polymer composite.

5.0 g of PPTA (Mw 4900, 42 mequiv. amide unit) and 2.4 g of sodium hydride (50% suspension in mineral oil, 50 mmol) were reacted in 100 ml of anhydrous DMSO at 333°K for 2 h under nitrogen atmosphere to give metalated PPTA. 50 g of NBR dissolved in 400 ml DMF was mixed with a metalated PPTA solution at room temperature with stirring. Both component polymers in a mixed solvent in a state of isotropic solution were recovered by coagulation in a large amount of acidic water and washed by hot water to remove the inorganic salt which was produced by regeneration of PPTA. The isolated product was dried at 323°K for 96 h in vacuo.

Processing

Blend rubber was milled on a two-roll open mill and compounded according to the standard formulation shown in Table 1. Compounded rubber was cured at 423 K by compression molding for 60 min. Curing was monitored by Curastometer (Japan Synthetic Rubber Co., Ltd.), which indicated the optimum cure conditions.

Testing

Tensile properties of cured samples were tested parallel to the milling direction. At least four samples were tested in each case. Dumb-bell shape specimens were employed for tensile strength tests (JIS K6301, No. 3), trousers type specimens for tear strength test (JIS K6301, B type), and ring shape specimens for testing temperature dependence of tensile strength and modulus. Stress-strain curves were obtained by Autograph DSS-2000 (Shimazu Co.) at 250%/min strain rate for the dumb-bell specimen, 200%/min for the ring specimen in the tensile test and 667%/min in the tear test.

Dynamic storage modulus, E', dynamic loss modulus, E'', and tan δ were measured by Rheovibron DDV-IIB (Toyo Baldwin Instruments Co., Ltd.) under nitrogen atmosphere at 11 Hz.

Table 1. Compounding Recipe.

	Parts By Weight
NBR	100.0
Zinc Oxide	3.0
Sulfur	1.5
Stearic Acid	1.0
Antioxidant *1	1.0
Vulcanization Accelerator *2	0.7
PPTA	Varied

*1) IPPD
    N-Phenyl-N'-isopropyl-P-phenylendiamine
*2) TBBS
    N-tert-Butyl-2-benzo-thiazolesulfenamide

Swelling test was conducted by immersing cured polymer composite specimens (20 mm x 20 mm x 1 mm) in Fuel A (iso-octane), Fuel B (iso-octane/toluene:70/30 by volume), Fuel C (iso-octane/-toluene:50/50 by volume), and dichloromethane according to JIS K6301. Oil resistance was evaluated from the volume increase at equilibrium swelling.

RESULTS AND DISCUSSION

Formation of Polymer Composite

N-sodium PPTA was synthesized by the metalation of PPTA with sodium hydride in DMSO [3]. NBR dissolved in DMF. Both solutions were mixed. A mixed solvent of DMSO and DMF forms a homogeneous, isotropic solution of N-sodium PPTA and NBR. Blend polymer was precipitated by coagulation without macroscopic phase separation. N-sodium PPTA was regenerated rapidly at the time of coagulation by acidic water according to the following reaction scheme.

$$
\left(\text{HN} - \boxed{\bigcirc} - \text{NHOC} - \boxed{\bigcirc} - \text{CO}\right)_n \underset{\underset{H^+ \quad Na^+}{}}{\overset{NaH/DMSO}{\rightleftharpoons}} \left(\overline{N} - \boxed{\bigcirc} - \overline{N} - \text{OC} - \boxed{\bigcirc} - \text{CO}\right)_n \quad Na^+ \quad Na^+
$$

By metalation of PPTA, the solubility parameter of the PPTA derivative approaches that of NBR and thus, a homogeneous and isotropic polymer solution was obtained. After coagulation with acidic water, regenerated PPTA was finely and homogeneously dispersed in the composite of PPTA/NBR.

Morphology

Figure 1(a) shows a polarized optical micrograph under crossed nicols of the cured film of the polymer composite PPTA/NBR (10/100 by weight). Figure 1(b) shows the micrograph under the same condition of NBR blended with chopped commercial PPTA fiber of 1 denier to compare the dispersed state of the reinforcement in the matrix. The polymer composite in Figure 1(a) showed no grain boundaries but a wholly birefringent image which indicates that the dispersed state of PPTA in the polymer composite is fine and uniform. To clarify the dispersed state of PPTA in the NBR matrix, the composite was extracted with DMF to remove NBR from the uncured sample and the remaining PPTA was inspected by transmission electron microscope as shown in Figure 2. It is seen that PPTA was dispersed in a form of microfibrils whose diameter is 10 ~ 30 nm, corresponding to 10^{-6} denier. This diameter of microfibrils agrees

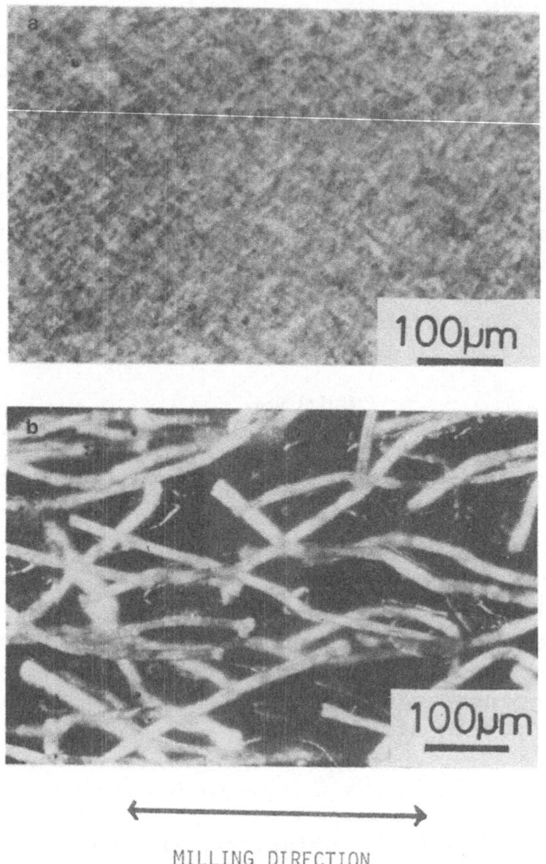

Fig. 1. Polarized optical micrographs under crossed nichols for
         (a)PPTA and NBR polymer composite, and (b)1 denier chopped
         PPTA fiber-reinforced NBR composite.

Table 2. Tensile properties of composites.

AN content in NBR(wt%)	PPTA *1	Young's modulus MPa	Tensile stress at 100% elongation MPa	Tensile stress at 300% elongation MPa	Tensile strength at break MPa	Elongation at break %
40	------------	2.0	1.2	2.1	4.7	540
	PPTA 2.5 phr	4.0	2.0	5.9	15.4	530
	5.0	6.9	4.0	12.1	17.3	420
	10.0	12.1	11.1	----	24.4	210
	Carbon Black	3.1	1.6	4.3	11.3	500
	PPTA Fiber	11.4	6.1	----	6.6	240
	Carbon Fiber	7.1	4.1	5.3	9.0	590
35	------------	2.0	1.0	1.6	5.4	600
	PPTA 2.5 phr	3.5	1.6	4.7	8.8	460
	5.0	5.8	3.3	9.2	13.8	390
	10.0	10.9	7.1	13.1	19.8	320
	Carbon Black	3.1	1.4	3.8	10.6	480
26	------------	1.3	0.8	1.3	2.5	570
	PPTA 2.5 phr	3.1	1.6	3.9	6.3	410
	5.0	4.5	2.3	7.0	12.0	520
	10.0	7.6	4.2	13.3	15.7	350
	Carbon Black	2.3	1.1	2.2	10.7	620

*1 Molecular weight of PPTA is 21900.

with those found for PPTA in the polymer composite of PPTA and
nylon [2], PVC [6] or ABS rein [7]. PPTA microfibrils in the
polymer composite might be statistically aligned in the direction
of milling during processing similarly to the macroscopic fibers as
shown in Figure 1(b). Further discussion will be made with respect
to the anisotropy in viscoelasticity and swelling behavior.

Stress-Strain Behavior

    Table 2 shows the tensile properties of NBR with different AN
content and the polymer composites with varying amount of PPTA.
Tensile modulus and strength increased with increasing amount of
PPTA. Young's modulus of the polymer composite of NBR (40 wt%
AN-comonomer fraction) reinforced with 6.4 vol% (10 phr)-PPTA was
raised to more than 10 MPa, whereas that of black stock with high
loading (40 phr of SRF carbon black) was only 5.1 MPa. To clarify
this remarkable effect of PPTA in the polymer composite, tensile
properties of NBR filled with chopped fiber of PPTA or carbon
fiber, were examined with the same reinforcement content. Although
tensile moduli of the chopped fiber-reinforced composites were
increased owing to the fibrous structure of the reinforcements,
their tensile strengths were decreased by 27% for PPTA chopped
fiber and by 37% for chopped carbon fiber from that of the PPTA
polymer composites.

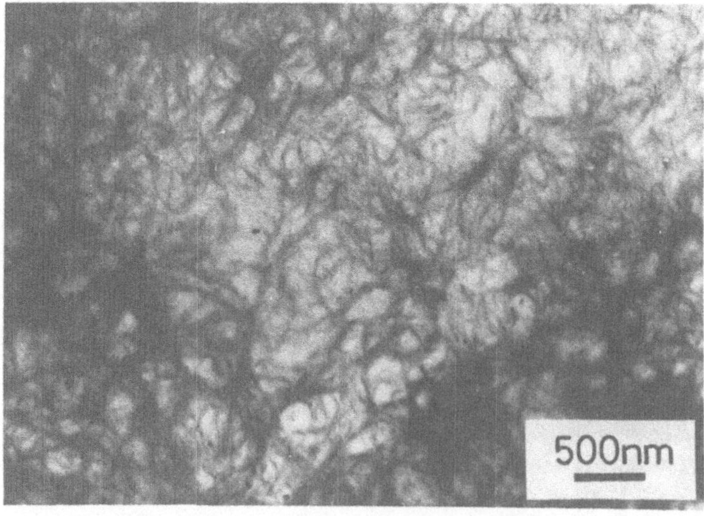

Fig. 2.   Transmission electron micrograph of PPTA microfibrils re-
          maining after extraction of NBR from the uncured polymer
          composite employing PPTA with M_w=4900.

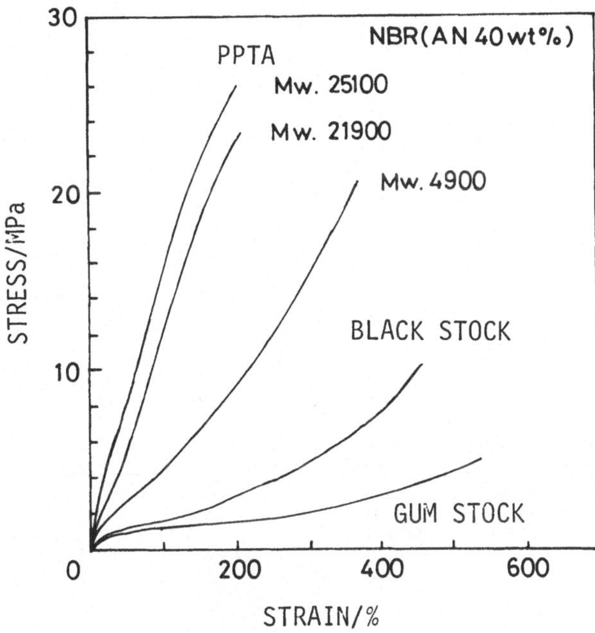

Fig. 3.   Stress-strain curves of NBR gum stock, black stock and
          PPTA polymer composite with various molecular weights of
          PPTA. Volume fraction of reinforcement is 6.4%.

Reinforcing effects of PPTA on NBR with various contents of
AN comonomer are also shown in Table 2. Relative modulus (the ratio
of modulus of polymer composite to modulus of matrix NBR) and relative tensile strength (the ratio of tensile strength of polymer
composite to tensile strength of matrix NBR) increased with
increasing AN comonomer content in NBR. This indicates that the
content of AN in NBR plays an important role in reinforcement. NBR
with high AN content has a strong affinity for the surface of PPTA
fibrils owing to the solubility parameter of NBR being closer to
that of PPTA with higher AN content. Intermolecular hydrogen bond
between -CN group in NBR and -NH in PPTA can be conceivably more
easily formed with increasing AN content in NBR. Such a phenomenon
was ascertained in the system of benzanilide as a model compound of
PPTA and poly(vinyl chloride) (PVC). Shifts of IR bands of C=O and
N-H were detected with increasing PVC, indicating the progress of
dissociation of hydrogen bonds of N-H and O=C in PPTA [8].

Figure 3 shows the stress-strain curves of the polymer composites of various molecular weights of PPTA and NBR with 40 wt% AN
comonomer content. The result indicates that the reinforcing
effect in polymer composites depends on the molecular weight of
PPTA. PPTA with a molecular weight less than 5000 showed a larger
reinforcing effect than the same volume fraction of reinforcing

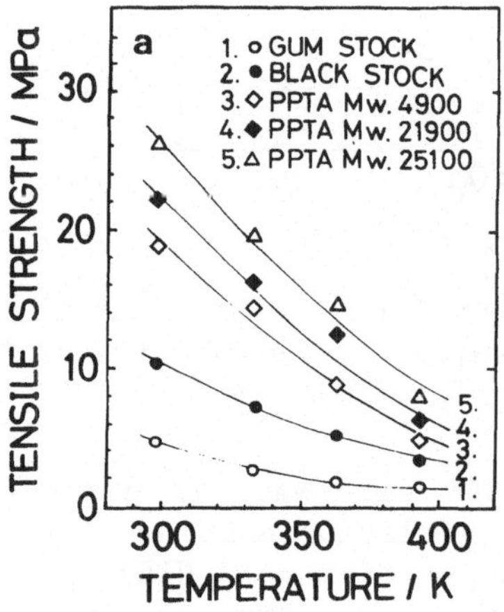

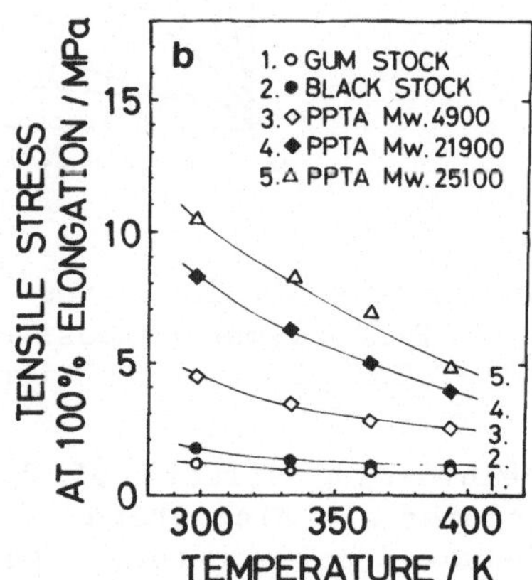

Fig. 4.   Temperature dependence of tensile strength (a)and 100%
modulus (b)for NBR gum stock, black stock and PPTA polymer
composites with 6.4 vol% of reinforcement.   AN Comonomer
content in NBR is 40 wt%.

carbon black (refer to the curve "Black Stock"). Tensile strengths
of the polymer composite with PPTA molecular weights greater than
20000 were comparable with that of highly loaded carbon black composites. However, the high tensile modulus of the PPTA composite
could not be achieved with a particulate-filled composite. In
spite of low molecular weight PPTA in the polymer composite (as low
as 4900), the tensile modulus and strength of the polymer composite
were superior to those of the chopped PPTA fiber-filled composites. As seen in Figure 1(b), the macroscopic fiber-reinforced
composite has a heterogeneous texture and the unreinforced region
of NBR allows the growth of cracks when a force is applied. On the
other hand, the texture of polymer composite shown in Figure 1(a)
impedes the growth of cracks in presence of the microfibrillar
PPTA.

Figures 4(a) and 4(b) show the temperature dependence of
tensile strength and modulus (stress at 100% elongation) of the
polymer composites, respectively. These mechanical properties
decreased with increasing temperature, but their order did not
change. Temperature dependence of tensile strength was comparable
to that of high-loaded black stock, and that of tensile modulus was
superior to the black stock (not shown). The polymer composite can
achieve excellent mechanical properties which are retained at elevated temperatures in the presence of even a small fraction of the
PPTA microfibrils (6.4 vol% or 10 phr).

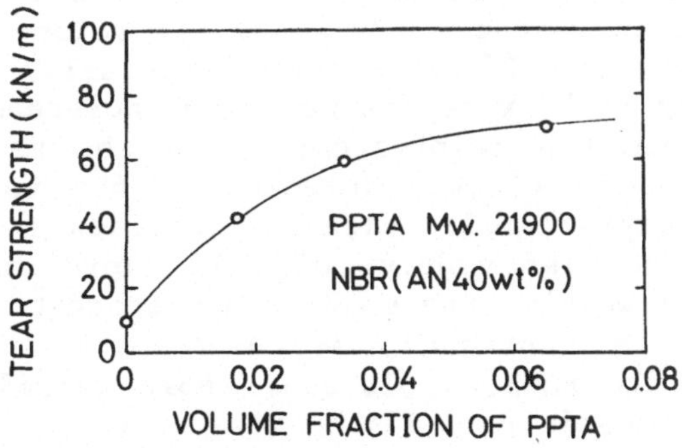

Fig. 5. Tear strength of PPTA polymer composite as a function of
volume fraction of PPTA with molecular weight 21900.

Figure 5 shows the tear strength of the polymer composite with various volume fractions of PPTA. Tear strength increased with increasing volume fraction of PPTA similar to the cases of tensile modulus and strength. The tear strength of the polymer composite with 6.4 vol% of PPTA reached to 70.3 kN/m, whereas that of black stock with the same volume fraction of reinforcement was 15.6 kN/m. This remarkable reinforcing effect can be explained by the crack propagation which is impeded by strong and fine PPTA microfibrils aligned across the stress direction. Thus, the advantages of fiber-reinforced composites in macro-scale are more effectively realized in a molecular composite. The applied stress is more uniformly distributed in reinforced texture and microfibrillar reinforcement effectively impedes the micro-crack formation and its growth.

Dynamic Mechanical Properties

Figures 6(a), 6(b) and 6(c) show the dynamic storage modulus, E', dynamic loss modulus, E'', and tan δ along the milling direction for NBR gum stock and the polymer composites with various molecular weights of PPTA (6.4 vol%) as a function of temperature, respectively. The peak of tan δ of the polymer composite did not shift to the higher temperature side by blending of PPTA with different molecular weights. This means that the PPTA microfibrils do not exert a noticeable effect on the motion of matrix (NBR) molecules in the bulk. This behavior accords with that of carbon black in NBR [9]. At present, however, it can not be concluded that the restricted molecular motion at the interfacial layer is entirely absent, since the evidence for the existance of an immobilized interfacial layer has been reported by several authors, according to which the bound rubber molecules were proved by using pulse NMR method [10]. The E' of the polymer composites in the rubbery plateau was above 10 MPa even for those employing low molecular weights of PPTA. The E' of black stock with the same volume fraction of the reinforcement showed about 3 MPa in that region. With increasing molecular weight of PPTA and its volume fraction in the polymer composite, the E' increased, especially in the rubbery region. The E'' in rubbery region was also increased. Although the interaction between the reinforcement and the matrix in polymer composites can not be clearly detected at 6.4 vol% PPTA, finely dispersed PPTA microfibrils clearly contribute to the rubber modulus through the high modulus of PPTA microfibrils. With increasing molecular weight, the modulus and strength of microfibrils are expected to be increased, which results in the increased modulus level in the rubbery plateau. More detailed quantitative analyses will be made in Part II.

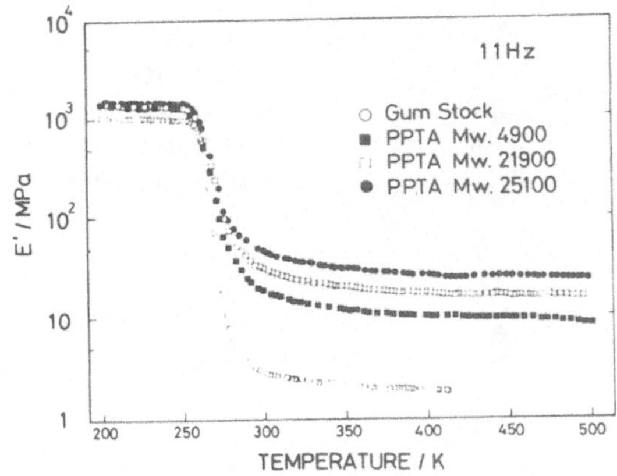

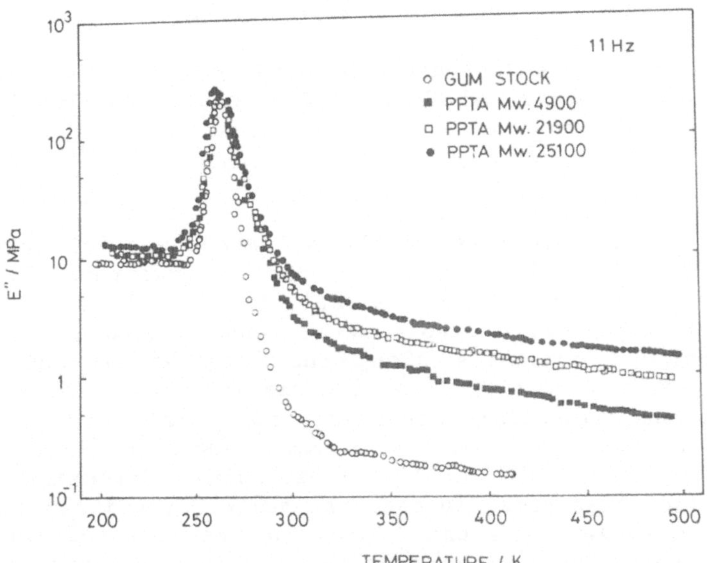

Fig. 6. Temperature dependence of dynamic mechanical properties of gum stock and PPTA polymer composites with 6.4 vol% for (a) E', and (b) E".

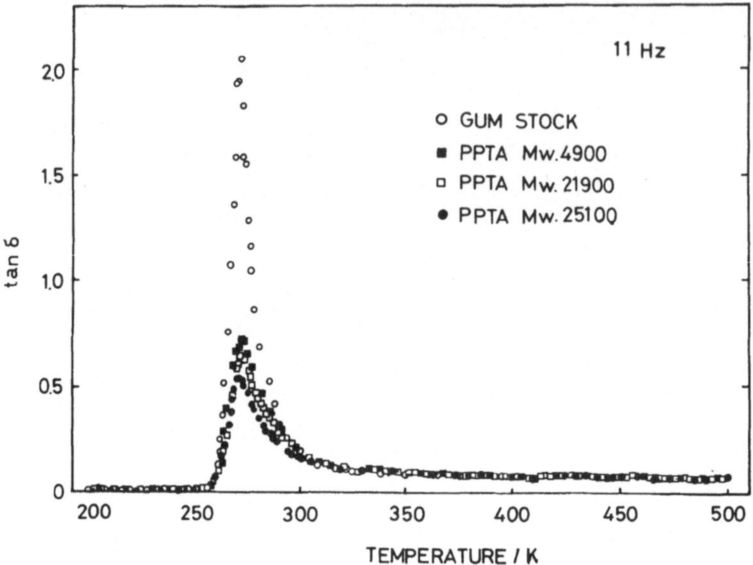

Fig. 6.   Temperature dependence of dynamic mechanical properties
          of gum stock and PPTA polymer composites with 6.4 vol%
          for (c) tan δ.

     Since PPTA molecules in the polymer composite form microfi-
brils, PPTA microfibrils are assumed to preferentially oriented
along the milling direction similarly to the fibers in macro-fiber
reinforced composite. Anisotropy in dynamic mechanical properties
of the polymer composite along the longitudinal and transverse
directions to the milling direction was measured. Figures 7(a) and
7(b) show the temperature dependence of dynamic mechanical properties, E', E'' and tan δ, of the polymer composite along both directions. The E' and E'' in the longitudinal direction maintained
higher values than those in the transverse direction over the whole
temperature range. The tan δ peak in the longitudinal direction
was lower than that in the transverse direction as seen in Figure
7(b). Table 3 shows the anisotropy in the dynamic storage modulus
of the polymer composites with various molecular weights of PPTA
and the macro-fiber composite. The anisotropic ratio (the ratio of
modulus along the longitudinal to the transverse direction) increased with increasing molecular weight of PPTA. This effect of
molecular weight means that the PPTA microfibrils with higher molecular weights tend to orient more easily along the milling
direction. PPTA with high molecular weight could form long and
strong microfibrils, which results in increased orientation along
the milling direction. Dynamic storage modulus of the polymer composites with low molecular weight (as low as 4900) was even larger

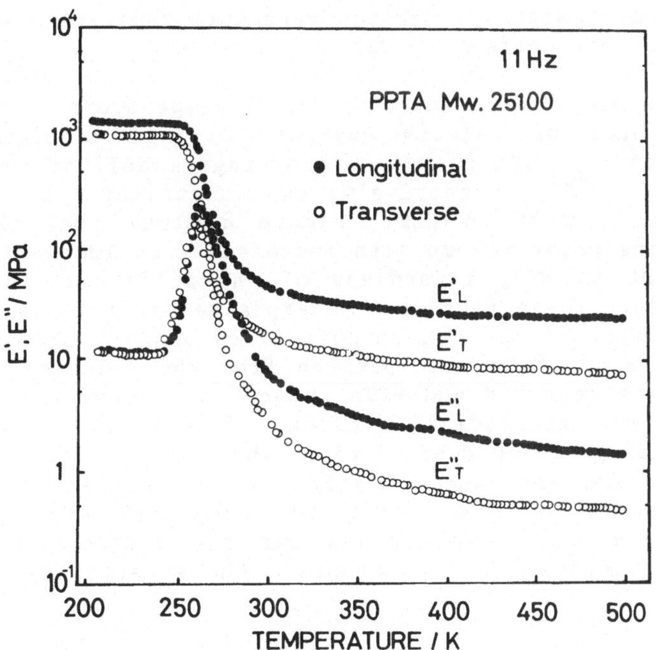

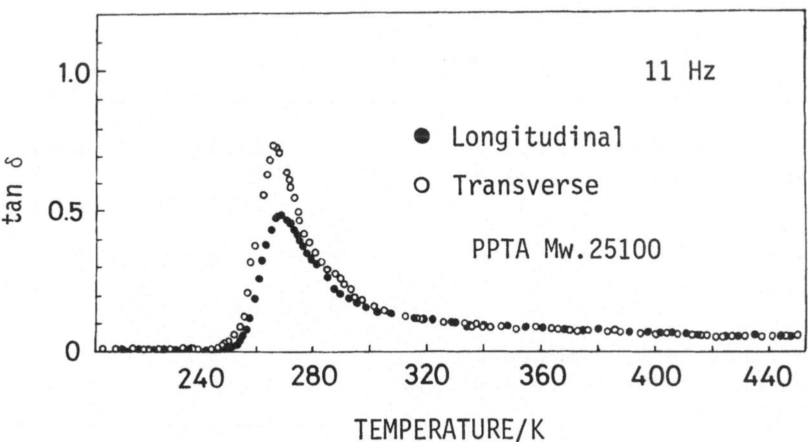

Fig. 7. Anisotropic effect of polymer composite in dynamic mechanical properties for (a) E' and E'' and (b) tan δ.

than that of the macro-fiber reinforced composite.

The reinforcing effects of the polymer composite observed in the dynamic mechanical properties were more remarkable with increasing AN comonomer content in NBR.

Temperature (represented by the temperature difference of T-Tg)-dependence of relative dynamic storage modulus, E'_c/E'_m where E'_c and E'_m are the dynamic storage moduli of the composite and the matrix NBR, respectively, was investigated by varying the AN comonomer content in NBR. Figure 8 shows that the relative modulus of the polymer composite increased with increasing AN comonomer content in NBR, regardless of the difference in Tg of the three samples. These facts can be explained as follows. The solubility parameter of NBR with higher AN comonomer content is closer to that of PPTA. When precipitated from the solution, the strong interaction between NBR and PPTA induces the formation of microfibril with high interfacial adhesion. This tendency is increased with increasing AN content, resulting in a higher degree of dispersion. The increased interfacial interaction at the microfibril surface of PPTA with high AN comonomer content results in the increase of relative modulus for the system. The relative modulus is a measure for representing the strength of interfacial adhesion.

Table 3. Anisotropies of Dynamic Storage Modulus for Polymer
          Composite and Macro-fiber Composite

	Mw. of PPTA	Storage modulus at 300 K (MPa)		Ratio
		Longitudinal	Transverse	
Polymer	4900	18.7	13.3	1.40
Composite	21900	31.2	14.9	2.09
	25100	47.0	16.6	2.83
Macro-Fiber Composite*		16.5	8.6	1.92

*Chopped fiber is 0.5 mm long.
 PPTA content is 10 phr (6.1 vol%).

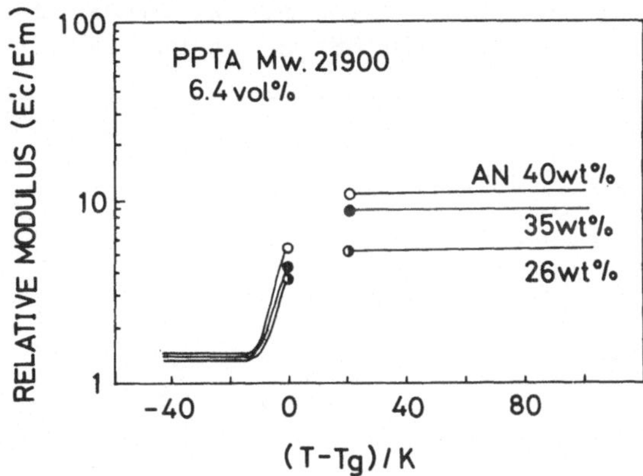

Fig. 8. Reinforcing effect of polymer composites on AN comonomer
content in NBR. Relation between relative modulus and temperature (T-Tg) for NBR with various AN comonomer content.

Swelling Behavior

One of the important properties of NBR is oil resistance. By
compounding the reinforcement and NBR, swelling ratio is much decreased by the increase in interaction between the rubber and the
reinforcement. The PPTA polymer composite has excellent mechanical
properties compared with black stock because the network of microfibrillar PPTA is developed uniformly in the matrix. Such a structure is expected to improve the oil resistance of the polymer composite, since the PPTA microfibrillar network is indifferent to oil
and restricts the adhering matrix rubber to be swollen by oil.
Table 4 shows the swelling behavior of gum stock, black stock and
polymer composites in fuels composed of iso-octane and toluene.
Fuels A, B and C are iso-octane, iso-octane/toluene = 70/30 (volume
ratio), and iso-octane/toluene = 50/50 (volume ratio), respectively. The restriction effect on swelling by the reinforcement
was decreased with increasing aromatic component in fuels, according with the value of the polymer-solvent interaction parameter. These tendencies agree with the effect of carbon black in
black stocks. Table 4 also shows the restriction effect of the molecular weight of PPTA in the polymer composites on their swelling
behavior. Higher comonomer concentration of AN in NBR gives a
stronger restriction effect on swelling in organic solvent.

Table 4. Volume increase of gum stock, black stock and polymer composites of PPTA/NBR after swelling.

	Fuel A[*1] %	Fuel B[*2] %	Fuel C[*3] %
AN 26 wt%			
Gum Stock	22.1	90.0	148.2
Black Stock [*4]	21.2	82.4	136.1
PPTA Composite [*5]	14.6	68.3	117.7
AN 35 wt%			
Gum Stock	11.4	52.3	89.4
Black Stock [*4]	9.5	48.7	77.3
PPTA Composite [*5]	6.0	44.5	71.9
AN 40 wt%			
Gum Stock	1.2	36.4	55.3
Black Stock [*4]	0.8	34.5	50.5
PPTA Composite [*5]	0.6	33.2	46.6

*1 iso-octane
*2 iso-octane/toluene = 70/30 (volume ratio)
*3 iso-octane.toluene = 50/50 (volume ratio)
*4 ISAF carbon black, 6.4 vol%, is employed.
*5 Molecular weight of PPTA in composite is 21900; 6.4 vol% PPTA is employed.

The relative volume fraction of rubber in the swollen gel was measured to analyze the restriction effect on swelling quantitatively. Figure 9 shows the plots of relative volume fraction of rubber vs. $\phi_2/(1-\phi_2)$ in dichloromethane where ϕ_2 is the volume fraction of the reinforcement. According to Kraus [11], the restriction relationship was represented by:

$$V_r/V_{ro} = 1 - m[\phi_2/(1 - \phi_2)]$$

where V_r is the volume fraction of rubber in the swollen rubber phase of the composite, V_{ro} is the same quantity of pure gum stock, and m is a parameter. The m values in dichloromethane for molecular weights of PPTA, 25100, 21900 and 4900, were 7.4, 6.1 and 5.0, respectively. On the other hand, that of black stock was evaluated as 1.3. In PPTA macro-fiber filled composite, V_r/V_{ro} was slightly increased with increasing volume fraction of fiber. The m value of the macro-fiber filled composite was evaluated as -1.5,

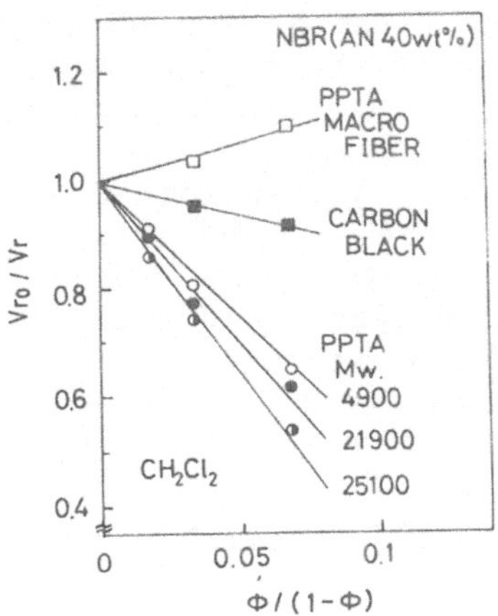

Fig. 9.   Restriction effect of reinforcement on swelling. Relation
of relative volume fraction of rubber in swollen gel to
$\phi/(1-\phi)$, a function of volume fraction of reinforcement.

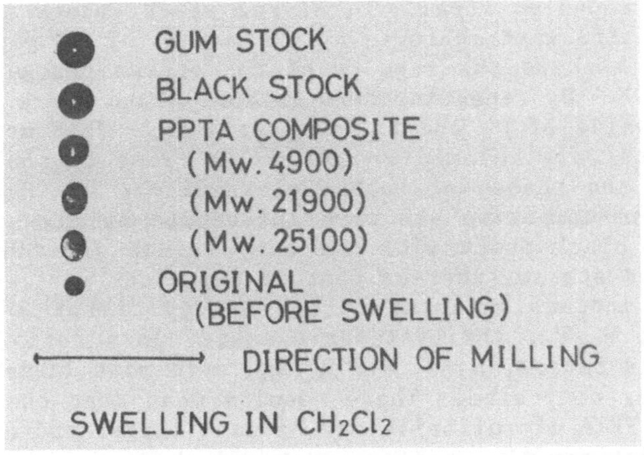

Fig. 10.   Scanning electron micrographs of tensile fractured surfaces for (a)gum stock, (b)black stock and (c)PPTA polymer composite. Volume fraction of reinforcement is 6.4%
and M_W of PPTA is 21900.

which means that a solvent invades into small empty space generated by clumping macro-fibers.

Anisotropic swelling behavior was observed in the PPTA polymer composite, especially in the strong swelling condition of dichloromethane as shown in Figure 10. Gum stock and black stock swelled isotropically. The anisotropy of swelling of PPTA polymer composite was increased with increasing molecular weight of PPTA. The strong anisotropy of swelling behavior found in the polymer composites reflects the texture of the polymer composite, in which PPTA with high molecular weight could form compact and strong microfibrillar networks with preferential orientation along the milling direction. The existence of PPTA microfibrils preferentially oriented along the direction of milling was confirmed not only by swelling behavior but also by dynamic mechanical properties. Detailed discussion will be given in Part II.

Mullins Effect

The softening phenomenon followed by the first stretching, so-called "Mullins effect", has been observed more remarkably in the black stock with increasing volume fraction and reinforcing effect. This effect is caused by breaking the loose or weak chains connecting the reinforcement to the matrix elastomer. The reinforcing effect can be evaluated by the magnitude of this effect. The input strain energy, W, and the hysteresis energy loss, H, at the constant repeated strain, λ, of gum stock, black stock and the polymer composite were measured by the method of Hirakawa and Urano [12]. Table 5 shows the results of the energy change at repeated strain (λ=37%). By repeating deformation, W and H are converged to equilibrium value of W_∞ and H_∞, respectively. This means that the network chains were relaxed enough to be close to the ideal state predicted by the theory of rubber elasticity. The Mullins effect of the polymer composite was more influenced by the reinforcement than that of black stock with the same volume fraction at a constant repeated strain, whereas that of gum stock was least affected even by the increasing input strain energy. That is, the input energy ratio, W_∞/W_1, the hysteresis energy loss ratio, H_∞/H_1, and the hysteresis ratios, H_1/W_1 and H_∞/W_∞, were most largely increased in the polymer composite. These results mean that the active surface area of PPTA microfibrils may be much larger and/or the number of linkages between the reinforcement and the matrix may be greater than those of carbon black for the samples employed in this paper. Such a softening effect is also preferable to the fracture properties from the viewpoints of energy dissipation. Remarkable Mullins effect of the polymer composite is in close relation to its superior mechanical properties.

Table 5. Hysteresis Behavior

	Gum Stock		Black Stock	PPTA Composite
W_1 (MJ/m^3)	$6.62\text{x}10^{-2}$*	1.27	$2.04\text{x}10^{-1}$*	$4.93\text{x}10^{-1}$*
H_1 (MJ/m^3)	$1.03\text{x}10^{-2}$	$2.26\text{x}10^{-1}$	$4.69\text{x}10^{-2}$	$2.15\text{x}10^{-1}$
W_∞/W_1	0.742	0.760	0.667	0.543
H_∞/H_1	0.580	0.365	0.397	0.255
H_1/W_1	0.156	0.178	0.230	0.437
H_∞/W_∞	0.116	0.085	0.105	0.204

*The same repeated strain, = 37%. NBR containing AN 40 wt% is employed. Volume fraction of reinforcement is 6.4 vol%. Molecular weight of PPTA is 21900.

Fractography

Tensile fractured and tear fractured surfaces were observed by scanning electron micrographs. Figures 11(a), 11(b) and 11(c) show the tensile fractured surfaces of pure gum stock, highly loaded black stock and the polymer composite, respectively. In the pure gum stock in Figure 11(a), the fractured surface was very smooth. The fractured surface of the PPTA polymer composite in Figure 11(c) was composed of four different fracture patterns: nucleation of fracture, mirror zone, mist zone and rough zone. Surface energy required for fracture is conceived to be very large in the polymer composite. In the black stock in Figure 11(b), the fractured surface is intermediate between both images in Figures 11(a) and 11(c).

Figures 12(a), 12(b) and 12(c) show the tear fractured surfaces of pure gum stock, black stock and polymer composite, respectively. Tear lines were branched at the ends of crack for the samples with high tear strength. This tendency is increased in the order of Figures 12(a), 12(b) and 12(c). In the PPTA polymer composite, microfibrils of PPTA act more effectively in impeding the crack growth due to bifurcation or deflection of crack. The fractography supports the superiority in tensile and tear strengths of the PPTA polymer composite.

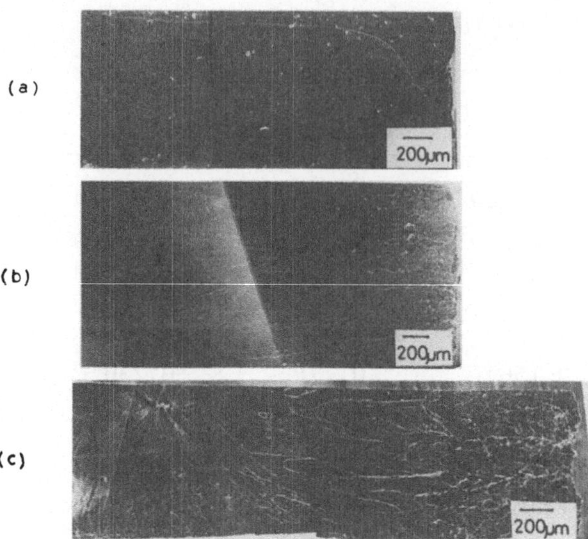

Fig. 11.   Scanning electron micrographs of tear fractured surfaces
           for (a) gum stock, (b) black stock and (c) PPTA polymer
           composite.   Volume fraction of reinforcement is 6.4% and
           M_W of PPTA is 21900.

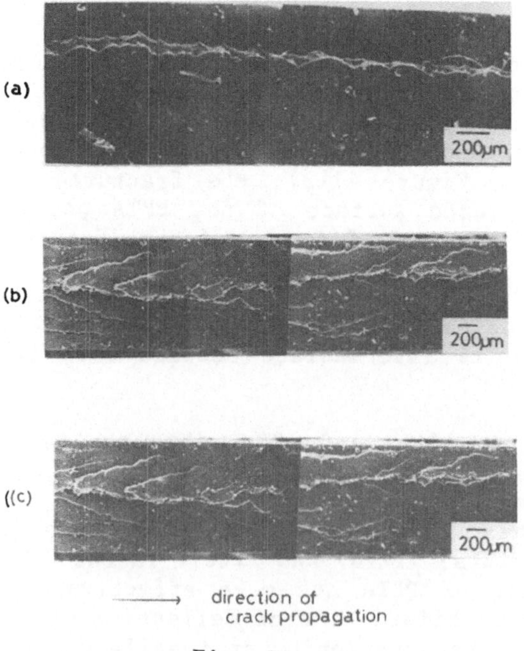

Fig. 12.

CONCLUSIONS

N-sodium PPTA in DMSO and NBR in DMF formed an isotropic solution. By regenerating with acidic water, polymer composite of PPTA and NBR was prepared. Regenerated PPTA formed microfibrils with a diameter of 10 ~ 30 nm in the matrix of NBR. Cured polymer composites had superior mechanical properties, such as high modulus, high tensile strength and high tear strength, with even a small fraction (6.4 vol%) of the reinforcement. PPTA microfibrils in the polymer composite preferentially oriented along the milling direction, resulting in anisotropies in dynamic viscoelastic properties and swelling with organic solvents. Reinforcing effects were ascribed to strong interfacial interaction between the reinforcement and the matrix, which were supported by large Mullin's effect. Rough fracture surface of the polymer composite observed by SEM also supported the improved mechanical properties. The reinforcing effect of PPTA in the polymer composite was superior to that of carbon black at the same volume fraction of reinforcement. A new type of high performance NBR was prepared based on the molecular composite concept.

ACKNOWLEDGMENT

The authors express thanks to Dr. H. Ikeda and his group in Tokyo Research Laboratory of Japan Synthetic Rubber Co., Ltd. for their collaboration in various engineering processes such as compounding and curing of samples.

REFERENCES

1. A. Kelly, "Strong Solids," 2nd edition, Clarendon Press, Oxford (1973).
2. M. Takayanagi, T. Ogata, M. Morikawa, and T. Kai, J. Macromol. Sci.-Phys., B17, 591 (1980).
3. M. Takayanagi and T. Katayose, J. Polym. Sci., Polym. Chem. Ed. 19, 1133 (1981).
4. T. I. Bair, P. W. Morgan and F. L. Killman, Macromolecules, 10, 1396 (1977).
5. M. Arpin and C. Strazielle, Polymer, 18, 591 (1971).
6. M. Takayanagi, K. Goto and K. Yamada, Rep. Prog. Polym. Phys. Japan, 24, 311 (1981).
7. M. Takayanagi and K. Goto, submitted J. Appl. Polym. Sci.
8. M. Takayanagi, Reprints of the 2nd Japan-Korea Joint Symposium on Polymer and Technology, Kyoto, p. 61 (1980).
9. S. Nohara, Kobunshi-Kagaku, 12, 47 (1955).
10. K. Fujimoto and T. Nishi, Polymer Preprints in Japan, p. 789, (1970).
11. G. Kraus, J. Appl. Polym. Sci., 7, 861 (1963).
12. K. Hirakawa and F. Urano, Rubber Chem. & Tech., 51, 201 (1971).

POLYMER COMPOSITES OF POLY(p-PHENYLENE TEREPHTHALAMIDE) AND NITRILE
BUTADIENE RUBBER: (II) CALCULATION OF ANISOTROPIC MODULI BASED ON
MICROFIBRILLAR LATTICE MODEL

Motowo Takayanagi and Kohei Goto*

Dept. of Applied Chemistry, Faculty of
Engineering, Kyushu University
Hokozaki, Higashi-ku,
Fukuoka, 812 JAPAN

*Japan Synthetic Rubber Co., Ltd.
Higashi-Yurigaoka, Asao-ku
Kawasaki, 214 JAPAN

ABSTRACT

Preparation and properties of polymer composites of poly-
(p-phenylene terephthalamide) (PPTA) and nitrile butadiene rubber
(NBR) were reported in the previous paper. Morphological observation proved that PPTA is uniformly dispersed in a form of microfibrillar network in NBR matrix. The diameter of the microfibrils
was about 10 ~ 30 nm. The milling process induced the anisotropy
in modulus of the polymer composites owing to the preferential
orientation of PPTA microfibrils along the milling direction. With
increasing molecular weight, the anisotropy was increased due to
the stress concentration onto the high modulus PPTA microfibrils
with high molecular weight. Anisotropy of modulus with respect to
the milling direction was calculated by using the quasi-three-dimensional lattice model. Lattice parameters of the unit cell were
evaluated by the aid of anisotropy of the swollen rubber
composite. It was necessary to take into account the decreasing
modulus with decreasing molecular weight of PPTA by introducing a
very small, low modulus fraction of the matrix into each lattice
element of the model. The temperature dependence of moduli of the
polymer composites with various molecular weights of PPTA and
various degrees of anisotropy were successfully calculated by the
lattice model over the whole temperature range from glassy to
rubbery state.

INTRODUCTION

A series of polymer composites is in exploration by one of
the authors [1,2]. A basic idea of polymer composites is to extend
the principle of reinforcing in a macro-fiber reinforced composite
to the molecular level. This means that rigid rod-like molecules
such as aramid are dispersed uniformly and finely in the flexible
molecules of the matrix to provide the polymer composites thus
obtained with excellent modulus and strength. For preparation of
polymer composites, a common solvent of rigid and flexible molecules is selected and the polymer composite is precipitated from
the solution of both components. It is preferable to prevent
liquid crystal formation of rigid molecules in the solution to
achieve uniform dispersion of rigid molecules in the matrix
polymer. For this purpose, the concentration of polymer in
solution is selected to be lower than the critical concentration
for liquid crystal formation. Another important point, to prepare
the polymer composites is in the process of coagulation of the
solution with non-solvent to avoid aggregation of rod-like
molecules.

In the previous paper [3], the molecules of poly(p-phenylene
terephthalamide) in presence of NBR were found to form a microfibrillar network with a fibril diameter of 10 to 30 nm. The matrix
NBR molecules filled the intermicrofibrillar space of PPTA. Vulca-
nization of the polymer composite of PPTA and NBR provided excellent mechanical performance in comparison with the black stock of
NBR. The polymer composite was processable as in the compound for
black stock. The difference between both cases was in the anisotropies of modulus and swelling behavior, which were remarkable in
the polymer composite. Microfibrils of PPTA were preferably
oriented along the milling direction. In this paper, the anisotropy of modulus of the polymer composite vulcanizates with various
molecular weights of PPTA will be interpreted by using the microfibrillar lattice model, which is deduced from morphological observation of the texture.

EXPERIMENTAL

NBR sample was obtained from Japan Synthetic Rubber Company,
JSR N220S, the acrylonitrile content of which was 40 wt%. PPTA was
prepared by low temperature polycondensation of terephthaloyl
chloride and p-phenylenediamine in solution. The molecular weights
of the samples employed were 4900, 21900, and 25100. A homogeneous
solution of N-sodium PPTA in DMSO [3] and a solution of NBR in DMF
were blended to form a homogeneous, isotropic solution, which was
poured into acidic water to coagulate and regenerate PPTA from
N-sodium PPTA. The coagulated polymer composite was washed and
dried. The compound of NBR prepared by the standard recipe was

milled with an open roll. According to the conditions found by JSR curastometer (Japan Synthetic Rubber Co.), the polymer composites were cured at 423°K for 60 min with a press. PPTA content in the composite was 6.4 vol% [3].

Dynamic complex tensile moduli of the polymer composites and pure rubber vulcanizate of NBR were measured at 11 Hz by using Rheovibron DDV-IIB (Toyo Baldwin Co.) in a nitrogen atmosphere. The degree of swelling was evaluated by the change of the size of the specimen after equilibrium is attained by immersing in dichloromethane at 20°C. The sheet of specimen for the swelling test was 2 cm square and 1 mm thick.

RESULTS AND DISCUSSION

Polarization optical micrograph revealed that the texture of the polymer composites was homogeneously birefringent and indicated that the PPTA was uniformly dispersed in the rubber matrix. The PPTA residue remaining after extraction of NBR with DMF from the uncured polymer composite was observed by transmission electronmicrograph, which showed the microfibrillar shape of PPTA [3]. Based on these observations, a three-dimensional lattice model was proposed to explain the anisotropy of moduli of the composites. The diameter of the fibrils was about 30 nm and denoted by "r". An isotropic lattice model is constructed as shown in Figure 1, the length of the unit lattice of which was denoted by "a". Figure 2 shows the equivalent mechanical model for the modulus calculation. It was assumed that only the fibril elements along the stress direction contribute to the composite modulus with their highest modulus value along the molecular axis.

The complex dynamic modulus of the polymer composite, E^*_c, is given by the following equation.

$$E^*_c = \frac{A}{a^2}\, E^*_f(||) + \frac{(a-r)^2}{a^2}\, E^*_r + \frac{2ar-r^2-a}{a^2}\left[\frac{r/a}{E^*_f(\perp)} + \frac{(a-r)/a}{E^*_r}\right]^{-1} \qquad (1)$$

where $E^*_f(||)$ and $E^*_f(\perp)$ are the complex moduli in the longitudinal and transverse directions of the PPTA microfibril, respectively, and the complex modulus of rubber matrix vulcanized is denoted by E^*_r. A is the cross-sectional area of fibril, $\pi r^2/4$. The volume fraction of PPTA microfibrils, V_f, is:

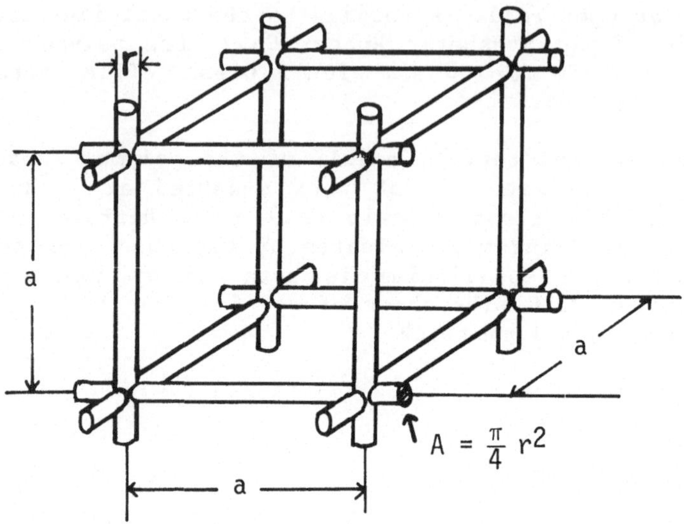

Fig. 1.   Ideal three dimensional lattice model representing isotropic microfibrillar composite. Interfibrillar space is filled by the matrix polymer.

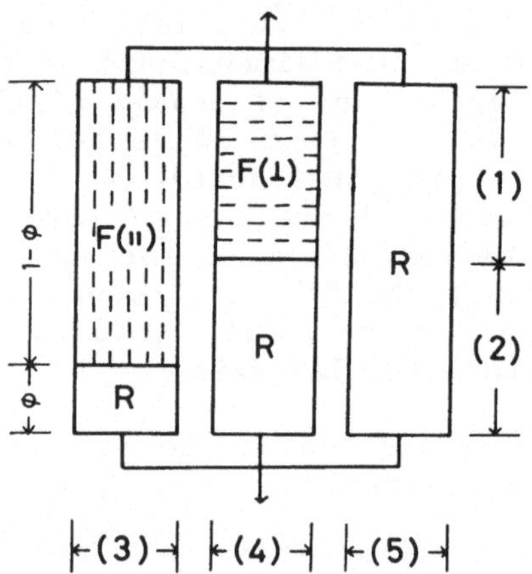

	(1)	(2)	(3)	(4)	(5)
a-axis direction	$\dfrac{r}{a}$	$\dfrac{a-r}{a}$	$\dfrac{A}{bc}$	$\dfrac{r(b+c)-r^2\cdot A}{bc}$	$\dfrac{(b-r)(c-r)}{bc}$
b-axis direction	$\dfrac{r}{b}$	$\dfrac{b-r}{b}$	$\dfrac{A}{ac}$	$\dfrac{r(a+c)-r^2\cdot A}{ac}$	$\dfrac{(a-r)(c-r)}{ac}$
c-axis direction	$\dfrac{r}{c}$	$\dfrac{c-r}{c}$	$\dfrac{A}{ab}$	$\dfrac{r(a+b)-r^2\cdot A}{ab}$	$\dfrac{(a-r)(b-r)}{ab}$

Fig. 2.   Mechanical model of the isotropic ideal three dimensional lattice model.

$$V_f = \frac{3\,r^2}{4a^2} \tag{2}$$

Substituting the observation data of r=30 nm and the content of PPTA microfibrils being 6.4 vol% into equation (2), the lattice length, "a", is evaluated as 182 nm. The modulus along the direction perpendicular to the molecular axis of PPTA filament assumes the value 1/10 or 1/100 of the modulus along the filament axis according to the crystal structure of PPTA [4]. $E_f^*(\perp)$ is assumed to be 1/10 $E_f^*(||)$ by taking into account E_f^* being not necessarily large enough to be comparable to that of a Kevlar filament. Figure 3 shows the temperature dependence of the absolute value of dynamic complex modulus of NBR, $|E_r^*|$. Strict calculation of equation (1) should be made by separating the complex modulus E* into real and imaginary parts, but in the case of very low tan δ or low loss modulus, E", the absolute value of E_r^* is very close to E´, and it is mathematically permissable to substitute it into equation (1). Only in the transition region, where tan δ amounts to a fairly large value (refer to Figure 6 in Part I), is this substitution not allowed. Fortunately, our discussion is made on the behavior in the glassy region and the rubbery region, and the transition region over a limited temperature range is not important. A similar assumption as in E_r^* is made for E_f^*, since the tan δ value of Kevlar filament was very small over the whole temperature range studied [5]. Thus, for simplicity, the symbol E* is used instead of |E*|. Later on, the absolute value of modulus is employed for calculation of the complex modulus of equation (1).

Figure 4 shows the temperature dependence of E_c^* calculated by using the data of E_f^* in Figure 3 and the $E_f^*(||)$ assumed to range from 5 to 50 GPa. The catalogue value of E_f for Kevlar 49 is cited in du Pont´s catalogue as 130 GPa. Depending on the molecular weight of PPTA, fineness of the fibrils, spinning and annealing conditions, the fibril modulus of PPTA might be varied, and at present it is assumed as an adjustable parameter. The modulus of the composite, E_c^*, in the rubbery zone is 20 ~ 30 MPa as seen in Figure 3 even for molecular weight of PPTA amounting to 25100. On the other hand, the calculated value of E_c^* in the rubbery zone showed a very high value of 100 MPa even for a very low modulus value of 5 GPa of PPTA as shown in Figure 4.

There are some problems in adopting the isotropic lattice model as shown in Figure 1 : one is the model being incapable of describing the modulus anisotropy with respect to the milling direction and the other is the ideal lattice with its element being continuous without interruption through the specimen. A more acceptable model is the one in which the lattice constants along

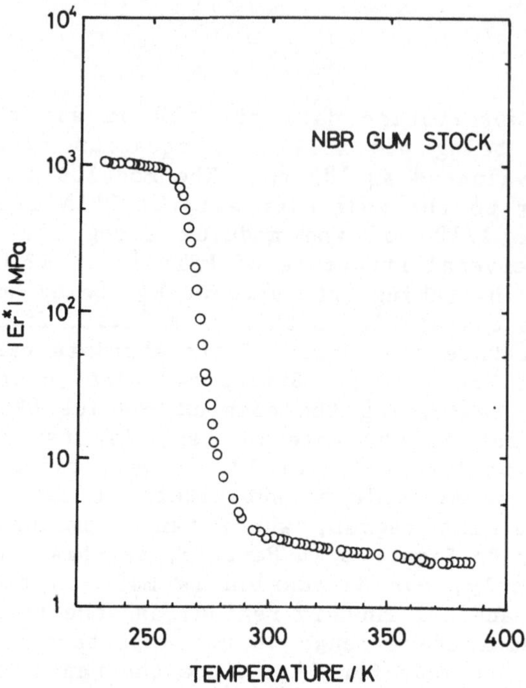

Fig. 3.   The temperature dependence of the absolute value of the
dynamic complex modulus of vulcanized NBR, $\left|E^*_r\right|$.

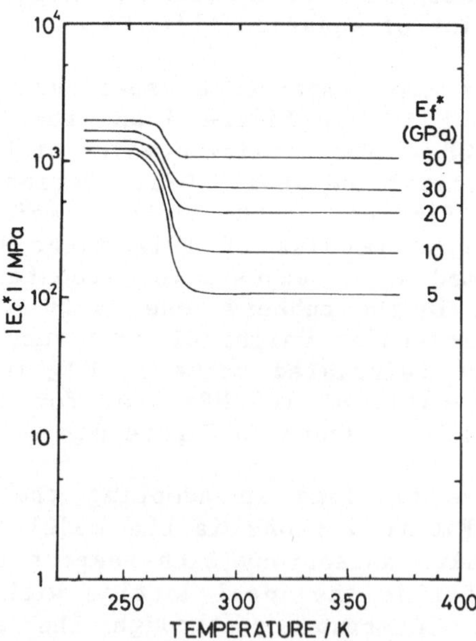

Fig. 4.   The temperature dependence of the absolute value of $\left|E^*_c\right|$
calculated by using the $\left|E^*_r\right|$ in Fig. 3 and E^*_f as indi-
cated in the figure.

the external Cartesian coordinates are different to take into account the anisotropy in modulus and the continuity of the lattice element of microfibrillar PPTA is interrupted by inserting a small fraction of the matrix NBR vulcanizate. The modulus of the micro-fibrils is also dependent upon various factors such as molecular weight, formation conditions of fibrils and others. With increasing molecular weight, PPTA fibrils tend to develop in more perfect shape with higher modulus and strength owing to the decrease in defects associated with chain ends of PPTA.

Based on the above considerations, a quasi-3-dimensional microfibrillar lattice model is constructed as shown in Figure 5, in which the lattice lengths of construction unit along and perpendicular to the milling direction and the thickness direction are denoted by "a", "b" and "c", respectively. Figure 6 represents the equivalent mechanical model which responds to the stress applied to the lattice model in Figure 5 along the "a", "b" or "c" direction. The complex moduli, $E^*_F(a)$, $E^*_F(b)$ and $E^*_F(c)$ are expressed by the following equations, respectively.

$$E^*_c(a)=\frac{A}{bc}E^*_F(\|) + \frac{(b-r)(c-r)}{bc}E^*_r + \frac{[r(b+c)-r^2-A]}{bc}\left[\frac{r/a}{E^*_F(\perp)} + \frac{(a-r)/a}{E^*_r}\right]^{-1} \tag{3}$$

$$E^*_c(b)=\frac{A}{ac}E^*_F(\|) + \frac{(a-r)(c-r)}{ac}E^*_r + \frac{[r(a+c)-r^2-A]}{ac}\left[\frac{r/b}{E^*_F(\perp)} + \frac{(b-r)/b}{E^*_r}\right]^{-1} \tag{4}$$

$$E^*_c(c)=\frac{A}{ab}E^*_F(\|) + \frac{(a-r)(b-r)}{ab}E^*_r + \frac{[r(a+b)-r^2-A]}{ab}\left[\frac{r/c}{E^*_F(\perp)} + \frac{(c-r)/c}{E^*_r}\right]^{-1} \tag{5}$$

where $E^*_c(a)$, $E^*_c(b)$ and $E^*_c(c)$ represent the moduli along the a-, b- and c-directions of the composite, respectively, and E^*_F is represented by the following equation.

$$E^*_F =\left(\frac{1-\phi}{E^*_f} + \frac{\phi}{E^*_r}\right)^{-1} \tag{6}$$

where ϕ is the fraction of the matrix rubber vulcanized NBR being connected in series with the PPTA fibril. When ϕ is very small, the volume fraction of microfibril, V_f, is represented by:

$$V_f \simeq \frac{A(a+b+c)}{abc} \tag{7}$$

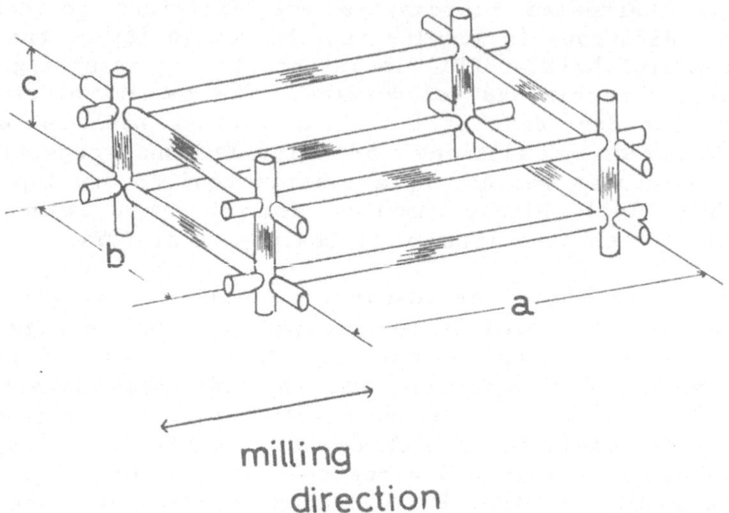

Fig. 5. Quasi-three-dimensional lattice model. The shadowed portion represents the interrupted region by the matrix rubber.

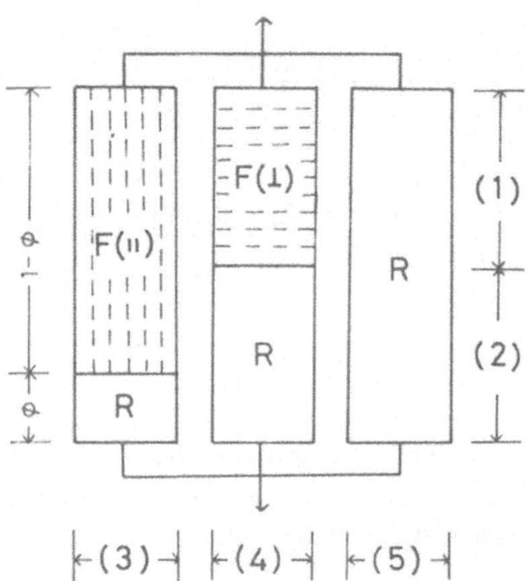

	(1)	(2)	(3)	(4)	(5)
a-axis direction	$\dfrac{r}{a}$	$\dfrac{a-r}{a}$	$\dfrac{A}{bc}$	$\dfrac{r(b+c)-r^2-A}{bc}$	$\dfrac{(b-r)(c-r)}{bc}$
b-axis direction	$\dfrac{r}{b}$	$\dfrac{b-r}{b}$	$\dfrac{A}{ac}$	$\dfrac{r(a+c)-r^2-A}{ac}$	$\dfrac{(a-r)(c-r)}{ac}$
c-axis direction	$\dfrac{r}{c}$	$\dfrac{c-r}{c}$	$\dfrac{A}{ab}$	$\dfrac{r(a+b)-r^2-A}{ab}$	$\dfrac{(a-r)(b-r)}{ab}$

Fig. 6. Mechanical models representing along the a-, b- and c-axes directions of lattice model in Fig. 5.

Preliminary calculations done by assigning conceivable values to the parameters in equations (3) to (7) revealed that the modulus value of PPTA, $E_f(||)$, is very influential on the modulus of the composite in its glassy state and the fraction of the matrix rubber, ϕ, is influential on the modulus of the composite in its rubbery state. Referring to such relations, the values of $E_f(||)$ and ϕ were approximately evaluated.

For calculation of the anisotropy in the modulus of the composite, the different values of lattice constants, "a", "b" and "c" are necessary to be evaluated. The ratio of a/b was rather easily evaluated from the moduli of the composite in a rubbery state. In a rubbery state, the relations of $E_f^*(||) \gg E_r^*$ and $E_f^*(||) \gg E_f^*(\perp)$, hold and equations (3), (4) and (5) are simplified by:

$$E_c^*(a) \simeq \frac{A}{bc} E_F^* \tag{8}$$

$$E_c^*(b) \simeq \frac{A}{ac} E_F^* \tag{9}$$

and

$$E_c^*(c) \simeq \frac{A}{ab} E_F^* \tag{10}$$

The anisotropic ratios of the composite moduli in a rubbery state are given by:

$$\frac{E_c^*(a)}{E_c^*(b)} \simeq \frac{a}{b} \tag{11}$$

and

$$\frac{E_c^*(b)}{E_c^*(c)} \simeq \frac{b}{c} \tag{12}$$

The left side in equation (11) is experimentally obtained and the ratio of a/b can be determined. The modulus of the composite along the thickness direction, $E_c^*(c)$, is rather difficult in direct measurement with high precision.

For this problem, the result of swelling tests along the thickness direction were employed. Detailed explanation is as follows. In the previous paper [3], anisotropy in modulus was found to be in close relation to the anisotropy in degree of swelling. Along the milling direction, the modulus is highest and the degree of swelling is minimum compared with those along the transverse and thickness directions.

Table 1. The change of specimen sizes by swelling with dichloromethane.

		Milling Direction (a_s/a)	Transverse Direction (b_s/b)	Thickness Direction (c_s/c)	Anisotropy ratio	
					$\dfrac{a_s/a}{b_s/b}$	$\dfrac{b_s/b}{c_s/c}$
	GUM Stock	2.02	2.02	2.20	1.000	0.918 (1.000)
Polymer Composite	4900	1.81	1.99	2.27	0.910	0.877 (0.955)
Polymer Composite	21900	1.52	1.92	2.23	0.792	0.861 (0.938)
Polymer Composite	25100	1.44	1.91	2.30	0.754	0.830 (0.904)

Dichloromethane was employed for swelling test of NBR. Table 1 shows the swelling ratios of specimen sizes along three directions: a_s/a for milling direction, b_s/b for the transverse direction and c_s/c for the thickness direction, where the subscript s means the size of swollen specimen. The anisotropy in swelling ratio is also cited as the anisotropy ratio. Gum stock (pure rubber vulcanizate) swells isotropically along the a- and b-direction, but the swelling ratio along the thickness direction is somewhat larger than the others. The same tendency was observed for the polymer composites. To extract the effect of microfibrillar PPTA in polymer composites on the swelling ratios along the b- and c- directions, the anisotropy ratio of $(b_s/b)/(c_s/c)$ for gum stock was taken as the reference and the corresponding anisotropy ratios for the polymer composites was divided by it. Such values are cited in Table 1 in the last column, being bracketed.

Figure 7 shows both the logarithmic plots of the anisotropic ratio of the modulus and the anisotropic ratio of swelling ratio along the a- and b- directions. In the range of experiment, linear relations hold between both logarithmic quantities. The anisotropic ratio of swelling, $(b_s/b)/(c_s/c)$, is in the range of linearity, which enables one to evaluate the anisotropic ratio of modulus along the b- and c- directions. By use of these relations, the relative ratios of sizes of lattice unit were evaluated, substitution of which into equation (7) enables the calculation of absolute values of lattice unit constants, a, b and c.

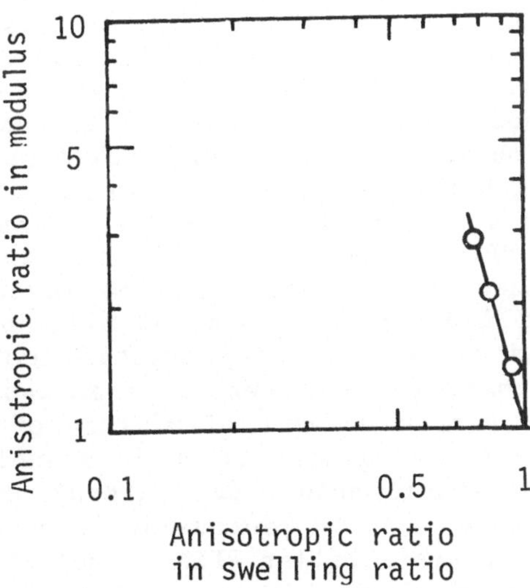

Fig. 7.   Anisotropic ratio in modulus vs. anisotropic ratio in swelling ratio for polymer composites with different molecular weight of PPTA.

Table 2.   The lattice unit constants (a, b and c), the modulus of
           microfibril (E_f) and the matrix fraction in microfibril
           (ϕ) for polymer composite.

Molecular weight of PPTA	a (nm)	b (nm)	c (nm)	E_f^* (GPa)	ϕ (vol %)
4900	237	169	154	7	0.52
21900	330	158	131	12	0.40
25100	446	158	113	13	0.27

     Table 2 summarizes the lattice unit constants of a, b and c,
the modulus of PPTA microfibril E_f^*, and the fraction of matrix NBR
interrupting the PPTA microfibril, ϕ, for various molecular weights
of PPTA. Under the assumption of constancy in modulus of PPTA microfibril [5], the temperature dependence of modulus of polymer
composite can be calculated by using the observed matrix modulus,
E_r^* given in Figure 3. With increasing molecular weight of PPTA
from 4900 to 25100, the "a" value increases and the b- and c-
values decrease. This means that the number of PPTA microfibrils
per unit cross sectional area along the milling direction increases
due to the decrease of the area of the bc plane. The milling process of polymer composite with a roll gives rise to the preferential orientation of microfibrils along the milling direction. Microfibrils composed of high molecular weight PPTA form highly developed filaments and receive larger shear stress at the interface
between the PPTA surface and NBR gum stock along the fibril axis,
which results in generation of stronger orientational force on the
microfibrils. The moduli of $E_f^*(||)$ are 7, 12 and 13 GPa for mole-
cular weights of PPTA being 4900, 21900 and 25100, respectively.
These modulus values of PPTA are very low in comparison with the
catalogue value of 130 GPa for Kevlar 49 filament. Microfibril of
PPTA is considered to be formed by crystallization of PPTA molecules during the coagulation process from an isotropic solution of
PPTA and NBR in a common solvent of mixture of DMSO and DMF. On
the other hand, Kevlar 49 aramid fiber is spun from a liquid crystalline dope in sulfuric acid under winding stress and well annealed. The difference in the processing is enough to explain the
low modulus of the microfibrillar PPTA in polymer composites. The
fraction of the matrix interrupting the fibril continuity, ϕ, is
increased with decreasing molecular weight of PPTA. The
microfibrils formed with low molecular weight PPTA might be
defective and fragile, which allows the matrix to penetrate into
the defects of fibrillar structure. Thus, ϕ, is 0.52% for PPTA
with molecular weight 4900, while ϕ is 0.27% for M = 25100.

Figures 8, 9, and 10 show the temperature dependence of the absolute value of complex modulus, $|E^*_c|$, for polymer composites with molecular weights of PPTA being 4900, 21900 and 25100, respectively. The solid lines represent the experimental curves given in our previous paper [3] and the open circles represent the calculated values by using the parameter values cited in Table 2 and the $|E^*_r|$ given in Figure 3. The agreement is fairly good.

In conclusion, PPTA molecules dispersed in NBR matrix form microfibrils. The polymer composite texture is well modelled by the incomplete lattice of microfibrillar PPTA and the intermicrofibrillar space is filled by the NBR gum stock. The continuity of PPTA microfibrils in a model is necessary to be interrupted by a small fraction of the matrix. The anisotropies in modulus and swelling ratio with respect to the milling direction can be described by the quasi-3-dimensional lattice model. Lattice parameters can be evaluated by the help of close correlation between swelling ratio and modulus ratio, which enables one to calculate the anisotropic moduli over the whole temperature range from glassy to rubbery state of polymer composites of PPTA and NBR.

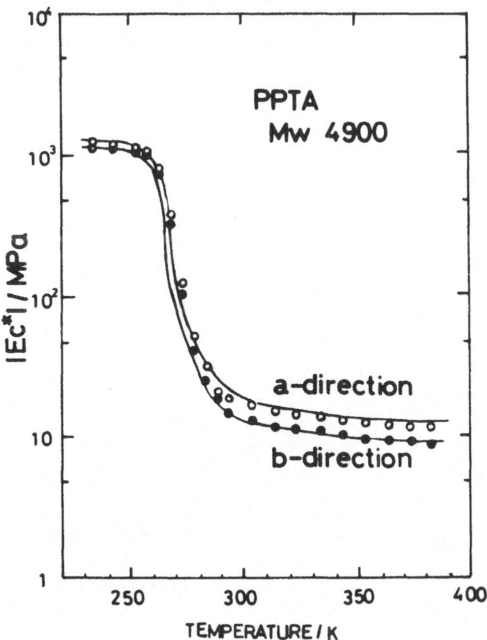

Fig. 8.   The temperature dependence of the absolute values of the modulus for polymer composite with molecular weight of PPTA, 4900. Solid lines represent the experimental curves and open circles represent the calculated values along the milling (a) and the transverse (b) directions.

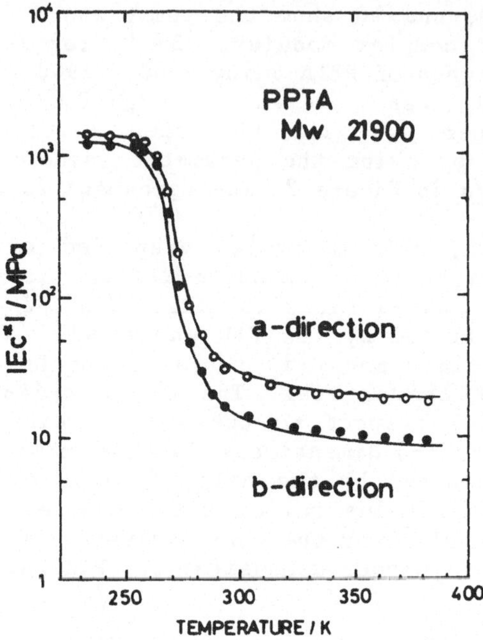

Fig. 9.   The same as in Figure 8 except the molecular weight of
         PPTA being 21900.

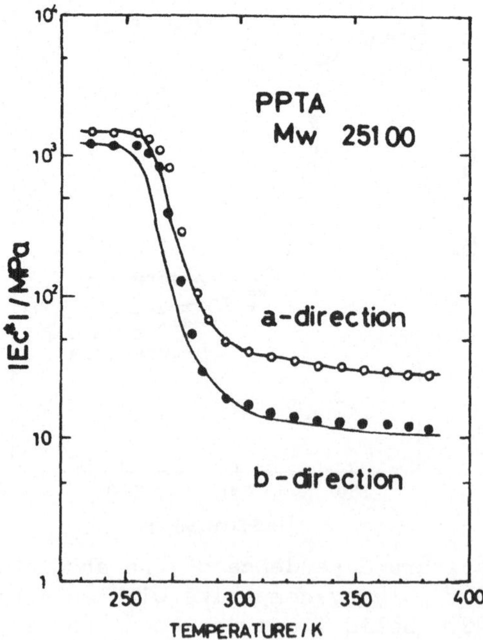

Fig. 10.   The same as in Figure 8 except the molecular weight of
          PPTA being 25100.

REFERENCES

1. M. Takayanagi, T. Ogata, M. Morikawa, and T. Kai, J. Macromol. Sci.-Phys., $\underline{B17}$(4), 591 (1980).
2. M. Takayanagi, Pure & Appl. Chem., $\underline{55}$, 819 (1983).
3. M. Takayanagi and K. Goto, "Polymer Composites of Poly(p-phenylene terephthalamide) and Nitrile Butadiene Rubber (I), Preparation and Properties," in this series.
4. S. Manabe, S. Kajita, and K. Kamide, Sen-i Kikai Gakkaishi, $\underline{33}$, T93 (1980).
5. K. Haraguchi, T. Kajiyama, and M. Takayanagi, Sen-i Gakkaishi, $\underline{32}$, T535 (1977).

THERMOSTIMULATED CREEP STUDY OF THE INTERFACE OF GLASS

BEAD-REINFORCED EPOXY COMPOSITES

J.P. Bayoux, C. Pillot, D. Chatain* and C. Lacabanne*

Materials Department
INSA
69621 Villeurbanne Cedex, France

*Solid State Phys. Lab
UPS
31062 Toulouse Cedex, France

ABSTRACT

The anelastic behavior of two different types of model composites is studied using the thermostimulated creep. This original method has a high resolving power and gives us accurate information about the distribution of the relaxation times in the matrix of an epoxy resin reinforced with glass beads having on their surface either a coupling agent like a silane or a releasing agent like silicone. From the analysis of spectra, we propose two different mechanisms to explain the reinforcement process through a modification of the matrix at its interface with the filler.

INTRODUCTION

The mechanical properties of composite materials have received considerable attention over recent years [1]. Measurements in shear on glass beads reinforced epoxy resin composites are fairly common in the literature [2-4] and experimental data have been compared with the moduli calculated from the various theoretical predictions [1-5]. In the glassy region, there is a good agreement but above the glass transition temperature, the material is reinforced much more than suggested by the theories. Some evidence is given for a difference in glass transition temperature between filled and unfilled materials [4]. In fact, the influence of the matrix interphase on the mechanical properties is not clearly understood [6]. The aim of this work was to precise that

point, using a new technique--the thermostimulated creep (TSC)
[7]--which enables the experimental resolving of the complex
spectra generally recorded.

EXPERIMENTAL

<u>Materials</u>

We used two different types of model composites based on
epoxy resin CY 208 from CIBA-GEIGY hardened with 50 per hundred
resin (phr) of HY 905 CIBA-GEIGY anhydride and 5 phr of N-benzyl-
dimethylamine. Reaction of curing (30 minutes at 60°C) is followed
by a 24 h stabilization at 90°C and a 12 h annealing at 120°C.
SOVITEC glass beads (4-44 μm) are covered either with a 0,5%
trichloroethylene solution of QZ 13 silicon from PROCHAL for weak
adhesion or with γ-glycidoxypropyltrimethoxysilane (A 186 silane
from UNION CARBIDE) for strong adhesion.

<u>Thermostimulated Creep (TSC)</u>

In thermostimulated creep experiments, the sample is placed
in a torsional balance as shown on Figure 1. The pendulum is placed
under helium atmosphere, so that the temperature of the sample can
be controlled from liquid helium temperature to 100°C.

The upper end of the strip-shaped sample can be submitted to
a given shear stress during the loading program. The recovery of
the sample is followed by a mirror. This TSC cell has been previously described [7].

THEORY

The principle of TSC experiments is as follows [8]:

a. A shear stress is applied to the sample at a temperature
$T_{\sigma ON}$ for 2 min.

b. The temperature is decreased to $T_{\sigma OFF}$ for freezing the
mobile units that one wishes to consider. Then, the stress is cut
off.

c. The recovery of the sample is stimulated by increasing
the temperature in a controlled manner. The strain γ and the rate
of change of the strain $\dot{\gamma}$ are simultaneously recorded as a function
of time.

TSC spectra obtained in composites are complex: the great
advantage of this technique is to allow their experimental resolution by applying fractional stresses.

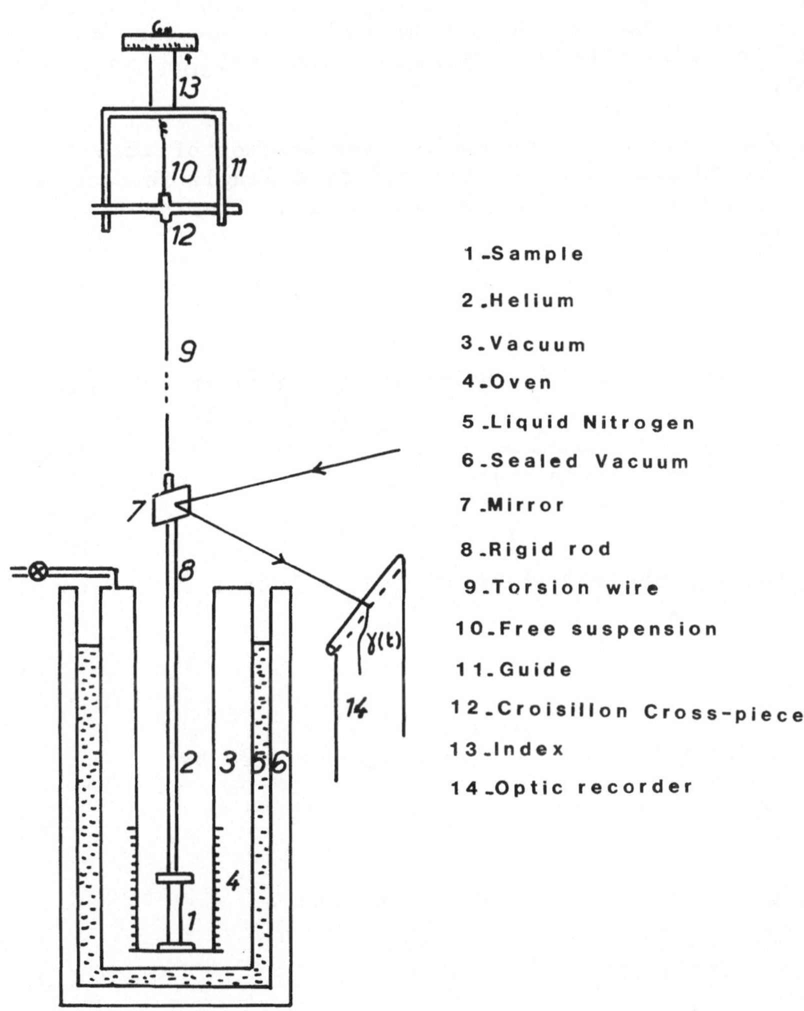

Fig. 1.   Experimental device for thermostimulated creep.

The principle of the experimental resolution of TSC spectra is as follows:

The stress σ is applied at $T_{\sigma ON}$ for 2 min. so that the units having a retardation time $\tau(T_{\sigma ON}) < 2$ min will be practically oriented while the others remain distributed at random. The temperature is lowered up to $T_{\sigma OFF} = T_{\sigma ON} - 10°$ and the stress is cut off while the temperature is maintained constant for 2 min. allowing the units with a retardation time $\tau(T_{\sigma OFF}) < 2$ min. to randomize.

The TSC spectra obtained after application of such fractional stresses can be practically described by a single retardation time $\tau_M(T)$. The law followed by the deformation after application of a stress σ is:

$$\gamma = \gamma_o \ (1-e^{-t/\tau_M}) \tag{1}$$

When the stress is cut off, the return to equilibrium follows:

$$\gamma = \gamma_o \ \exp \left[-\int_0^t \frac{dt}{\tau[T_M(t)]} \right] \tag{2}$$

By measuring γ_0, the variation of the real part of the compliance ΔJ is obtained from:

$$\gamma_o = \Delta J \sigma \tag{3}$$

The rate of change of the deformation is defined by:

$$\dot{\gamma} = \frac{d}{dt} \gamma \tag{4}$$

The mechanical retardation time is given by:

$$\tau_M(T) = \gamma(T)/\dot{\gamma}(T) \tag{5}$$

So the temperature dependence of the mechanical retardation time will be deduced from the knowledge of γ and $\dot{\gamma}$.

In polymers, two laws have been proposed for fitting the experimental points: the Arrhenius law,

$$\tau(T) = \tau_{oa} \exp E_a/kT \tag{6}$$

where τ_{oa} is the pre-exponential factor, k is the Boltzmann constant, and E_a is the activation energy; the Vogel law:

$$\tau(T) = \tau_{ov} \exp [\alpha(T - T_\infty)]^{-1} \tag{7}$$

where τ_{ov} is the preexponential factor, α is the thermal expansion coefficient of the free volume, and T_∞ is the critical temperature at which the retardation time becomes infinite.

For each i elementary process, the variation ΔJ_i of the real compliance (equation 3) and the temperature dependence of the retardation time τ_{Mi} (T) (equation 5) will allow us to calculate the real J' and imaginary J'' part of the compliance versus temperature and frequency:

$$J' - J_u = \sum_{i=1}^{n} \Delta J_i \frac{1}{1 + \omega^2 \tau_{Mi}^2} \tag{8}$$

$$J'' = \sum_{i=1}^{n} \Delta J_i \frac{\omega \tau_{Mi}}{1 + \omega^2 \tau_{Mi}^2} \tag{9}$$

where J_u (u for unrelaxed) is the instantaneous elastic compliance. By substituting continuous spectra, we can write [11]:

$$J' - J_u = \int_{-\infty}^{+\infty} [L/(1 + \omega^2\tau^2)] \ d \ \ell n\tau \tag{10}$$

$$J'' = \int_{-\infty}^{+\infty} [L\omega\tau/(1 + \omega^2\tau^2)] \ d \ \ell n\tau \tag{11}$$

where $L(\tau)$ is the retardation time spectrum, $L(\tau) \ d(\ell n\tau)$ defining the contributions to the creep compliance associated with retardation times whose logarithms lie in the range between $(\ell n\tau)$ and $(\ell n\tau + d\ell n\tau)$.

So, the retardation time spectrum can be deduced from the
storage or the loss compliance. The iteration method of Ninomiya
and Ferry [9] using J´ has been adopted for this work.

RESULTS

The purpose of the study was to obtain a better knowledge of
the interactions between filler and matrix in composites. For our
investigation, we have chosen to take two very different kinds of
filler concerning their reactivity toward the matrix.

The comparative TSC study of unfilled and filled epoxy resin
has shown the existence of an intense TSC peak around the glass
transition temperature i.e. at room temperature. Silicone-treated
beads does not modify significantly the peak of the unfilled resin;
on the contrary, silane-treated beads shifts the TSC peak by 4°.
All those peaks are complex; they have been experimentally resolved
into four elementary processes represented on Figure 2. For each
peak, the temperatures $T_{\sigma OFF}-T_{\sigma ON}$, precising the program of fractional stresses have been indicated on the figure. For silicone
treated glass beads, each elementary TSC peak has the same temperature position as for the unfilled matrix. Considering now the
intensities of the peaks, for the matrix with silane treated beads,
the most intense peak appears in the section isolated for
$T_{\sigma OFF}-T_{\sigma ON}=30-40°C$; for the unfilled matrix and the matrix with
silicone-treated glass beads, it is observed in the section
20-30°C.

This shows an increase in the glass transition temperature
already visible on the global distribution and in agreement with
dynamic mechanical measurements [10].

From the recordings of $\gamma(t)$, $\dot{\gamma}(t)$ and $T(t)$, eq. 5 gives the
temperature dependence of the retardation time of each elementary
TSC peak. As shown on the Arrhenius diagram of Figure 3, two different behaviors are observed: on the low temperature side of the
complex TSC peak, i.e. below the glass transition temperature, T ,
the retardation times follow the Arrhenius equation (eq. 6); on
the high temperature side of the complex TSC peak, i.e. above the
glass transition temperature, τ follows the Vogel equation (eq.
7). Retardation times are not much different for the unfilled
matrix and the matrix with silicone treated beads whereas they
differ when the filler reacts with the matrix (silane treated
beads).

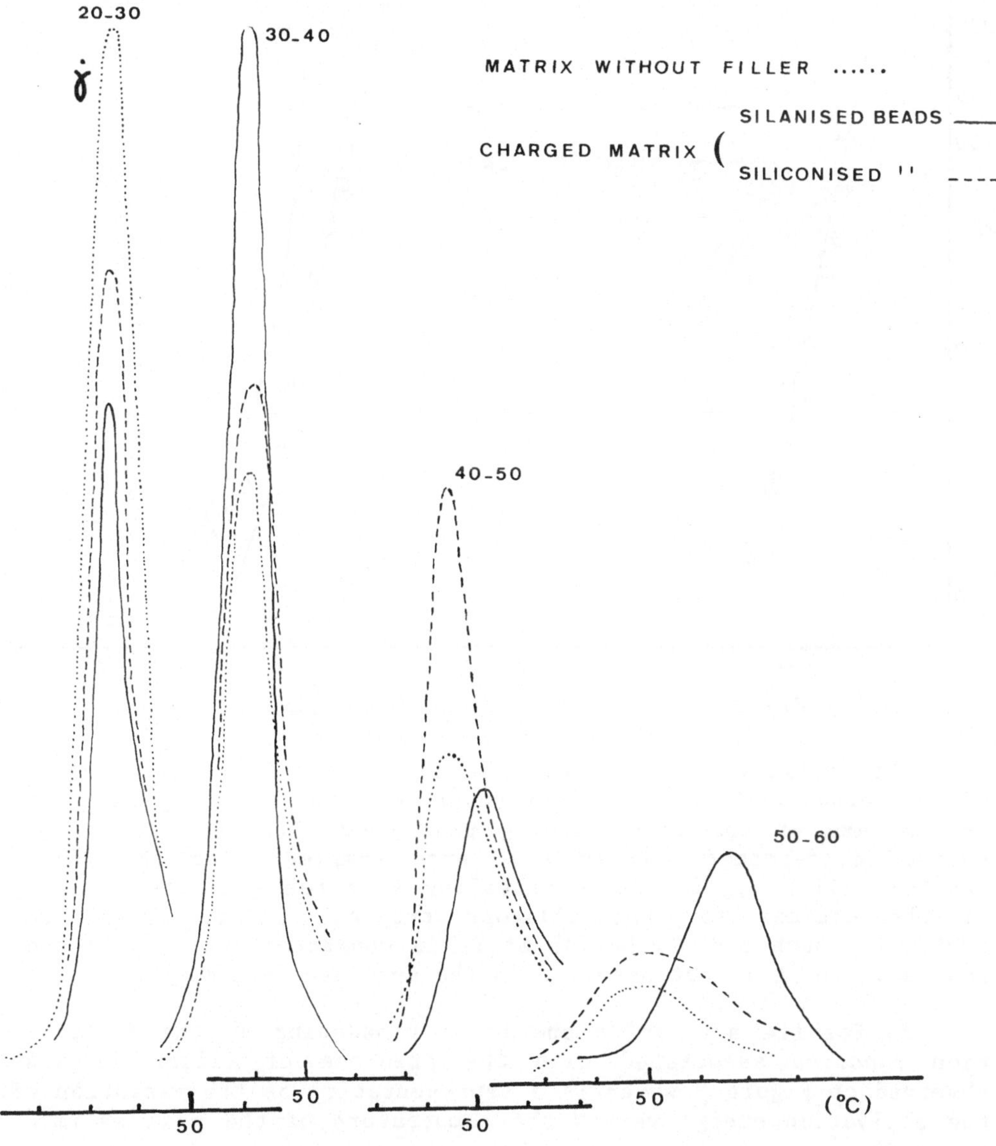

Fig. 2.   TSC spectrum of the samples.

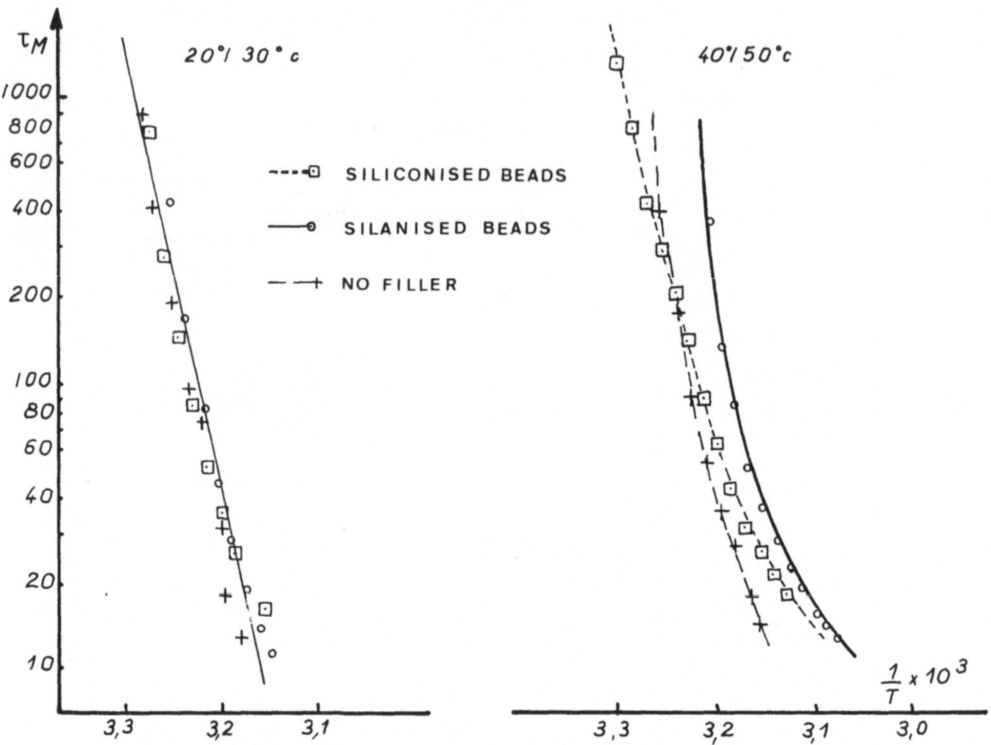

Fig. 3.   Analysis of retardation times.

The critical temperatures, T_∞, that linearize the variations of $\ell n\tau$ versus $1/(T-T_\infty)$ have been reported versus the temperature of the maximum of the corresponding peak above T_∞. This plot shows serious differences between the three samples. For the silane treated filler, T_∞ increases revealing a decrease in the mobility of the chains; for the silicone treated filler, it is the contrary. Such a distribution of T_∞ is characteristic of the composite since it is not observed in the unfilled matrix.

As for the activation energy, a broadening of the distribution function associated with the presence of filler is also observed on Figure 5 which is a representation of the variation of the activation energy versus the temperature of the peak maximum, below Tg.

With the TSC technique, one can go further and calculate the retardation time spectrum $L(\tau)$ at a given temperature as shown before. Figure 6 represents $L(\tau)$ at 20°C. For the unfilled matrix and the matrix filled with silicone treated beads, two maxima are observed: one is situated at 10^3 sec and the other, slower, at $10^{4.5}$ sec. For the case of a filler treated by a silane as a coupling agent, two additional peaks appear at 10^2 sec and 10^7 sec.

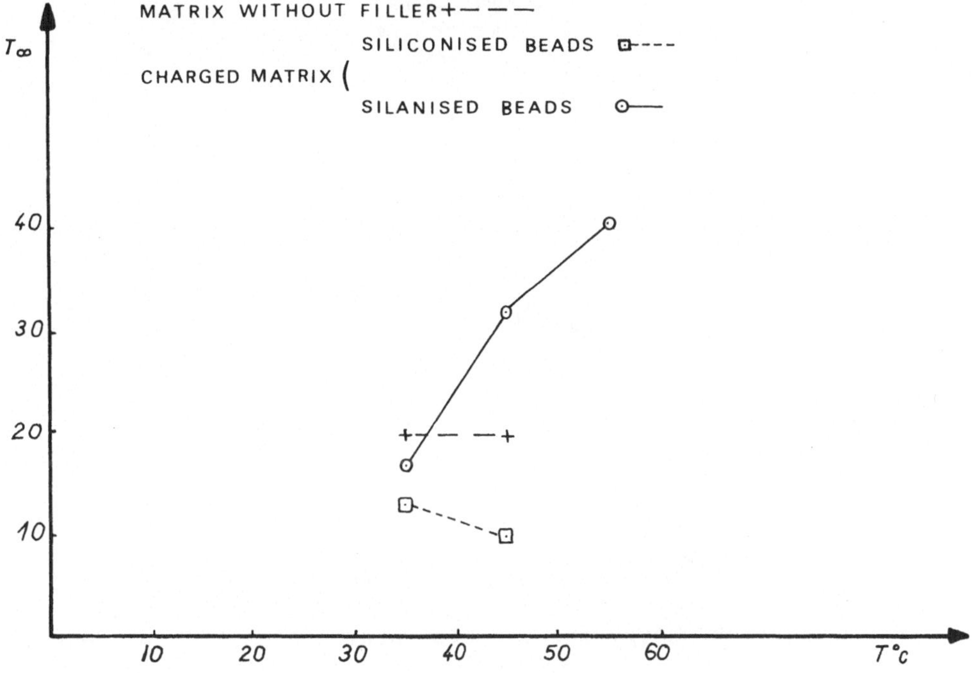

Fig. 4.   Variation of T_∞ from the Vogel equation.

The discussion of their molecular origin is presented below.

DISCUSSION

Ferry thought that the slower relaxation corresponds to dangling segments constituted of a few monomers fixed at one end only. A statistical theory of polycondensation [12] considers the existence of such structures.

In our case, it appears that the relative proportion of dangling chains is greater when the filler is silicone treated. In this case, indeed, the hydroxyl functions of the raw glass are concealed and chains only bound at one end become "dangling" chains. For the case of a filler treated by a silane as coupling agent, simple models could explain the two additional retardation modes.

With the very same skeleton for the network of the matrix the differences in behavior can be related to the presence of the dangling chains. With a silane treated filler, these chains are drawn toward the reactive surface as it is the case for carbon black and rubber [13]. These chains are bound to the surface but the bonds are reversible allowing a relaxation of the stress when the matrix slips on the filler's surface. This phenomenon is rather slow and the relaxation appears at about 10^6 sec, that is to

　　　　　　　　　　　　　　　　　　　　　J. P. BAYOUX ET AL.

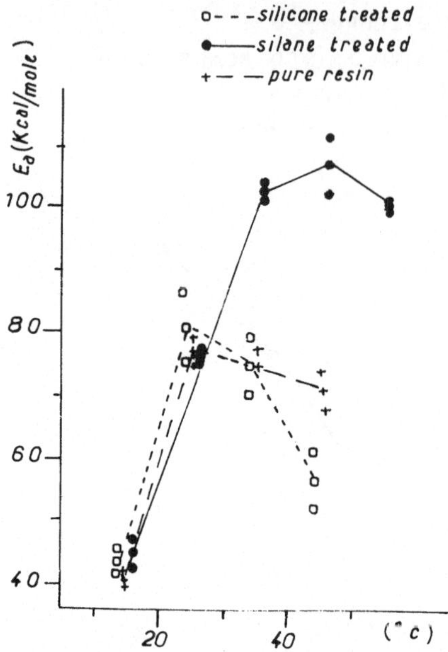

Fig. 5.　Activation energy for the elementary relaxation processes.

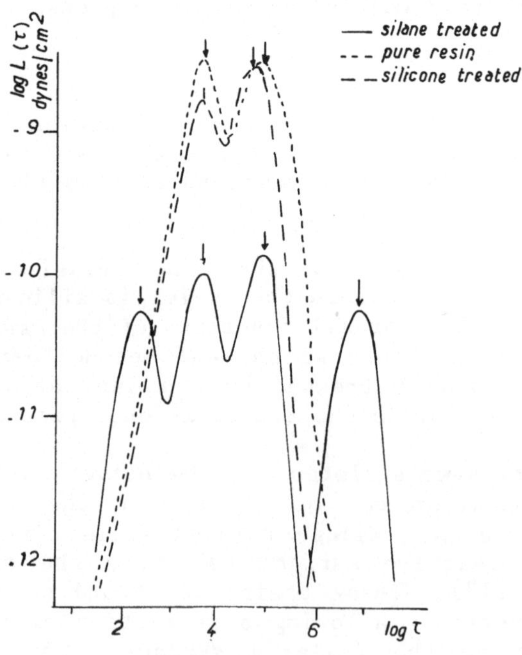

Fig. 6.　Retardation spectrums.

say, 2.5 decades above that of the "free" dangling chains. Around the filler a region of the network is "cleaned" of its dangling chains and a supplementary relaxation appears characteristic of this region at 10^2 sec, which is lower by a factor of 10 from the network with dangling chains. For the rest of the composite, the matrix keeps its two main relaxation peaks as most of the network is rather far from the filler itself.

CONCLUSION

The thermostimulated creep technique allows a new approach of the concept of boundary interface in composite mechanics [14]. The relaxation mechanisms in reinforced epoxy resins have been investigated.

The matrix itself exhibits two relaxations, the faster corresponds to the network and the slower to the dangling chains.

The addition of a silicone treated filler increases the relative proportion of dangling chains.

The addition of a silane treated filler generates two more relaxation peaks, one at a long relaxation time corresponding to the dangling chains creeping from one reactive site to the other at the surface of the filler, the other at a shorter relaxation time corresponding to the network cleaned of its dangling chains which have been "attracted" by the filler.

REFERENCES

1. J. A. Manson, L. H. Sperling, "Polymer Blends and Composites," Plenum Press, New York (1976).
2. J. H. Speake, R. G. C. Arridge, G. J. Curtis, J. Phys. D7, 412 (1974).
3. T. B. Lewis, L. E. Nielsen, J. Appl. Polym. Sci. 14, 1449 (1970).
4. R. J. Crowson, R. G. C. Arridge, J. Mat. Sci. 12, 2154 (1977).
5. B. E. Read, G. D. Dean, "The Determination of Dynamic Properties of Polymers and Composites, Adam Hilger Ltd, Bristol (1978).
6. G. C. Papanicolaou, S. A. Paipetis, P. S. Theocaris, Colloid and Polym. Sci. 256, 625 (1978).
7. J. C. Monpagens, Ph.D. Thesis, University of Paul Sabatier, Toulouse (1977).
8. J. C. Monpagens, D. Chatain, C. Lacabanne, P. Gautier, J. Polym. Sci.-Phys. Ed. 15, 767 (1977).
9. K. Ninomiya, J. D. Ferry, J. Coll. Sci. 14, 36 (1959).
10. J. P. Bayoux, Ph.D. Thesis, University of Claude Bernard, Lyon (1980).

11. J. D. Ferry, "Viscoelastic Properties of Polymers," John Wiley,
        New York (1970).
12. H. Lee, K. Neville, "Handbook of Epoxy Resins," McGraw Hill,
        New York (1957).
13. F. Bueche, J. Rubber Chem. Tech. $\underline{32}$, 1259 (1959).
14. H. Ishida (to be published).

APPLICATIONS OF SOLID-STATE MAGIC ANGLE NMR SPECTROSCOPY TO FIBER

REINFORCED COMPOSITES

A. M. Zaper, A. Cholli and J. L. Koenig

Department of Macromolecular Science
Case Western Reserve University
Cleveland, Ohio 44106

ABSTRACT

The solid-state NMR techniques of cross polarization (CP), high power proton decoupling and magic angle sample spinning allow one to study the binding of coupling agents to silica surfaces. γ-mercaptopropyltrimethoxysilane (γ-MPTS), γ-aminopropyltriethoxy-silane (γ-APS), γ-glycidoxypropyltrimethoxysilane (γ-GPS), γ-methacryloxypropyltrimethoxysilane (γ-MPS) and vinyltriethoxysilane (VTES) were the organosilanes utilized on high-surface-area silica. High resolution solid-state ^{13}C NMR spectra were obtained for coupling agents adsorbed on silica surfaces and for the corresponding coupling agents condensed as bulk polymers. Chemical shifts and line widths of the resonances of the chemically modified silicas are compared to those resonances arising from the bulk organosilanes. The spectra confirm chemical binding of the coupling agent to the silica surface. A graphite-filled epoxy is characterized in terms of molecular motion. In this study ^{13}C spin-lattice relaxation times in the rotating frame of reference ($T_{1\rho}$) were obtained for the composite.

INTRODUCTION

Reinforced polymers are high performance materials with many uses. The properties of composites depend on the fiber, coupling agent/fiber interface, coupling agent/polymer interface and on the polymer. The investigation into the nature of these components on the molecular level is useful in optimizing mechanical properties. Because of the insolubility of composites, FT-IR [1] has been one of the few spectroscopic techniques used to characterize these

systems. ^{13}C NMR spectroscopy has principally been used to characterize chemical systems in the liquid or solution state and only
recently has it become possible to study solid materials with this
method. The difficulty with solid-state ^{13}C NMR studies in the
past was associated with excessive dipolar broadening, low natural
abundance and long relaxation times of ^{13}C nuclear spins. However,
recent developments in three techniques; high power proton decoupling, cross polarization (CP) [2] and magic angle sample spinning
(MAS) [3-5] result in high resolution ^{13}C NMR spectra for insoluble
polymers [5] and organosilane moieties on silica surfaces [6-8].
The techniques also provide a method for obtaining new information
about the solid-state structure of materials and the surface
chemistry of chemically modified fillers. In surface studies ^{13}C
NMR has the particular advantage of avoiding interference effects
from glass, a problem that may arise in other surface study
methods. The usefulness of solid-state ^{13}C NMR (CP/MAS) spectroscopy is shown for the characterization of the different regions of
a composite. ^{13}C CP/MAS NMR studies of silane coupling agents on
silica surfaces were undertaken. An epoxy composite was also
investigated. In this report the application of ^{13}C CP/MAS experiments of γ-mercaptopropyltrimethoxysilane (γ-MPTS), γ-aminopropyltriethoxysilane (γ-APS), γ-glycidoxypropyltrimethoxysilane (γ-GPS),
γ-methacryloxypropyltrimethoxysilane (γ-MPS) and vinyltriethoxysilane (VTES) on silica surfaces are discussed. ^{13}C relaxation
times were used to monitor changes in a graphite-filled epoxy
composite.

EXPERIMENTAL

A. Samples

The silane coupling agents were obtained from Petrarch
Systems, Inc. and used without further purification. The fumed
silica, obtained from Cabot Corp., has a surface area of 390 m^2/g.
The silica was heated at 100°C in a vacuum oven for a day prior to
use. The coupling agents subsequently applied (γ-MPTS, γ-APS,
γ-GPS, γ-MPS and VTES) were first hydrolyzed in a 2% aqueous solution where the pH of the treating solution was adjusted to approximately 4. For the case of γ-APS, the aqueous solution was at
natural pH. The silica was added to the aqueous solutions for 5
minutes with stirring, then washed with distilled water and vacuum
filtered. The samples were dried overnight at 80°C under vacuum.
The corresponding condensed coupling agents were prepared by
driving off the water from similar aqueous solutions as those used
for treatment. These samples were also dried overnight at 80°C.
The epoxy resin used was a diglyceride of biphenol A (EPON 828,
Shell Co.) cure with nadic methyl anhydride (Fisher Scientific).
The curing accelerator was benzyldimethylamine (Fisher
Scientific). The graphite fibers used in the epoxy composite,

which contained 45% by weight fibers, were obtained from Union
Carbide.

B. Spectroscopy

The ^{13}C NMR experiments were performed at 37.7 MHz on a
Nicolet NT-150 spectrometer. Cross polarization and magic angle
spinning were used to obtain all the spectra and also for the
relaxation measurements. The rotors used to spin the samples in
the case of the surface study measurements were machined from poly-
chlorotrifluoroethylene for the barrel and polyoxymethylene for the
base and were spun at approximately 2-2.5 kHz. For the case of the
epoxy samples, a spinner made entirely of polyoxymethylene was used
and spun at 3.5 kHz. Radiofrequency field strengths between 50 and
60 kHz were obtained. The ^{13}C CP/MAS NMR spectra of the condensed
coupling agents are the time average of 5000 transients, while the
number of transients averaged to 20,000 for each of the treated
silica samples. In all cases the contact time was 1 msec while the
delay between successive sequences was 2 sec.

I. SURFACE STUDIES

The ^{13}C CP/MAS NMR spectra of the representative surface-
modified silicas and the corresponding condensed silane coupling
agents are shown in Figures 1 to 5. Chemical shift differences
between the peaks of the carbons of the polymerized organosilanes
and the corresponding organosilanes on silica surfaces are listed
in Table 1. The chemical shifts were measured relative to tetra-
methylsilane (TMS). For most cases, the trend appears to be an upfield chemical shift for the propyl carbon resonances of the coupling agents on the silica when compared to the same carbons of the
bulk organosilanes. Chemical shift differences will be analyzed.
Line widths at half resonance height are listed for all the carbon
peaks of the silanes in Table 2. Significant line narrowing and
broadening will be discussed.

A. γ-Mercaptopropyltrimethoxysilane

Figure 1 shows spectra of samples prepared with γ-mercaptopropyltrimethoxysilane coupling agent. The top spectrum (Figure
1A) is that of the condensed organosilane and the spectrum below it
(Figure 1B) is that of the same coupling agent adsorbed onto a
silica surface. The ^{13}C CP/MAS NMR spectra show two well resolved
peaks. The resonance at 13.8 ppm in the spectrum of the polymerized coupling agent can be assigned to the methylene carbon
attached to the silicon atom and the resonance at 29.1 ppm can be
assigned to the remaining two methylene carbons. In the ^{13}C NMR
spectrum of γ-MPTS on silica, the α-carbon peak is found at 11.6
ppm and the β and γ-carbons at 27.2 ppm. Maciel et al. [6] have

Table 1. ^{13}C Chemical Shift Differences (in ppm relative to TMS)
(Condensed Coupling Agent - Coupling Agent on Silica)

Coupling Agent	Respective Carbons of Coupling Agent						
	a	b	c	d	e	f	g
γ-MPTS	2.2	1.9	1.9				
γ-APS	1.1	-2.4	-.1				
γ-GPS	1.0	.8	.6	.0	.3	-1.4	
γ-MPS	.6	.5	.5	-1.1	.8	.1	.6
VTES	1.6	2.2					

Table 2. Line Widths at Half Height (in Hertz)

Coupling Agent	Respective Carbon Peaks of Coupling Agent						
	a	b	c	d	e	f	g
γ-MPTS condensed	147.1	77.6	77.6				
γ-MPTS on silica	84.3	70.4	70.4				
γ-APS condensed	139.9	213.7	176.7				
γ-APS on silica	197.0	226.7	142.9				
γ-GPS condensed	97.2	46.0	81.0	81.0	6.5	19.5	
γ-GPS on silica	76.1	40.1	132.6	132.6	137.4	137.4	
γ-MPS condensed	26.8	42.8	39.6	35.0	40.1	73.9	39.7
γ-MPS on silica	86.0	72.1	91.8	60.0	36.1	100.4	86.9
VTES condensed	96.7	134.3					
VTES on silica	114.4	152.9					

also reported the ^{13}C NMR spectrum of the same silylating agent and
have found the two resonances within 1 ppm of the above assignments. If the two peaks on the spectrum of the condensed coupling
agent (Figure 1A) are compared to the same two peaks on the
spectrum of the coupling agent bound to the silica surface, a
chemical shift of the resonance peaks is observed. When the coupling agent is attached to the silica, the peaks corresponding to
the propyl chain are shifted upfield approximately 2.0 ppm as
compared to the bulk polymer. These carbon are in close proximity
(especially the α-methylene carbon which shows the larger upfield
shift) to the chemical binding sites which may lead to steric hindrance of the silanes causing the chemical shift [9]. An
additional peak which appears at approximately 44 ppm arises from
residual glycine in the sample spinner. Glycine is the material
used for optimizing peaks for the setting of the magic angle.

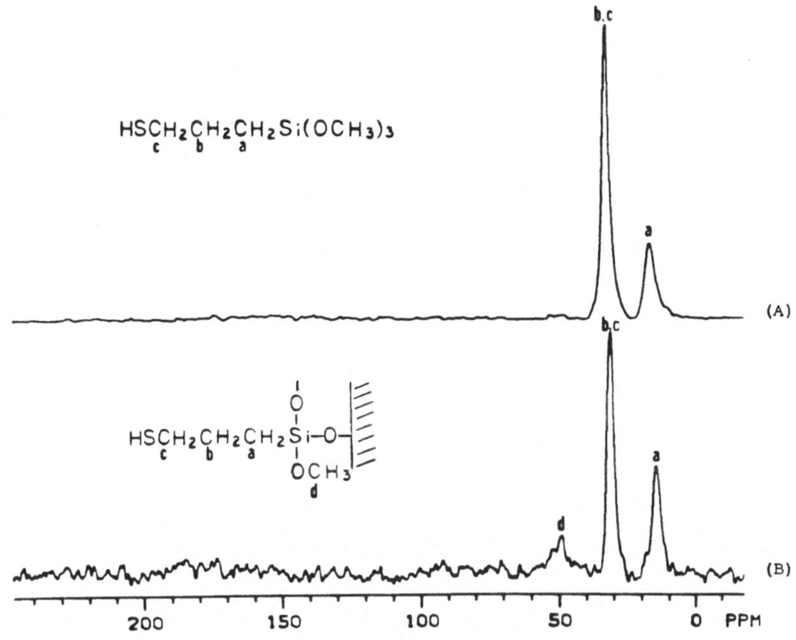

Fig. 1.   (A) CP/MAS ^{13}C NMR spectrum of polymerized γ-mercaptopro-
          pyltrimethoxysilane (γ-MPTS). The spectrum was recorded
          at 37.7 MHz with a single contact time of 1.0 msec during
          cross-polarization and a recycle time of 2.0 sec. 5000
          transients were collected during accumulation. Assign-
          ments of the resonance peaks are indicated. (B) CP/MAS
          ^{13}C NMR spectrum of adsorbed γ-MPTS on a silica surface.
          20,000 transients were collected during accumulation. The
          chemical shift differences of spectra A and B can be found
          in Table 1.

At this time it is not possible to quantitatively compare peak intensities between the two spectra because they are a function of the effectiveness of the proton-carbon cross polarization, i.e. polarization efficiency must be the same for each type of carbon involved, a condition which is not always reached for this type of analysis. It has been observed that there is a linewidth change when comparing the two spectra. The two resonances of the organosilane are narrower for the spectrum of the treated surface, 84.7 and 70.4 Hz for the α and β, γ-methylene carbon peaks, respectively, when compared to the spectrum of the same condensed coupling agent, 147.1 Hz and 77.6 Hz for the α and β, γ-resonances, respectively. This reduction in linewidth can be attributed to a decrease in molecular motion which may be caused by the steric hindrance of binding to the silica. It was also noticed that there was a greater linewidth reduction for the resonance arising from the methylene unit bound to the silicon atom than for the resonance due to the β and γ-methylene carbons. When there is a reduction in molecular motion one would expect an appreciable narrowing of resonances in ^{13}C CP/MAS NMR spectra. The reduction in motion leads to a narrower distribution of isotropic chemical shifts. This is in contrast to liquid NMR spectra which show very narrow peaks in spite of great mobility of liquids. Because of their rapid molecular motion, liquids have fewer terms in their nuclear spin Hamiltonians, which govern linewidths, than do solids. A more detailed explanation can be found elsewhere [10].

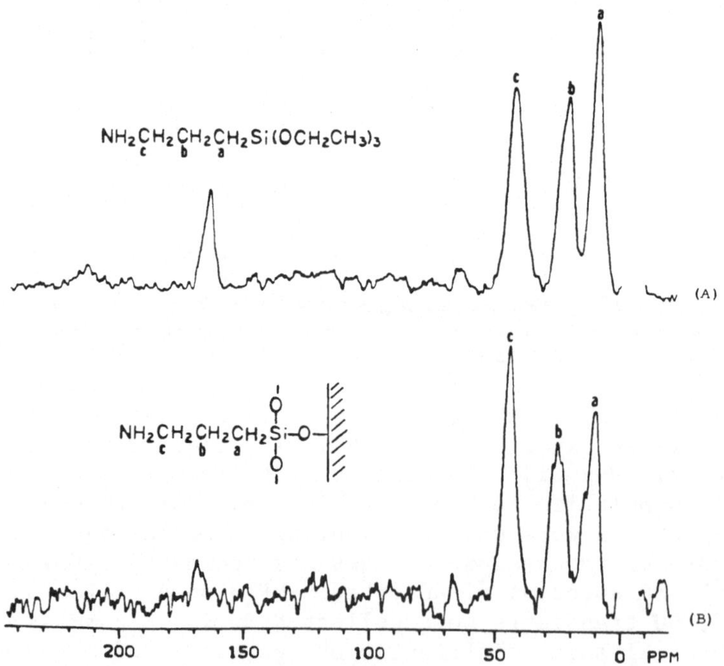

Fig. 2. CP/MAS ^{13}C NMR spectra of γ-aminopropyltriethoxysilane (γ-APS) (A) polymerized, (B) on a silica surface.

B. γ-Aminopropyltriethoxysilane

Condensed γ-aminopropyltriethoxysilane and the same coupling agent treated onto a silica surface produce well resolved ^{13}C CP/MAS NMR spectra as can be seen in Figure 2. The strong resonances appearing at 11.7 ppm, 22.9 ppm and 43.9 ppm in Figure 2A are readily assigned to the α, β, and γ-carbons of the methylene units of the propyl chain, respectively. The absence of ethoxy group resonances confirms that the γ-APS is completely hydrolyzed both in the solution and on the silica surface. The chemical shift of the α-methylene carbon is slightly shifted on the spectrum of the organosilane-treated silica which may be a result of steric hindrance arising from close proximity to the chemical binding sites.

The resonance of the γ-methylene carbon is narrower (142.9 Hz) in the spectrum of the coupling agent on the silica surface than in the spectrum of the condensed material (176.7 Hz). A narrower peak implies restricted flexibility or restricted freedom of the molecule. One of the factors for restricted molecular motion may arise from chelate ring formation [11]. Hydrolyzed γ-APS on a silica surface has been proposed to exist in two structural forms: a chelate ring form and an extended form. Intramolecular hydrogen bonding of amine groups and unreacted silanols on the silica leads to formation of a six-membered chelate ring. Heat treatment can disrupt hydrogen bonding leaving the aminopropyl group in chain form. It is possible that the duration and temperature of heating used during sample preparation were not extensive enough to destroy all of the intramolecular hydrogen bonding on the silica surface. The methylene unit bound to the amine group is affected to the greatest extent by this bonding and this is indicated by the narrower resonance for this carbon. A broader peak in this case would have indicated an increase in flexibility due to a loss of hydrogen bonding.

The spectrum of condensed γ-APS has an additional prominent peak arising at 164.0 ppm. The chemical shift value of 164 ppm corresponds well to that expected for the carbon of a carbonyl group. If this peak is attributed to a carbonyl carbon then the carbonyl is most probably part of the aminebicarbonate salt which is formed by a reaction with carbon dioxide in air [12]. This reaction can occur with both bulk γ-APS and γ-APS on the surface. The aminebicarbonate salt is unstable to heat. With 100°C heat treatment, the carbon dioxide should be removed from the silica surface while 150°C is necessary to remove the carbon dioxide from the bulk polymer. The 80°C heat treatment applied during sample preparation was not adequate to destroy the salt in the condensed material. A slight carbonyl peak also appears at 163.7 ppm on the

spectrum of γ-APS on silica, probably due to residual aminebicarbonate salt (most was probably removed during heat treatment). The
carbonyl peak in the spectrum of condensed γ-APS was integrated
with respect to the center methylene carbon peak and the percentage
was found to be between 50 and 55%. This value corresponds well to
results obtained for percentage salt formation in a sample of this
material. Koenig and coworkers have found that when glass mats
treated with γ-APS are dried in air, 54% of the primary amine of
the coupling agent were interacting with carbon dioxide [12].

C. γ-Glycidoxypropyltrimethoxysilane

Solid-state ^{13}C NMR spectra were also obtained for condensed
γ-glycidoxypropyltrimethoxysilane (Figure 3A) and γ-GPS on a silica
surface (Figure 3B). The five resonances in the spectrum of the
condensed coupling agent are readily assignable to the six
different carbons of the hydrolyzed coupling agent. The peaks at
11.3 ppm and 25.2 ppm in Figure 3A are attributed to the α and
β-methylene carbons of the propyl chain and are both shifted upfield by approximately 1 ppm on the spectrum of the coupling agent
on the silica surface. In addition to being shifted upfield on the
spectrum for adsorbed coupling agent, these two resonances are also
narrower. In the spectrum of the coupling agent on silica the   α

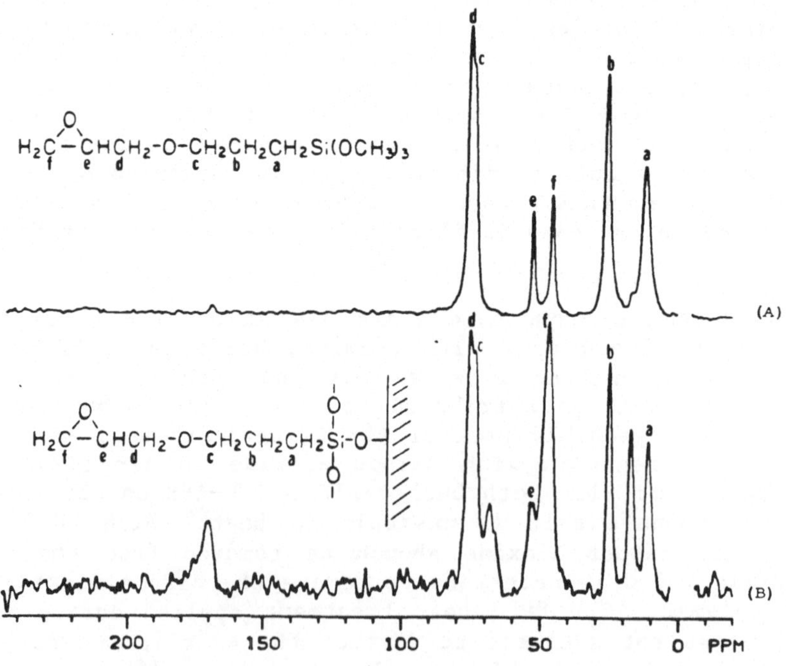

Fig. 3. CP/MAS ^{13}C NMR spectra of γ-glycidoxypropyltrimethoxysilane (γ-GPS) (A) polymerized, (B) on a silica surface.

and β-methylene resonances have line widths of 76.1 Hz and 40.1 Hz, respectively, while in the spectrum of the condensed organosilane, the line widths are 97.2 and 46.0 Hz, respectively. This is due to the increased rigidity of these methylene units when the coupling agent is attached to the surface. The two oxymethylene chain carbon peaks coincide and appear at 75.0 and 73.8 ppm on the spectrum of polymerized γ-GPS while the methine and methylene carbons of the epoxide ring are well resolved and show resonances at 52.4 and 45.3 ppm, respectively.

Significant differences appear in the spectrum of γ-GPS on the silica surface (Figure 3B). The resonances corresponding to the carbons of the epoxide ring (e and f) are considerably broader. In Figure 3B, the resonances for the e and f carbons essentially form one peak with a line width of 137.4 Hz while two resonances appear in the spectrum of the condensed coupling agent with line widths of 6.5 Hz and 19.5 Hz, respectively. One possible explanation for the increase in intensity of the peak at approximately 46 ppm is that is is also comprised of a contribution from unhydrolyzed methoxy groups of the coupling agent. The broadness of the combined resonances of the epoxide ring carbons may also indicate a possible functional group interaction with the silica surface. It is feasible that the oxygen of the epoxide ring is interacting with the surface causing a chemical shift distribution at different sites on the surface which in turn appear as appreciably broadened resonances. Additional factors contributing to the line broadening may originate from surface impurities, steric hindrance and other inhomogeneity factors [13,14].

In Figure 3B, the spectrum of γ-GPS on silica shows two new resonances, one peak at 16.6 ppm and another at 67.8 ppm. During the preparation of this sample the organosilane was hydrolyzed to a silane triol in an acidic (pH 4.0) aqueous solution. The acidic medium may have catalyzed a cleavage of the epoxide ring. Several reactions can occur following the ring opening [15]. A likely reaction in this type of medium is one with water that would form primary and secondary alcohols. Products with these types of structures would have additional methylene units and carbon-oxygen bonds. The resonances for carbons bonded to oxygens can appear in the 67 ppm chemical shift range and methylene carbon resonances may appear at approximately 16 ppm. A reaction of the epoxide ring in the acidic solution would explain the presence of the two additional peaks appearing at 16.6 ppm and 67.8 ppm.

D. γ-Methacryloxypropyltrimethoxysilane

Figure 4 shows the ^{13}C CP/MAS NMR spectra for condensed γ-methacryloxypropyltrimethoxysilane and γ-MPS adsorbed onto a silica surface. Resonances for all of the carbons of the coupling agent

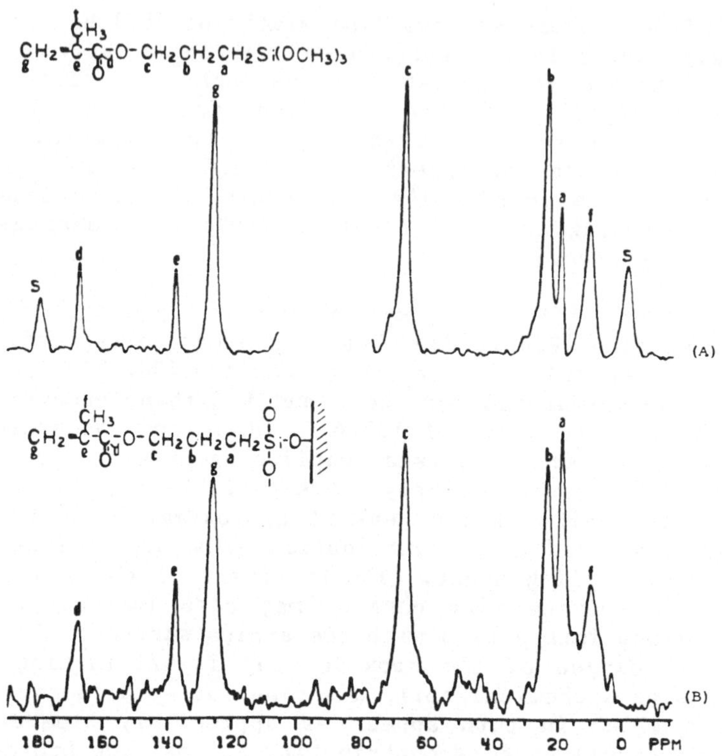

Fig. 4. CP/MAS ^{13}C NMR spectra of γ-methacryloxypropyltrimethoxy-
         silane (γ-MPS) (A) polymerized, (B) on a silica surface.

appear clearly on both spectra. The peaks appearing at 19.1 ppm,
23.5 ppm and 67.3 ppm correspond to the α, β, and γ-methylene
carbons of the propyl chain, respectively on the CP/MAS spectrum of
the condensed coupling agent (Figure 4A). These three resonances
are shifted to 18.5 ppm, 23.0 ppm and 66.8 ppm, respectively, on
the spectrum of γ-MPS treated silica (Figure 4B). This upfield
shift may be a result of steric hindrance of the silanes which
arises from the close proximity of the carbons to the chemical
binding sites. The vinyl group carbons appear at 126.2 ppm and
137.8 ppm (g and e) on the spectrum of the condensed coupling
agent, and at 125.6 ppm and 137.0 ppm, respectively, on the
spectrum of the treated silica.

A functional group interaction effect is observed when comparing the spectra of condensed γ-MPS to that of γ-MPS on a silica
surface. The carbonyl carbon resonance at 167.4 ppm in the top
spectrum appears at 168.5 ppm in the γ-MPS treated silica
spectrum. In this case the carbonyl carbon resonance is
broadened. This broadening may be due to the presence of two types

of carbonyls − a free carbonyl and a carbonyl which is interacting
with the surface, possibly through hydrogen bonding [16].   This
interaction causes a chemical shift distribution which is reflected
by a broadened asymmetric resonance peak.   The formation of a
hydrogen bond to the surface has an electron withdrawing effect
which causes deshielding of carbons. Deshielding is reflected by a
downfield chemical shift as was observed for the carbonyl carbon.
This functional group interaction may also explain the slight upfield chemical shift for the resonances of the vinyl carbons of the
coupling agent on the silica surface.   Hydrogen bonding of some of
the carbonyl groups of γ-MPS with the hydroxyl groups of silica may
lead to steric hindrance of neighboring carbons resulting in the
upfield chemical shift.

The methyl carbon peak of the methacryloxy functional group
of γ-MPS appears at approximately 9.9 ppm in both spectra in Figure
4. There is an increase in the linewidth for the methyl resonance
when γ-MPS is adsorbed to silica.   In Figure 4B, peak f has a linewidth of 86.9 Hz while in Figure 4A this same resonance has a linewidth of 39.7 Hz. This results may be indicative of slightly hin-
dered rotation of group when the molecule is bound to a surface.

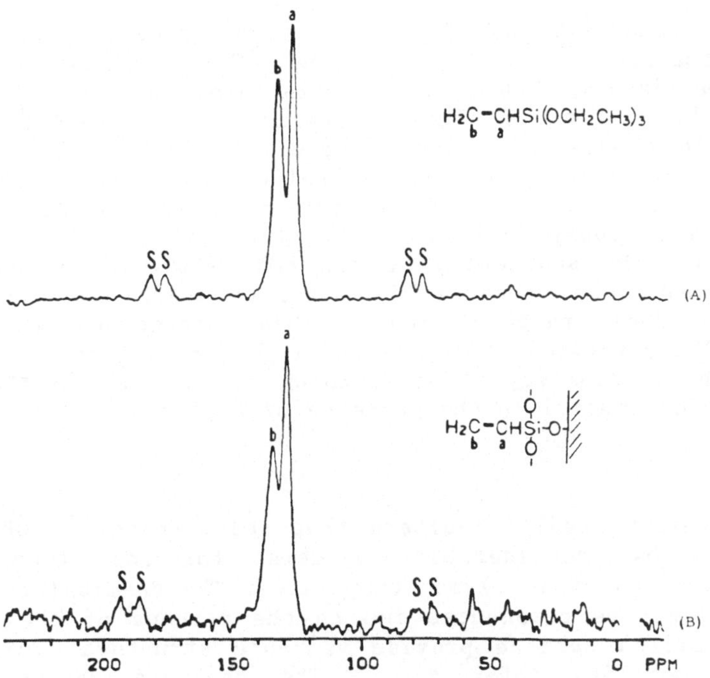

Fig. 5.   CP/MAS ^{13}C NMR spectra of vinyltriethoxysilane (VTES) (A)
polymerized, (B) on a silica surface.

E. Vinyltriethoxysilane

Solid-state ^{13}C NMR spectra were also obtained for condensed vinyltriethoxysilane and VTES on a silica surface as can be seen in Figure 5. Both spectra have two prominent resonances corresponding to the two vinyl carbons of the coupling agent. The organosilane is completely hydrolyzed as is indicated by a lack of resonances due to ethoxy carbons in both spectra. The methine carbon resonance appears at 132.3 ppm on the spectrum of the condensed coupling agent, and at 130.7 ppm on the spectrum of the coupling agent on the slica. The methylene carbon of the vinyl group has a resonance appearing at 138.2 ppm when the sample is condensed, and is shifted upfield to 136.0 ppm when the coupling agent is adsorbed. This upfield chemical shift appears to signify steric hindrance of the silanes which would occur as a result of chemical binding to the surface. The less intense resonances on either side of the vinyl carbon peaks on both spectra are spinning sidebands of the prominent resonances.

II. COMPOSITE SYSTEM

Graphite Filled Epoxy

Solid-state ^{13}C NMR $T_{1\rho}$ relaxations studies indicate the effect of the presence of fibers on the ^{13}C $T_{1\rho}$ relaxation times of an epoxy matrix (see Table 3). In principle, the ^{13}C $T_{1\rho}$ measurements should distinguish between regions of different motional character in composite polymeric materials [17]. This would be observed by multiple decays of the relaxation times or changes in the average value of the ^{13}C relaxation time. The data for the graphite-filled epoxy indicate a reduction in the ^{13}C $T_{1\rho}$ relaxation times. The shortening of the NMR relaxation times in the composite shows the presence of molecules which are highly constrained in their range of motion when located on or near the surface. The observed differences of $T_{1\rho}$'s for different nuclei may indicate the possibility of determining differences in the motion of the relevant nuclei in the presence of fillers.

SUMMARY

The present results indicate that solid-state ^{13}C CP/MAS NMR spectroscopy has considerable potential for the study of the different regions of a polymer composite. The chemical shifts and line widths of the resonances due to the carbons of the coupling agents on silica surface provide valuable structural information and local mobility information. The characterization of the composite and the effect of fillers on the relaxation parameters reveal more about the structure and changes in molecular motion in the interfacial regions.

Table 3. Carbon-13 $T_{1\rho}$'s of Graphite-Epoxy Composites

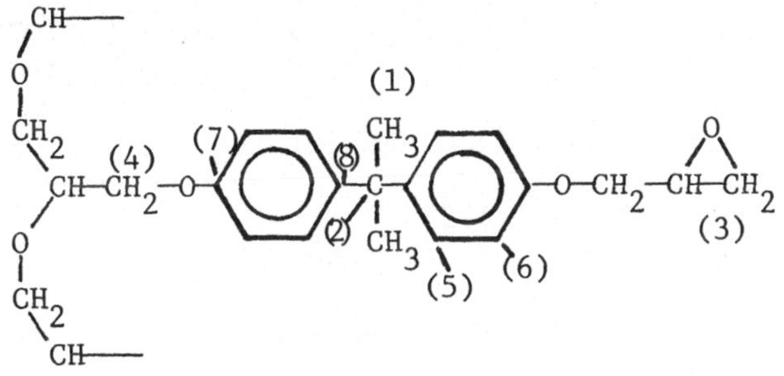

CARBON	EPOXY $T_{1\rho}$	COMPOSITE $T_{1\rho}$
Methyl (1)	18.1 msec	10.7 msec
Quaternary (2)	28.0	14.4
Methylene (4)	12.2	10.3
Protonated Aromatic		
(5)	22.4	9.7
(6)	7.8	8.1
Nonprotonated Aromatic		
(7)	29.8	22.1
(8)	24.6	22.6

ACKNOWLEDGEMENT

The authors wish to express their gratitude to the National Science Foundation for the support they provided in this research. Additionally, A. M. Zaper wishes to thank the Center for Applied Polymer Research (CAPRI) at Case Western Reserve University, as well as the support of the B. F. Goodrich Company; and A. Cholli expresses his gratitude to the Materials Research Laboratory of CWRU.

REFERENCES

1. H. Ishida and J. L. Koenig, Polym. Eng. Sci., 18, 128 (1978).
2. A. Pines, M. G. Gibby and J. S. Waugh, J. Chem. Phys., 59, 569 (1973).
3. E. R. Andrew, A. Bradburry and R. G. Eades, Nature, 182, 1659 (1959).
4. I. J. Lowe, Phys. Rev., 2, 285 (1959).
5. J. Schaefer and E. O. Stejskal, J. Am. Chem. Soc., 98, 1031 (1976).
6. G. E. Maciel, D. W. Sindorf and V. J. Bartuska, J. Chromotography, 205, 438 (1981).
7. G. R. Hays, A. D. H. Claque, R. Huis, G. Van Der Velden, Appl. Surf. Sci., 10, 247 (1982).
8. D. E. Leyden, D. S. Kendall, T. G. Waddell, Anal. Chim. Acta., 126, 207 (1981).
9. I. D. Gay, J. Phys. Chem., 78, 38 (1974).
10. R. G. Griffin, Anal. Chem., 49, 951A (1977).
11. C. Chiang, H. Ishida and J. L. Koenig, J. Colloid Interface Sci., 74, 396 (1980).
12. S. R. Culler, H. Ishida and J. L. Koenig, J. Colloid Interface Sci., accepted.
13. J. J. Chang, A. Pines, J. J. Fripiat and H. A. Resing, Surface Science, 47, 661 (1975).
14. K. Dusek, M. Ilavsky and S. Lunak, in "Cross-linking and Networks," K. Dusek, B. Sedlacek, C. H. Overberger, H. F. Mark and T. G. Fox, eds., John Wiley and Sons, New York (1975) p. 29.
15. G. Odian, "Principles of Polymerization," John Wiley and Sons, New York (1981) p. 512.
16. H. Ishida and J. L. Koenig, J. Colloid Interface Sci., 64, 565 (1978).
17. D. C. Douglass in "Polymer Characterization by ESR and NMR," A. E. Woodard and F. A. Bovey, eds., Amer. Chem. Soc., Washington (1980).

ACID-BASE INTERACTIONS AND SOME PROPERTIES OF COMPOSITES

H. P. Schreiber and Yongming-Li

Department of Chemical Engineering
Ecole Polytechnique, Montreal (Quebec)
H3C 3A7 Canada

INTRODUCTION

Understanding the dependence of processing and of mechanical properties in multi-component polymeric materials on the interaction forces among the system's components represents a means of tailoring the overall performance of such materials. The subject merits detailed study, the present paper contributing to that objective.

Inverse gas chromatography [1] (IGC) was the method used to establish the interaction character of polyethylene (PE), and polyvinyl chloride (PVC) polymers and of $CaCO_3$ and C black fillers. This characterization was based on the concept that acid-base interactions could play a significant role in the behavior of polymer systems [2,3]. The IGC route to interaction characteristics was modelled after earlier uses of gas chromatography to similar ends [4]. Specific retention volumes, V_g^o, by the solids for vapors with known acid-base tendencies [5] were used to define an Acid-Base Parameter (ABP),

$$ABP = (V_g^o)_a / (V_g^o)_b - 1 \qquad [1]$$

the respective vapors being n-propyl acetate (a) and n-butylamine (b). The characterized materials were then used as discussed below.

EXPERIMENTAL

Preparation of materials, mixing properties:

A Brabender Plasticorder was used to prepare PE, and PVC compounds with up to 40% (wt.) of the fillers, $CaCO_3$ (Winnifil "S"), and carbon black (Vulcan 3H). In blends using PE, the 1.2 melt index, low-density polymer was stabilized against thermal degradation by the addition of 0.1 wt.% Santonox-R^{TM}. PVC blends, based on a K-65 polymer, also involved 12.5 phr DIOP plasticizer and 3 phr lead tribase stabilizer. In a control study the fillers were used as received. Subsequently they were modified by surface treatments using a large volume microwave plasma (LMP) process [3,6]. The interaction characteristics of the fillers were altered by LMP treatments in n-butylamine and in n-butyl alcohol [3]; $CaCO_3$ was also treated in styrene and ethylene vapors to produce plasma-polymer coatings which would, in the case of ethylene treatment, resemble the structure of the PE host. Treatment in styrene vapor should produce a plasma-polymer with which PE may be expected to be very immiscible.

Blend preparation was done at an instrument T = 190°C and N = 50 rpm, the torque (τ) being allowed to attain a steady state. The mixing energy requirement, ε, was then computed from

$$\varepsilon = 2\pi N \int_o^t \tau.dt... \qquad [2]$$

Blend samples were microtomed and examined for fineness of dispersion with a Quantimet image analyser.

Mechanical Properties

The present report is based on the stress-strain properties of blends at high deformation. Those proved to be quite sensitive to variations in the interaction between polymer-filler pairs. Particular attention was placed on examining stress-strain properties at various draw rates, and an Instron tester was therefore operated at rates from 0.5-20 cm.min^{-1}. The retention of mechanical properties following controlled aging was also felt to be linked, potentially, with interaction behavior; accordingly, elongation and stress at failure were measured following the accelerated aging of blends under N_2, at 100°C, for up to 7 days.

Results

ABP parameters for the materials involved are given in Table I. PE can be considered inert, while PVC behaves as an acidic substrate. The $CaCO_3$ as received is mildly basic, the carbon acidic;

Table I. ABP Parameter Summary: All values at 30°C.

Material	ABP	Material	ABP
PE	0.05		
PS	0.11	$(CaCO_3)_a$	-0.38
PVC	-0.52	$(CaCO_3)_b$	1.02
$CaCO_3$	0.27	$(CaCO_3)_s$	0.08
C	-0.77	$(CaCO_3)_e$	-0.03
		$(C)_a$	-0.52
		$(C)_b$	0.64

Table II. Dispersion, Mixing Characteristics of Blends at 20%
Filler Content.

Parameter: System:	ε (arb. units)	$\overline{d}^{(1)}$ (μm)	$\overline{d}^{(1)}_{max}$ (μm)
PE			
+ $CaCO_3$	16.2	2.4	6.0
$(CaCO_3)_a$	18.3	2.8	5.7
$(CaCO_3)_b$	19.7	2.7	6.0
$(CaCO_3)_s$	19.8	3.1	7.5
$(CaCO_3)_e$	9.7	1.5	4.0
+ C	11.5	1.8	3.5
C_a	9.6	2.5	5.0
C_b	12.8	2.8	5.3
PVC			
+ $CaCO_3$	17.0	2.8	4.5
$(CaCO_3)_a$	26.5	5.5	6.0
$(CaCO_3)_b$	10.4	1.9	3.0
$(CaCO_3)_s$	19.1	3.3	4.2
$(CaCO_3)_e$	18.7	3.5	4.5
+ C	30.6	4.5	9.5
C_a	27.8	3.3	7.0
C_b	16.1	2.6	4.0

(1) $\overline{d}$ is arithmetic mean particle diameter.
d_{max} is diameter of particles in upper 10% of distribution.

LMP treatments exert appreciable influence, the respective treatments placing the solids into acid, base or near-inert categories. The effects of plasma-treatments, which deposit plasma polymers onto the filler substrates, are here used for illustration purposes. Specially designed LMP reactors will be needed to optimize the surface treatment of particulates such as those used.

The relevance of interaction variables to the mixing and dispersion behavior of filled blends is illustrated in Table II for PE and PVC at 20% filler load levels. The inert PE matrix is most readily blended with the $CaCO_3$ as surface modified by ethylene plasma treatment. The relatively mildly basic CaCO as received performs better than fillers modified to display pronounced proton donor or acceptor properties. The statement applies both to energy requirement for blending and to particle size attained. In the case of PVC, there is strong dependence of dispersion parameters on ABP; for $CaCO_3$, the strongest interaction $PVC-(CaCO_3)_b$ produces excellent mixing results, the others deteriorating in simple relation with the difference between polymer and filler ABP values. The inherently acidic carbon black does not readily disperse in the PVC matrix. Plasma treatment improves on the performance, when the filler surface is modified into the basic range. In this polymer matrix, carbon C_a also disperses more readily than the unmodified solid, presumably because its acidic properties have in fact been softened by LMP treatment in butanol vapor. Similar results to those in Table II were obtained for 10% and 40% filler loading. It is apparent that surface modification of fillers by plasma polymerization may be a fruitful approach to the improvement of mixing process economics, particularly when plasma reactors well suited for the treatment of particulates are in operation.

Earlier indications [3] that acid-base concepts are involved in developing the mechanical properties of filled polymers are amplified by present data. To normalize the comparison of interaction effects on the stress-strain behavior of filled compounds, a parameter, W, representing the work required to break a sample has been defined by

$$W = \int_{o}^{e_{max}} \tau \, de \qquad\qquad [3]$$

Here τ is the stress and e the strain. The sample-to-sample variation of W is shown in Figure 1 for PE and PVC compounds with 20% $CaCO_3$ filler, tested at a draw rate of 5 $cm.min^{-1}$. Very little significant change in the total strength of PE compounds occurs when the filler is surface treated in ethylene plasmas. We suggest that this is due to very strong coupling between host polymer and a plasma-polymer layer on the filler surface with which the host is highly miscible. The degree of property loss is most severe when acidic or basic properties of the filler are accentuated. This stands to reason when the host polymer itself is incapable of participation in specific interaction of this kind. LMP-modified filler in styrene vapor also fails to respond positively, arguably because of the well-known immiscibility of PE and PS - an immiscibility which is of steric origin.

The PVC case, illustrated in Figure 1, clearly displays the effect of acid-base interactions on mechanical properties. Increasing the difference between ABP values for polymer and filler, leads to increases in W; thus optimum performance is obtained for the amine-modified $(CaCO_3)_b$.

Figures 2 and 3 show the variation of W with draw-rate for PE and PVC compounds, respectively. Results follow patterns interpretable along lines entirely analogous to those stated above. Least sensitive PE compounds are those with $(CaCO_3)_e$ and the unmodified filler. Treatments rendering the filler strongly acid or basic, or sterically "incompatible" with the polymer (e.g. $(CaCO_3)_s$), result in major decreases of W with draw rates. The strength of polymer-filler linkages appears to be more important to the mechanical properties of compounds at higher rates of deformation. This statement applies to PVC compounds (Figure 3); in this series, the compounds containing carbon black are particularly noteworthy. The amine-modified carbon black is a vastly superior filler to the other versions used in this comparison. Interestingly, C_a is in fact less acidic than the carbon as received; its performance is in keeping with that alteration in the strength of specific interactions, as would be given by $\triangle$ABP.

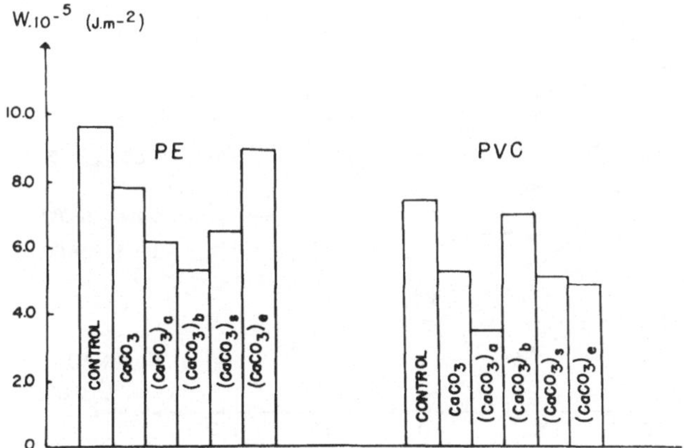

Fig. 1. Energy to Rupture PE and PVC Compounds: 20% $CaCO_3$, 5 cm. min^{-1} draw rate.

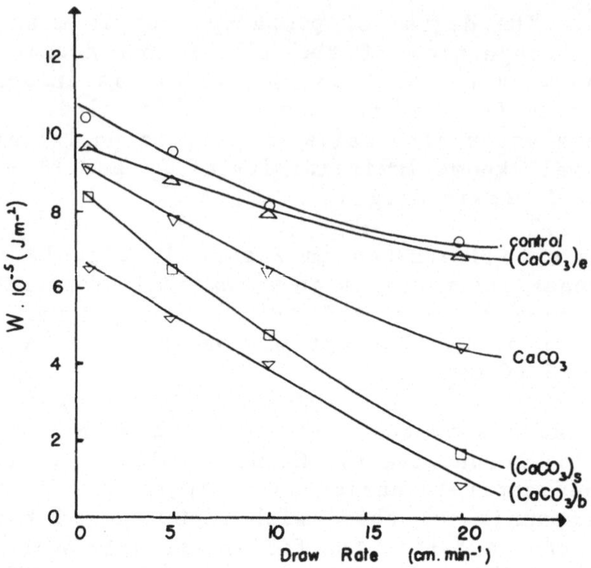

Fig. 2.   Response of PE Compounds to Extention Rate.

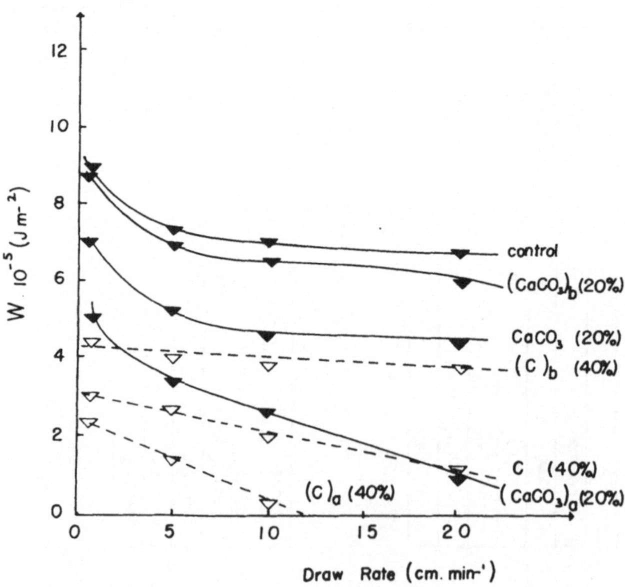

Fig. 3.   Response of PVC Compounds to Extention Rate.

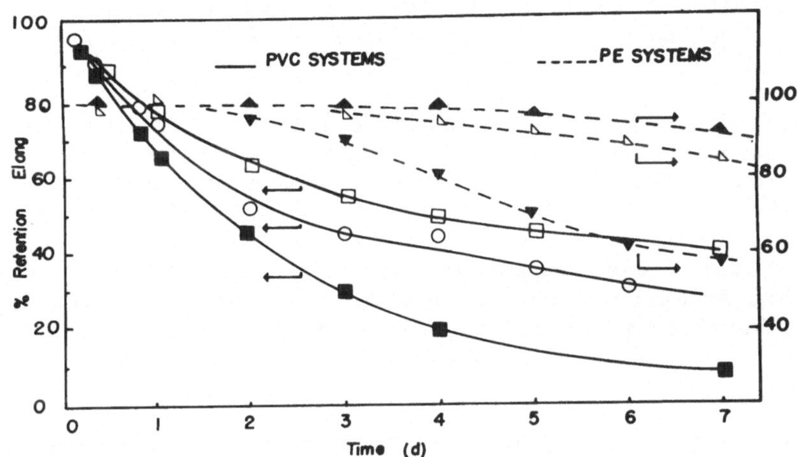

Fig. 4.   Retention of Elongation at Break of Filled PVC and PE:
40% loading, aging at 100°C.   PVC: □ $(CaCO_3)_B$ ○ $(CaCO_3)$
■ $(CaCO_3)_A$.   PE: ▲ $(CaCO_3)_E$ △ $(CaCO_3)$ ▼ $(CaCO_3)_B$

Finally, the relationship between property retention and ABP values
is illustrated for PE and PVC compounds in Figure 4. That relationship for PE systems is complex, but again, performance is optimized
when matrix and filler surfaces are chemically similar. Accentu-
ating the acid or base properties of the filler does not produce
benefits in mechanicals at high load. The PVC system again clearly
displays the strong response to acid-base interactions. Minimum
property loss upon exposure to T > Tg is emphatically seen in
compounds with large Δ ABP for polymer and filler, presumably
because of (thermodynamically) stable couplings at phase contacts.
The mechanical properties of C-filled polymers were less affected
by surface treatment. This may be due to the much smaller particle
size in the carbon, which would de-emphasize the importance of interfacial strength in the failure mechanism [7].

CONCLUSIONS

An acid-base balance parameter, ABP, computed from gas chro-
matographic experiments, can be used to characterize the acid-base
balance of polymers and fillers.

Dispersion effectiveness of fillers in (particularly) a non-
inert polymer matrix varies with the magnitude of polymer-filler
interactions.

Mechanical properties and their durability are functions of ABP, most notably when the polymer has pronounced acid or base character, and when filler particle size is significant (> 1-2 m).

Plasma-induced surface modifications of fillers lead to marked control of dispersion and mechanical property behavior; such modification methods may be viewed as new approaches to coupling agent technology.

This work was supported by the Natural Sciences and Engineering Research Council, Canada. We thank Mrs. Y. Sapieha and M. R. Wertheimer for aid in plasma experiments.

REFERENCES

1. O. Olabisi, L. M. Robeson and M. T. Shaw, Polymer Miscibility, Academic Press, New York (1979), Ch. 3.
2. F. W. Fowkes and S. Maruchi, Prepr. Am. Chem. Soc. Polym. Div., 37, 606 (1977).
3. H. P. Schreiber, M. R. Wertheimer and M. Lambla, J. Appl. Polym. Sci., 27, 2269 (1982).
4. H. G. Harris and J. M. Prausnitz, J. Chromatogr. Sci., 7, 685 (1969).
5. P. M. Cuckor and J. M. Prausnitz, J. Phys. Chem. 76, 598 (1972).
6. H. P. Schreiber, Y. Tewari and M. R. Wertheimer, J. Appl. Polym. Sci., 20, 2663 (1976).
7. A. N. Gent, G. S. Fielding-Russell, D. L. Livingston and D. W. Nicholson, J. Mater. Sci., 16, 949 (1981).

CHARACTERIZATION OF THE SURFACE AND THE INTERFACE OF THE CARBON FIBER

A. Ishitani

Physical Chemistry Laboratory
Toray Research Center, Inc.
1-1, 1-Chome, Sonoyama, Otsu
Shiga, Japan

INTRODUCTION

The surface of the carbon fiber and also the interface with the matrix resin plays an important role in the performance of the carbon fiber reinforced plastic. Extensive studies on the problem have been carried out in TORAY group utilizing mainly X-ray Photoelectron Spectroscopy (XPS). This article describes collections of the CF work carried out in our laboratory concerning the surface and the interface.

Difficulties encountered in the study come from very special features of the carbon fiber as an industrially used material. First, the chemical processes involved in the carbonization of the precursor (polyacrylonitrile, PAN) are not fully understood, although it is known that the residual elements and the functional groups, and their distribution are important factors. Second, the morphology as thin fibers of several micron diameter is unfavorable for the surface analysis as well as for depth profiling. Besides, there are good reasons to suspect that a fiber has heterogeneous structure in the direction of its radius. Third, the carbon fiber has intense light absorption covering from the ultraviolet to the far infrared regions due to the electronic transition. This makes the utilization of any kind of photon probes extremely difficult.

XPS has proven to be effective in circumventing these difficulties and in obtaining useful information as indicated below. However, use of FT-IR, Raman microprobe and solid-state high-resolution NMR are necessary to obtain more detailed structural infor-

mation, especially when the interaction between the CF surface and
the matrix polymer is to be examined. The effects of low concentration elements should be studied by SIMS (Secondary Ion Mass
Spectrometry).

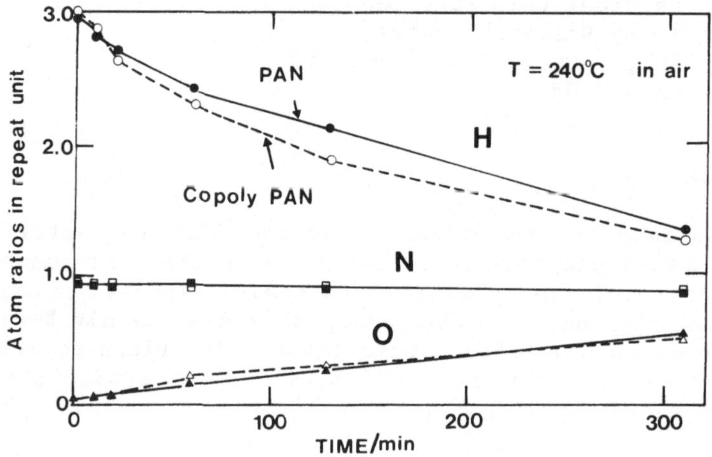

Fig. 1. The changes of elemental composition of PAN fibers in the
thermal stabilization process.

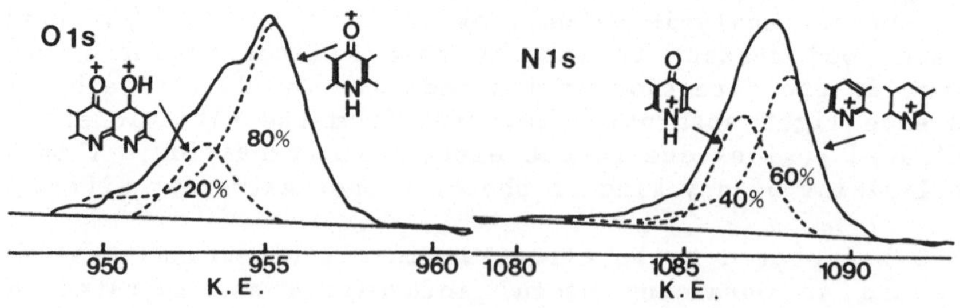

Fig. 2. ESCA O1s and N1s spectra of thermally stabilized PAN.

Chemical Structure Variation in the Carbonization Process

The detailed study of the chemical change in the whole carbonization process, from the precursor polymer to the final carbon fiber, is desirable to fully understand the surface of CF, because the residual elements and the functional groups in the surface area depend much on the process, although the major structure of the final product is graphite-like.

The thermal stabilization process of PAN homopolymer and a copolymer of PAN with 0.5 mol% hydroxyethylacrylonitrile at 240°C for up to 300 min is studied by the combined techniques [1] of elemental analysis, XPS [2] and FT-IR [3].

The change of the elemental composition of the homopolymer and the copolymer fibers in the stabilization process are compared in Figure 1. The dehydrogenation accompanies introduction of unsaturated bonds and oxygen atoms into the polymer chains. The hydrogen concentrations of the stabilized fibers are about 50% of that of the precursors. The reaction rate of dehydrogenation of the copolymer is larger than the homopolymer. Oxygen atom content increases to 0.5 atom per repeat unit of PAN at the final stage of the stabilization process. The concentration of nitrogen shows little change in the process for both precursors.

XPS spectra of O1s and N1s in the stabilized PAN are curve resolved and assigned in Figure 2. The O1s spectrum consists of two components, the higher kinetic energy component with 80% intensity is assigned to acridone type carbonyl and the other one with 20% intensity is assigned to normal carbonyl and alcohol groups. The N1s spectrum has two components which are assigned to nitrogen in acridone ring (60%) and the one in naphthyridine and hydronaphthyridine rings (40%).

The ladder structure shown in Figure 3 can be estimated as the product of the stabilization process of PAN and the copolymer from the above observation.

The IR spectral changes of PAN in the thermal stabilization process are shown in Figure 4. Major variation observed is decreased intensity of the nitrile band (2240 cm^{-1}) and appearance of three new bands (1725, 1660 and 1595 cm^{-1}). These three bands could be assigned as 1725 cm^{-1}; carbonyl group conjugated with double bonds, 1660 cm^{-1}; carbonyl group in acridone, 1595 cm^{-1}; C=C group in aromatic rings. There is no significant difference between the thermally stabilized fibers produced from PAN and the copolymer.

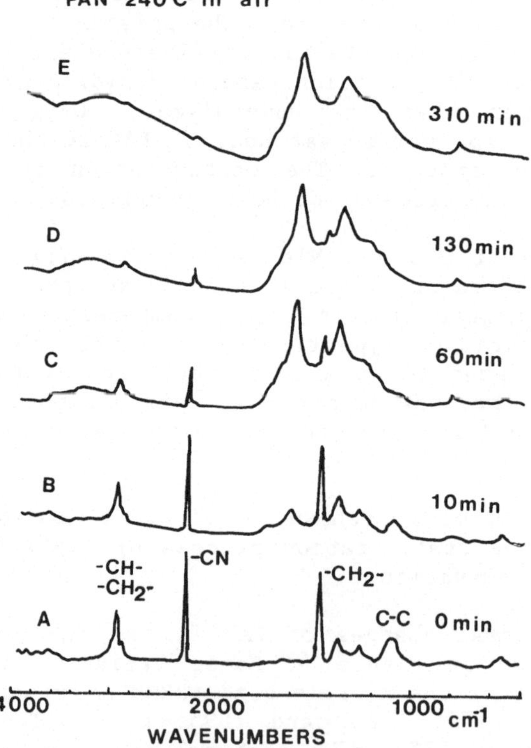

Fig. 3.   The chemical structure proposed for thermally stabilized
          PAN.

Fig. 4.   FT-IR spectrum change of PAN in the thermal stabilization
          process.

The kinetic study of the reaction was carried out using the band
intensities of nitrile, methylene and two kinds of carbonyl
groups. Examples for the carbonyl groups are given in Figure 5.
There is apparent difference in rate of generation of the carbonyl
group in acridone between PAN and the copolymer, while there is
none for the other carbonyl group. The reaction rate for acridone
generation was estimated to be twice as much as the plain ketone
assuming the first order reaction for thermal stabilization. This
indicates that copolymerization accelerates mainly the dehydrogenation reaction.

In summary, the majority of oxygen atoms taken into the thermally stabilized PAN are found to be in carbonyl group of acridone. The copolymerization is confirmed to accelerate the dehydrogenation reaction and also formation of acridone in the staboliza-
tion process.

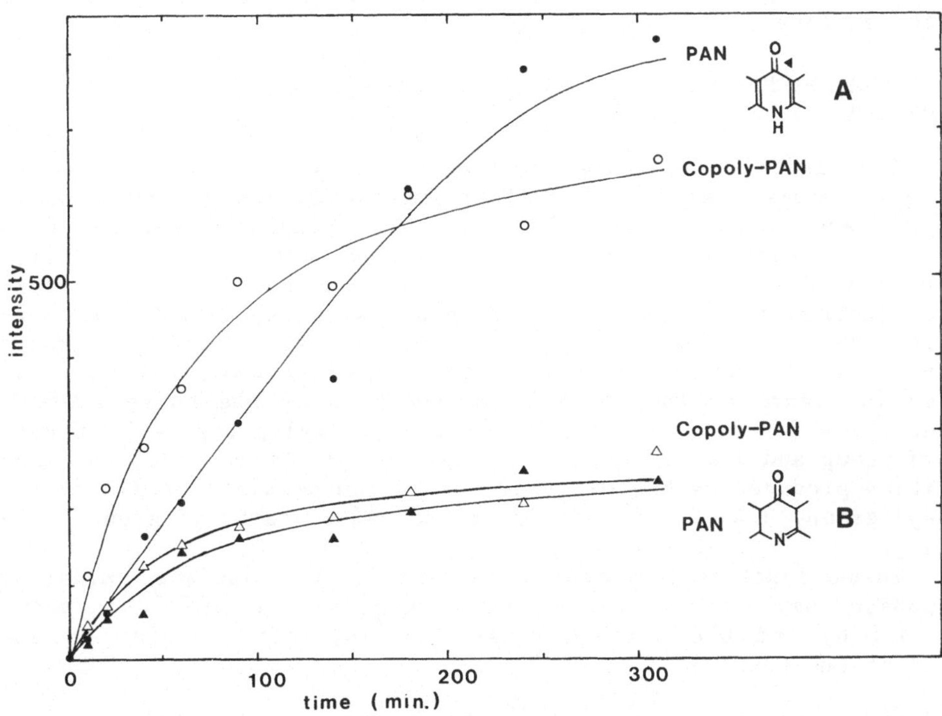

Fig. 5. Variation of band intensity due to two kinds of carbonyl
         groups in the thermal stabilization process for PAN and
         PAN copolymer.

Surface Composition and In Depth Profiles of the Carbon Fibers

XPS has started to be used to characterize surface composition of carbon fibers frequently [4,5,6,7]. Argon ion etching [4] can be used to reveal heterogeneous structure of the carbon fibers in the depth direction. Figures 6 and 7 show the results on the carbon fiber (CF) and the graphite fiber (GF). Here, CF is obtained by heat treatment up to about 1500°C, whereas GF is prepared by a condition over 2000°C. Both of CF and GF have substantial amount of oxygen which is analyzed to originate from alcoholic and also carboxyl groups without any surface oxidation. They are considered to be introduced onto surface by quenching of the free radicals which are produced in the inert gas of the furnace at high temperature, when the fibers are brought into air. The oxidized layers thus created in both of CF and GF are thin and removed quickly with the 0.5 h etching. The N1s peak due to the residual nitrogen of CF does not show much intensity variation, although the center of the peak shifts about 2eV to lower binding energy values by the etching, indicating the presence of more oxidized nitrogen on the surface.

Functional Groups Generated by the Surface Oxidation

The digital difference spectrum technique on XPS is used for the qualitative analysis of surface functional groups generated by the surface oxidation [8,9,10,11]. The difference spectra were obtained by subtraction of C1s spectrum of control CF and GF from that of oxidized CF and GF respectively. The obtained C1s difference spectrum of CF has three components with chemical shifts corresponding to hydroxyl groups ($-C-OH$; 286 eV), carbonyl groups ($\supset C=O$; 287 eV) and also carboxyl groups ($-C\underset{OH}{\overset{O}{\lessgtr}}$; 288.6 eV) as indicated in Figure 8. The chemical composition of the surface of CF after the moderate oxidation is 73% of hydroxyl group, 17% of carboxyl group and 10% of carbonyl group. On the other hand, the composition produced by the extensive oxidation consists of 24% of hydroxyl group, 22% of carbonyl group and 54% of carboxyl group.

These functional groups generated by the surface oxidation disappear completely with the thermal treatment at 1000°C in vacuo for 0.5 h, and the surface oxygen concentration returns to the level of the control.

The difference spectrum of GF in Figure 8 has two components. The higher binding energy component at 288.7 eV which disappears after thermal treatment at 1000°C is assigned to carboxyl group ($-C\underset{OH}{\overset{O}{\lessgtr}}$), and the binding energy peak which is left after the thermal treatment is attributed to disordering of graphite crystal lattice brought about by the surface oxidation.

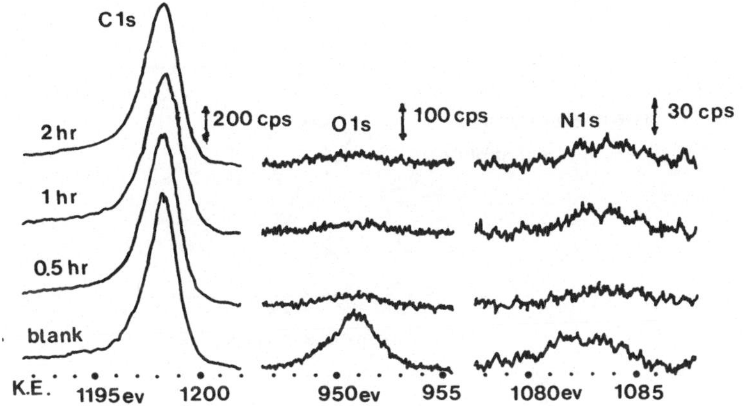

Fig. 6.   The etching effect on CF.

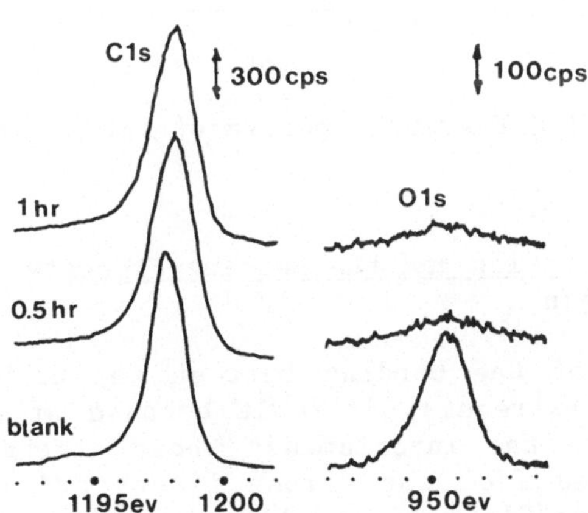

Fig. 7.   The etching effect on GF.

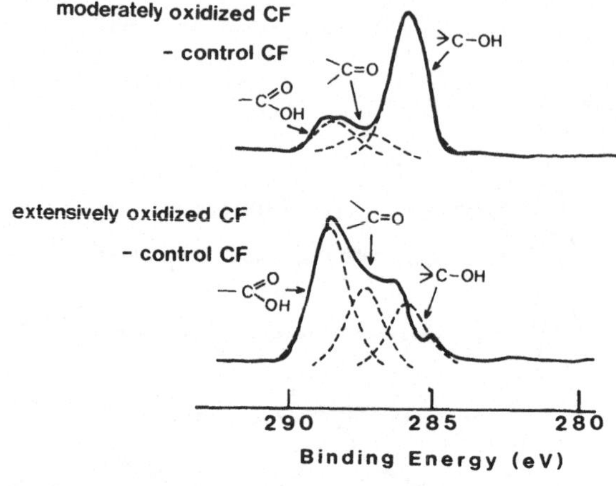

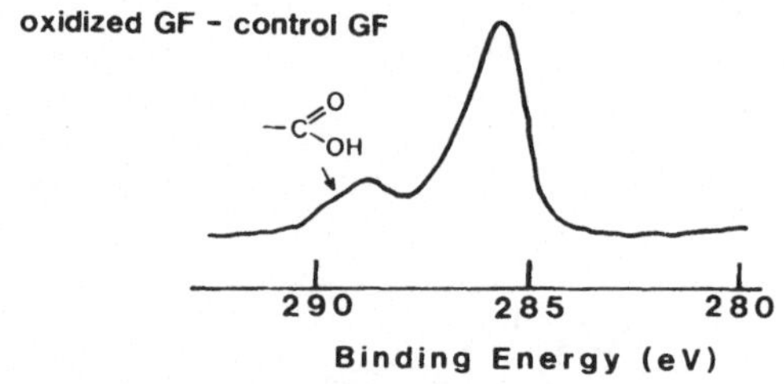

Fig. 8.   Digital difference spectra of the surface oxidized CF
          and GF.

The Surface Composition and the Bonding Property
to the Matrix Resin

Estimation of the bonding between the carbon fiber and the
matrix resin is extremely difficult because of the morphology of
the fibers. Here the interlaminar shear strength (ILSS) of the
composite prepared from the carbon fibers and an epoxy resin is
related to the O1s/C1s ratio of XPS which is well established [4]
to represent the amount of the functional groups on the carbon
fiber surface.

The surface oxidation increases the O1s/C1s ratio and ILSS
simultaneously both for CF and GF as shown in Figure 9. A good
linear relation is observed for CF. The thermal treatment in vacuo
decreases the O1s/C1s ratio and ILSS reversibly in CF. Therefore,

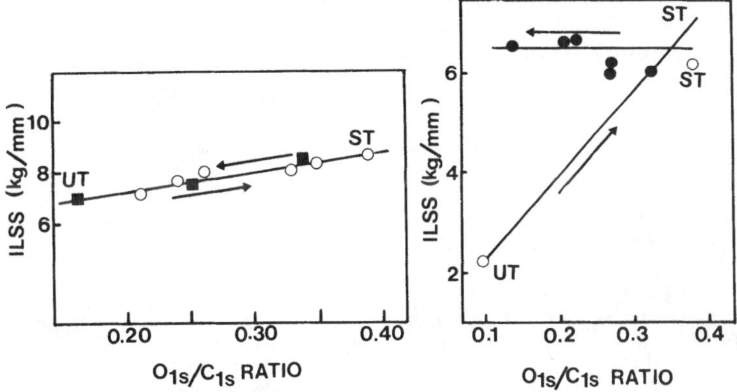

Fig. 9. Relation between ILSS and O1s/C1s ratio: UT: Control,
ST: Surface oxidized fibers, ●■: Fibers heated in vacuo.

introduction of the functional groups on the surface is a main factor for the bonding property. However, ILSS does not decrease in
GF with the thermal treatment although the O1s/C1s ratio
decreases. It indicates that physical perturbation such as roughness or irregularity on the lattice structure brought on the surface by the treatment determines the bonding property. This observation corresponds well with the behavior of the C1s line shapes
described in the previous section.

New Techniques for Surface and Interface
Characterization of Carbon Fiber

FT-IR: The high sensitivity and the extensive data processing
capability of FT-IR contributes much to CF work. A FT-IR reflection-absorption spectrum of the surface oxidized CF is shown in
Figure 10. A band at 1680 cm^{-1} is identified and assigned to the
carbonyl stretching vibration of the aromatic carboxyl group. A
FT-IR-ATR digital difference spectrum also reveals thermal oxidation of an epoxy polymer thin film on an extensively surface oxidized CF.

Raman: Raman spectroscopy is useful for the surface characterization of CF because of the small penetration depth (50nm) of
the exciting light. Further improvement of the surface sensitivity
is being tried by evaporating Ag on the surface in order to utilize
the Surface Enhanced Raman Scattering (SERS) effect. An example is
indicated in Figure 11. The remarkable enhancement is observed for
CF, whereas no significant effect is seen for GF probably due to
poor contact with the Ag thin film. Microprobe Raman is also one
of the promising techniques. The focused beam of 1μm diameter can
analyse and identify small inclusions and defects, and also reveal
heterogeneity of the lattice structure.

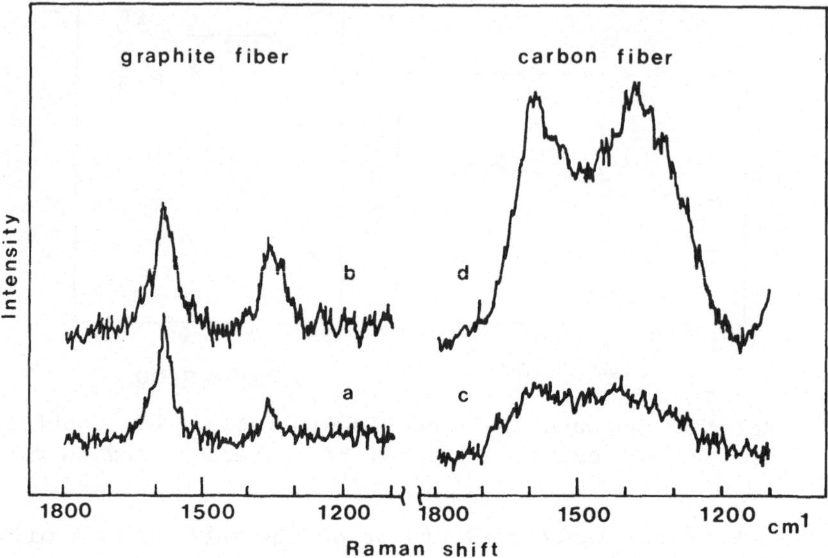

Fig. 10.   A FT-IR-RAS spectrum of a surface oxidized CF.

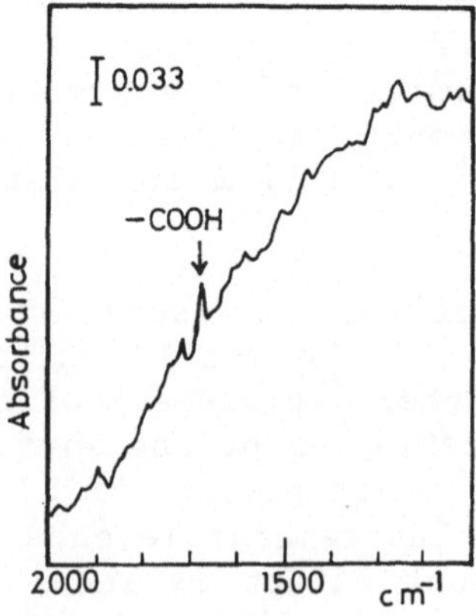

Fig. 11.   The SERS effect on the carbon fibers.

<u>High Resolution Solid NMR</u>: The Cross Polarization/Magic Angle Spinning (CP/MAS) and the other modes of the techniques are expected to observe the surface and the interface area selectively. The discussion on the interaction between the CF surface and the matrix resin will be possible because of the detailed structure information from NMR.

<u>Secondary Ion Mass Spectrometry (SIMS)</u>: Detection and imaging of the low concentration elements on the CF surface as well as their depth profiles are available from this technique.

REFERENCES

1.   T. Takahagi, I. Shimada, M. Fukuhara, K. Morita and A. Ishitani, Proceedings of International Symposium on Carbon, Toyohashi, Japan, 321 (1982).
2.   T. Takahagi, I. Shimada, M. Fukuhara, K. Morita and A. Ishitani, to be published in J. Polym. Sci., (1984).
3.   I. Shimada, T. Takahagi, M. Fukuhara, K. Morita and A. Ishitani, to be published in J. Polym. Sci., (1984).
4.   A. Ishitani, Carbon 19, 269 (1981).
5.   F. Hophgarten, Fibre Science and Technology 11, 67 (1978).
6.   F. Hophgarten, Fibre Science and Technology 12, 283 (1979).
7.   K. Waltersson, Fibre Science and Technology 17, 289 (1982).
8.   T. Takahagi and A. Ishitani, to be published in Carbon.
9.   A. Proctor and P. M. A. Sherwood, Anal. Chem. 54, 13 (1982).
10. A. Proctor and P. M. A. Sherwood, Surface and Interface Anal. 4, 212 (1982).
11. A. Proctor and P. M. A. Sherwood, Carbon 21, 53 (1983).

APPENDIX

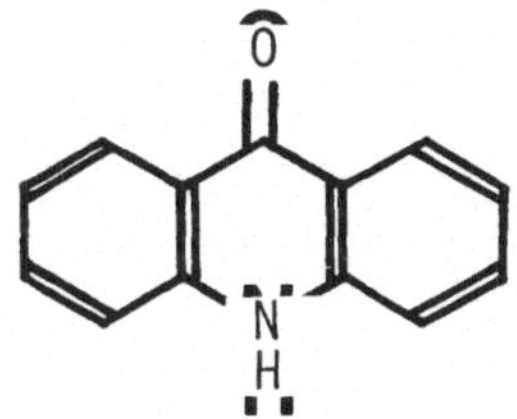

acridone

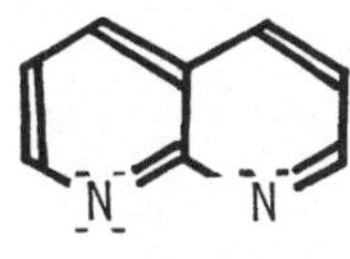

naphthyridine

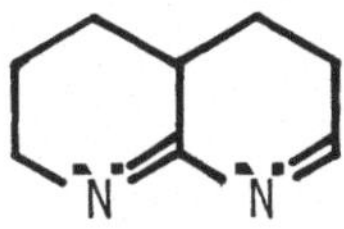

hydro-
naphthyridine

SURFACE CHEMISTRY AND BONDING OF PLASMA-AMINATED POLYARAMID

FILAMENTS

Ronald E. Allred*, Edward W. Merrill and
David K. Roylance

Massachusetts Institute of Technology
Cambridge, Massachusetts 02139

*Division 1812
Sandia National Laboratories
Albuquerque, New Mexico 87185

ABSTRACT

Thermomechanical performance of polyaramid-reinforced, resin-matrix composites often is limited by poor adhesion in the filament-matrix interphase region. This study describes a method to improve adhesion by forming covalent bonds across the interface through amine functional groups. Amine functionality has been introduced onto poly(p-phenylene terephthalamide), PPTA, filaments by exposure to ammonia or monomethyl amine RF glow discharge plasmas. Surface amine concentration rises rapidly upon plasma exposure and reaches a steady state in 30 to 60 sec. Weibull parameters for the filament strength distribution are unchanged by the plasma amination reaction. The amine groups are stable in air and water. They may be reacted directly with epoxide resins, or modified to functionalities that can react with other polymer matrix materials.

PPTA/epoxy laminates reinforced with aminated fabric have higher interlaminar tensile and peel strengths than laminates reinforced with untreated fabric. The failure mode changes from interphase dominated to a mixture of filament splitting and matrix cracking as surface amine concentration increases. Moisture absorption of untreated PPTA fabric/epoxy laminates occurs by a non-Fickian interfacial wicking mechanism. After amination, the absorption rate is reduced by a factor of three and occurs by a Fickian bulk diffusion mechanism. These results indicate that the

mechanical properties and environmental resistance of polyaramid-reinforced composites may be improved by covalent bonding at
the filament-matrix interface.

INTRODUCTION

The thermomechanical performance of aramid-reinforced composites often is limited by their low off-axis strength. Prior work
has shown that this is caused by poor adhesion at the filament-matrix interphase. Surface oxidation to improve wetting by
the liquid matrix resin improves bond strength by approximately 30
percent, but failures still are interphase controlled. Coupling
agents which do not chemically bond to both phases have little
effect on strength properties.

This study describes a method to create covalent bonds
between polyaramid filaments and an epoxy resin network to improve
interfacial adhesion. Reactive amine groups are incorporated into
the filament surface structure by exposure to ammonia or monomethyl
amine plasmas. The amine groups are stable, and plasma treatments
less than 600 sec long do not affect filament tensile strength.

Subsequent reactions of the surface amine groups with
epoxides form stable covalent bonds at the interface. Increased
interfacial adhesion is evident from changes in the composite
mechanical and moisture absorption behavior. After an introductory
discussion of polyaramid adhesion and ammonia plasma reactions,
these amination and characterization studies are discussed in
detail. Amination results with monomethyl amine plasmas are
similar to those with ammonia and will not be discussed.
Monomethyl amine results may be found in Ref. 1.

BACKGROUND

<u>Composite Interphase</u>

In this discussion, the filament-matrix interface is distinguished from the interphase region. The interface defines a sharp
boundary in chemistry and morphology between the filament and
matrix structures. The interphase region extends a finite depth
from the interface into the matrix. In the interphase, a gradual
transition in matrix morphology occurs from the interface structure
to that of the bulk structure between filaments.

The importance of interphase quality in determining structural integrity is generally recognized [2-6]. Virtually all composite properties are affected by interphase strength. Off-axis
(transverse tension, interlaminar shear) and compressive properties
are particularly sensitive to interphase strength. In addition,

interphase failure often determines the failure mode and resultant strength of the composite. As such, adequate filament-matrix adhesion is a necessary ingredient in useful composite systems. A quantitative definition of adequate adhesion depends upon the composite system, ply orientation, and particular application; however, for most applications, it is desirable to maximize interfacial adhesion. Major exceptions to this precept are pressure vessel and ballistic applications, where it is often desirable to minimize interfacial adhesion.

Interphase strength is determined by a large number of physical and chemical factors [7-10]. Physical factors include filament surface area and roughness, filament volume fraction, residual fabrication stresses (determined by filament and matrix thermal expansions and stiffnesses, cure shrinkage, etc.), and microvoid concentration. Chemical factors affecting interphase bond strength include wettability, chemical reactivity (primary, secondary or H-bonding), matrix morphology, filament and matrix cohesive energy densities, and environmental resistance. The interdependence of these factors, when combined with design constraints, such as operational environment, off-axis forces and required fracture toughness, dictates that a specific approach be taken to improve interfacial bonding in a given system.

Extensive research has been directed towards developing filament surface treatments to improve composite interfacial properties. The primary goal of those efforts has been to improve filament wettability with chemical bond formation as a secondary objective. Filament wetting by the matrix resin is the essential first step in forming an interfacial bond. Improper wetting reduces the contact area between the filament and matrix which causes microvoids at the interface that serve as stress concentration sites [11]. Good wetting is also required to attain the molecular level contact necessary to form chemical bonds. This approach is illustrated by the development of coupling agents for glass filaments [12] and oxidative surface treatments for graphite filaments [13]. Those modifications increase interfacial strengths by 2-3 times over untreated composites and also provide increased hygrothermal stability [12,13].

Similar improved bonding methods have not been developed for polyaramid-reinforced composite systems. Consequently, interface-sensitive properties are weaker in polyaramid containing systems than their glass or graphite counterparts. Representative data for these systems are compared in Table I. Transverse tensile strengths of polyaramid/epoxy are only 40 to 50 percent as strong as those of glass/epoxy or graphite/epoxy, and are characterized by bare, visually unaltered filaments with little epoxy adhering to them [14, 18, 23, 24]. This observation implies that the interphase

Table I. Off-Axis Strengths of Filament-Reinforced Epoxy Composites

Filament	Transverse Tension MPa	Interlaminar Shear MPa	References
Polyaramid	7 - 18	21 - 40	14 - 25
Glass	17 - 41	55 - 97	7,18,21
Graphite	26 - 38	85 - 110	7,13,18,26

rather than the cohesive strength of the filaments or matrix limits the off-axis strength of polyaramid/epoxy composites.

Compressive properties of polyaramid/epoxy laminates are also low compared to glass and graphite/epoxy [27,28]. The weak interphase may be a contributing factor to the poor composite compressive strength in addition to the filament morphology [28]. The polyaramid/epoxy interphase may also be more sensitive to moisture than the bulk filament and resin. Moisture diffuses more rapidly in the filament axial direction [29] and flexural strength is a strong function of composite moisture content [30,31], as is transverse tensile strength [24,25]. These results indicate that the interphase bonding is weakened by the presence of moisture. These mechanical response and environmental sensitivity observations clearly indicate that the performance of polyaramid-reinforced composites is limited by weak interfacial adhesion.

Polyaramid Adhesion Studies

The low off-axis properties of polyaramid-reinforced composites have stimulated numerous investigations to improve filament-matrix adhesion. Three types of approaches have been taken in these studies: (1) filament surface oxidation or etching, (2) matrix chemical modification, and (3) development of filament-matrix coupling agents.

Surface oxidation to improve transverse strength has been examined with thermal [19] and oxidizing/etching plasma treatments [18,32,33]. Such treatments increase off-axis composite strengths by about 30 percent. The gain in off-axis strengths is somewhat offset by losses in filament tensile strength. Increased off-axis strengths appear to be due to improved wetting and perhaps some covalent bond formation across the interface through residual surface free radicals on the treated filaments [32].

Matrix chemical modification to improve interfacial bonding has not been thoroughly examined for polyaramid/epoxy systems. Chiao and co-workers [15,20] observed that interlaminar shear and transverse tensile strengths are highly dependent upon the matrix resin system; however, no attempt was made to relate adhesion to matrix chemistry. Allred, Street and Martinez [18] demonstrated that increasing matrix toughness by the addition of a dispersed rubber second phase caused a corresponding 40 percent increase in the measured transverse tensile strengths of Kevlar 49/epoxy. The fracture zone with a toughened matrix was still characterized by bare, unbonded PPTA filaments. This result indicates that failure was interphase controlled and that increases in transverse strength were due to increasing the critical flaw size. It may also indicate the presence of microvoids at the interface as a consequence of poor wetting.

Surface chemical modifications of polyaramid filaments have been studied by numerous investigators. These approaches have aimed at improved wetting or providing filament—matrix coupling through the application of reactive sizes [19,34], plasma-polymerized coatings [32], or the creation of reactive surface sites [35]. While some of these methods have improved adhesion with other polymers, none have improved bonding with epoxy resins. Penn, Bystry and Marchionni [34] conclude that the polyaramid/epoxy bond is as good as it can be using only intermolecular forces, and that only chemical, covalent bonds across the interface will strengthen the Kevlar/epoxy bond.

The previous adhesion studies on polyaramid filaments lead to the conclusion that gains made by improving wettability or by bonding through residual free radicals are limited. Additional improvement in interfacial strength, thus, requires that the substituent phases be chemically bonded. It is further required that adhesion improvements should not be accompanied by a substantial loss in filament strength. It is expected that covalent bonding at the interface will also provide increased hygrothermal stability.

Covalent bonding may be attained by incorporating chemical groups into the filament surface that are reactive with the resin matrix. Primary amine, NH_2, groups are highly reactive with epoxides and may be introduced into polyaramid surfaces by exposing the filaments to amine plasmas.

Plasma Amination Concept

The concept of surface amination of polymers with amine plasmas was first presented by Hollahan, Stafford, Falb and Payne [36]. They demonstrated that amine surface functionality could be introduced into a wide variety of aliphatic polymers by exposure to ammonia, RF glow discharge plasmas. Heparin readily bound to the treated polymer surfaces which showed the surface amino groups were reactive. Improved bonding between epoxy and poly-tetrafluorethylene exposed to an ammonia glow discharge has also been demonstrated [37]. Ammonia plasma treatment of poly(ethylene terephthalate) filaments to improve adhesion to rubber was examined by Lawton [38]. He found that ammonia as well as many other plasma gases increased adhesion and did not noticeably affect the bulk fiber properties.

Graphite [39] and glass [40] filaments have also been successfully aminated in ammonia plasmas; however, the strength of the glass filaments was severely degraded. Evans and Kuwana [41] have recently studied the introduction of amino groups into graphite electrodes by exposure to ammonia, RF glow discharge plasmas. They found that surface amination occurred rapidly without causing structural damage to the bulk material.

These previous amination studies indicate that amine groups should be introduced into PPTA without substantial losses in filament strength by exposure to ammonia, RF glow discharge plasmas. The surface amine groups should be reactive with epoxides and form covalent bonds at the composite interface. Wertheimer and Schreiber [32] may have previously aminated PPTA with their microwave plasma experiments using ammonia. This is difficult to ascertain, however, since amines were not analyzed for, nor is the triazine resin used for their bonding studies reactive with amines.

AMINE PLASMA CHEMISTRY

<u>Plasma Characteristics</u>

The theory and technology of glow discharge plasmas has been presented by a variety of authors [42-46]. Only those concepts necessary for the present discussion will be summarized here. Glow discharge plasmas are initiated and sustained by the applied electromagnetic field. Free electrons are accelerated by the field, and partially ionize the gas molecules through inelastic collisions. The degree of ionization is on the order of 1 in 10^4-10^6 molecules [43]. Lower pressures reduce the impingement flux of the plasma on a substrate. Reduced impingement can be beneficial for surface modification of polymers in that substrate ablation rates and heating are correspondingly reduced.

Besides the relatively few ions, the plasma gas molecules exist as free radicals, or as various excited state species. Glow discharges are also abundant sources of ultraviolet radiation as well as the weak visible emissions which give them their characteristic color. Ion and neutral molecule temperatures remain near that of the reactor vessel, hence the designation low-temperature plasma. Electrons, because of their high mobility, are rapidly accelerated by the applied field and have energies (temperatures) one to two orders of magnitude higher than the plasma gas molecules. Glow discharge plasmas are, thus, not in a state of thermal equilibrium.

Plasma-Solid Interactions

Besides being a rich source of active species, plasmas simultaneously create active sites on polymer surfaces which can combine with the plasma species. Since the bulk of the substrate is maintained at a low temperature and unexposed to high-energy radiation, plasmas are ideally suited for polymer surface chemical modification. Plasma interactions with solid surfaces have been reviewed in depth by Hudis [47].

Reactive sites on the polymer surface are created by a variety of means including ultraviolet radiation, ion impact, meta-stable energy exchange and free radical interactions [47]. Surface hydrogen abstraction to create free radicals is the principal result of all these mechanisms. Concurrently, main chain bond scission leads to a continuous ablation of the polymer surface with time. The rate of material loss follows the general thermo-oxidative and ultraviolet stability trends based upon polymer composition and structure. Ablation rate is also a direct function of input power to the plasma, as is UV intensity [48] and ion formation.

All of the above mechanisms will contribute to the formation of surface free radicals on polyaramids. The strong absorption of aramids in the ultraviolet due to the conjugated carbonyl-aromatic chain structure [49] suggests that UV effects could be particularly important for creating reactive surface sites. The all-para structure of PPTA makes it more resistant to UV degradation than other aramids; however, significant strength losses are observed in PPTA upon UV exposure in air [49]. In vacuum, the strength of PPTA is unaffected by long UV exposures [49]. Ablation rates of nylon and poly(ethylene terephthalate) exposed to RF plasmas are low compared to most polymers [48], which indicates that PPTA is also likely to exhibit a low ablation rate.

In summary, glow discharge plasmas are rich sources of active species and radiation which can create active sites on a polymer surface. Recombination of surface radicals with plasma radicals can modify the polymer surface chemistry. This process is opposed by plasma ablation of the polymer surface. The relative rates of these processes are determined by the plasma input power, gas pressure, and flow rate, and the reactivity of the species in the plasma and on the polymer surface.

Ammonia – Polyaramid Reactions

A recent study of the plasma decomposition of ammonia has been reported by d´Agostino, et al. [50]. Their findings indicate that the decomposition process follows apparent zero-order kinetics through the following reaction sequence:

$$NH_3 \longrightarrow \cdot NH_2 + \cdot H \tag{1}$$

$$H_2 \longrightarrow \cdot H + \cdot H \tag{2}$$

$$\cdot H + NH_3 \longrightarrow \cdot NH_2 + H_2 \tag{3}$$

$$\cdot NH_2 \longrightarrow \cdot \overset{\cdot}{N}H + \cdot H \tag{4}$$

$$\cdot \overset{\cdot}{N}H + \cdot \overset{\cdot}{N}H \longrightarrow N_2 + H_2 \tag{5}$$

Reactions (1), (2) and (4) are nonequilibrium processes that occur through electron impact with ground state molecules, bimolecular dissociation of vibrationally excited molecules, or through a joint vibrational-impact mechanism as discussed by Capitelli and Molinari [51]. Reaction (5) occurs at a very high rate and is the main channel for $\cdot NH$ disappearance [50].

These observations suggest that $\cdot NH_2$ is the most likely species available for recombination with surface radicals. The PPTA amination reaction in an ammonia plasma may be idealized by Eqn. (6):

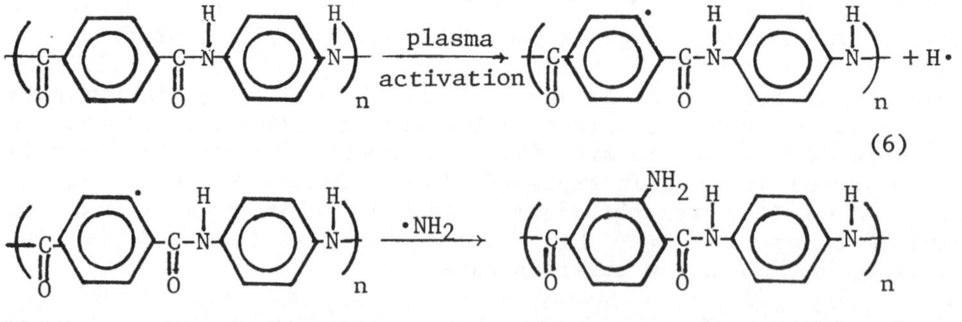

$$\tag{6}$$

Eqn. (6) is very idealistic for a variety of reasons. The surface chemistry of PPTA filaments differs considerably from the bulk. Published X-ray photoelectron spectroscopy (XPS) results show that the surface exists as a highly oxidized hydrocarbon layer with little nitrogen present [52]. Nitrogen species in the plasma other than $\cdot NH_2$ should lead to a variety of other surface groups as would reactions with the amide group. Competing oxidation reactions would also be expected. In addition, XPS spectra for plasmatreated polystyrene show a loss of shake-up structure from the aromatic groups that indicates extensive restructuring of the surface layers [46]. Eqn (6) is, nevertheless, a useful means for envisioning the PPTA amination process.

EXPERIMENTAL

Materials

Kevlar 49 (E. I. duPont de Nemours and Company) was chosen as the polyaramid substrate for the amination experiments because it has been used and studied extensively as a composite reinforcement. Kevlar 49 is a condensation polymer of terephthaloyl chloride and p-phenylene diamine which forms poly(p-phenylene terephthalamide), PPTA [53,54]. One lot of 1.3 m wide scoured, style 181 fabric (2 x 2 picks per mm, 380 denier, 8 harness satin; Ref. ASTM D 3318-76 "Woven Cloth from High Modulus Organic Fiber" [55] was used for all experimental procedures. The particular lot was identified as ref #20521, merge #6G004 and was woven by Clark-Schwebel Fiber Glass Corp. (White Plains, NY). Scouring processes are designed to remove the warp size applied during weaving. The polyvinyl alcohol size used in this lot was removed by a continuous dip process into a detergent solution held at 65-95°C [56]. A 15-20 min residence time for the fabric in the scouring solution has been shown by infrared analysis to remove 99.9 percent of surface extractables [56].

One hundred by two hundred-fifty mm swatches of fabric were dried in vacuum overnight at 110°C prior to plasma exposure. The filament specimens were in a woven fabric form to facilitate handling. Fabric is also particularly advantageous for determining amine content with wet chemical methods, and for fabricating composite specimens by hand lay-up procedures.

Ammonia gas (anhydrous, 99.99%) used in the glow discharges was used as received from Matheson Gas Products (East Rutherford, NJ). Diglycidyl ether of bis-phenol A (DER 332, Dow Chemical Co., Midland, MI) epoxy resin was cured with m-phenylene diamine, MPDA (99+%, Aldrich Chemical Co., Milwaukee, WI), to fabricate composite specimens. All other chemicals were reagent grade or better.

The model compounds 1,4-bis(N,N'-benzamido)benzene and 4-amino benzanalide were synthesized from p-phenylene diamine and benzoyl chloride in a pyridine solution. The 1,4-bis(N,N'-benzamido)benzene was formed by dripping a 10% excess of benzoyl chloride into the diamine-pyridine solution. 4-amino benzanalide was formed with a 10:1 excess of p-phenylene diamine. Both compounds were recrystallized from N,N-dimenthylformamide. .

<u>Plasma Treatment</u>

Plasma exposures were conducted in a Model #PM-310 (Branson International Plasma Corp., Hayward, CA) reactor, which is designed for batch process, oxidative removal of photoresists. The PM-310 is a capacitively coupled system driven at 13.56 MHz with a 127 mm diameter by 305 mm length reaction zone. Prior to plasma treatment, the fabric swatches were dusted with freon to remove any loose contaminates. The fabric swatches were held near the center of the plasma with a glass frame holder.

Exposure time, input power and pressure were the primary variables examined for the plasma amination process. Time was varied at intervals between 5 seconds and 2 hours. Input power (reflected to zero) was examined at 20, 50, 100 and 150 watts. Plasma pressure was varied between 0.5 and 3.0 torr.

The standard plasma treatment consisted of pre-evacuation of the reactor for 5 minutes, followed by bleeding in the plasma gas for 2 minutes before initiating the plasma. Evacuation of the reactor for 5 minutes reduced the pressure to 0.05-0.08 torr prior to the introduction of the plasma gas. Based upon stabilization of pressure and appearance of the emitted light, the plasma reached an apparent steady-state in approximately 5-10 seconds. At 1 torr, an ammonia plasma appeared a light purple to the eye. The plasma treated fabrics were stored in envelopes and desiccated to reduce potential adverse environmental interactions with UV radiation and moisture.

To examine the effect of oxygen on the amination reaction, some runs were conducted with a pre-evacuation to 0.4 torr. The effect of a post-bleed of the plasma gas on amine concentration was examined by leaving the fabric in ammonia for 5 minutes after termination of the plasma. Plasma treatments of model compounds were run on 10 mm diameter pellets made in an infrared spectroscopy pellet press (Perkin Elmer) or as crystals thinly spread on a clean petri dish.

<u>Surface Characterization</u>

Amine concentration was determined quantitatively by ion exchange with an azo dye, Ponceau 3R (Daiichi Pure Chemicals Co., Ltd., Tokyo, Japan). Ponceau 3R has two sodium sulfonate functional groups that will exchange with protonated amine groups (Fig. 1). The dye is subsequently removed from the filament surface and its concentration measured by visible light absorption at 499 nm. Dye concentration is then calculated from its known

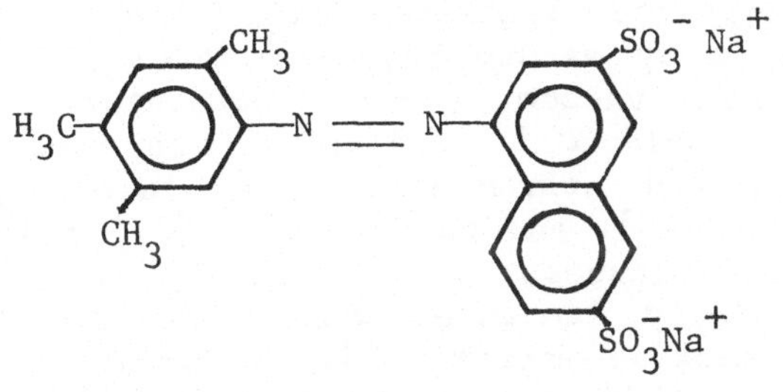

Fig. 1. Chemical Structure of Ponceau 3R.

optical density curve. This procedure was developed by Hendrick, Grant and Howard [57] to measure primary amine groups on proteins adsorbed on membranes. They proved the technique to be quantitative by comparison with isotope labeling experiments. Their procedure with minor modifications for handling fabric samples was followed closely. Details of the procedure used are given in Appendix A.

Scanning electron microscopy (SEM) was conducted on a model AMR 1000A (Amray, Inc.; Bedford, MA). XPS analyses were obtained with a Physical Electronics (Eden Prairie, MN) model 548 spectrometer digitally interfaced with an RT-11 (Digital Equipment Corp., Burlington, MA) disk storage and HP-2649C (Hewlett Packard Corp., Corvallis, OR) graphics terminal. All spectra were taken at 5 x 10^8 torr or less with a Mg $K_{\alpha 1,2}$ excitation source (1253.7 eV). A clean gold 4f line at 84.0 eV was used as an energy calibration. Low resolution spectra were taken at a 100 eV electron pass energy. High resolution spectra were taken at a 25 eV pass energy.

Composite Fabrication

T-peel specimens were fabricated by impregnating two fabric plies with 100g 332/15.5g MPDA and then laminating the fabric under pressure. The 332/MPDA system was formulated by heating the components to 70°C, combining them and mixing at 70° for 3 minutes prior to impregnation of the fabric. Those laminates were pressed to 0.46 mm stops in a hot press and heated to a 150°C cure temperature. After a 2-hour hold at 150°C, they were cooled to room temperature under pressure. Forty mm of poly(tetrafluoroethylene) film was placed between the fabric plies at one end to allow clamping into the Instron. The T-peel tests were run at a cross-head speed of 50 mm/min.

Twelve-ply composite laminates were fabricated with the 332/MPDA resin system by vacuum bag autoclave molding. The wet layup was placed under vacuum at room temperature and heated at 2.5°C/min to 110°C, held 2 h, and cooled to room temperature. Vacuum was maintained throughout the cure cycle. Both quasi-isotropic [0/90/±45]$_S$ and warp-aligned [0/90] laminates were fabricated. Cured thickness was nominally 2.5 mm. Volume fractions of the 12-ply laminates were determined by selective dissolution of the epoxide matrix [58]. Filament and void volume fractions were nominally 54 percent and less than 0.5 percent.

Mechanical Testing

Single filaments were removed at random from within the fabric swatches. Handling was minimized to reduce possible surface damage. The filaments were bonded to 76.2 mm long paper gaskets (Ref. ASTM D 3379-75 Tensile Strength and Young's Modulus for High-Modulus Single-Filament Materials [55]) with model airplane glue. After clamping into the Instron grips, the paper was cut and the filaments loaded at a cross-head speed of 10 mm/min.

Mechanical test specimens were machined from the composite laminates by diamond grinding. Reduced gage section tensile bars were taken from the [0/90] laminates. Specimen geometry was similar to ASTM D 638-75 Tensile Properties of Plastics [59]. Gage section dimensions were 6.35 mm wide by 25 mm length by the as-molded thickness. Strains were monitored with a 25 mm, 20 percent extensometer. Cross-head speed was 1.27 mm/min. All failures occurred in the gage section.

Interlaminar tensile strength was measured with 28.5 mm diameter plugs taken from quasi-isotropic laminates. The plugs were bonded to threaded stainless steel rods for mounting in the Instron. The stainless fixtures were sandblasted and the composite plugs etched for 30 min in an oxygen plasma (1 torr, 75 watt) prior to bonding. EC-2214-R (3M Co., St. Paul, MN) adhesive cured 16 h at 90°C was used to join the plugs to the stainless fixtures. The interlaminar tensile plugs were loaded at a cross-head speed of 1.27 mm/min. Strain was not measured. All failures reported occurred within the composite laminate.

Moisture Absorption

Moisture diffusion specimens were taken from the [0/90] ply stacking sequence laminates and ground on all surfaces to the nominal dimensions of 19.0 mm x 19.0 mm x 2.0 mm thickness. After machining, the specimens were polished smooth with 600 grit paper, washed with detergent followed by a distilled water rinse, and wiped with ethanol. They were then heated in a vacuum oven at 50° C until they reached a stable weight, slow cooled under vacuum, and stored in a desiccator.

Moisture weight gains were measured with a microbalance fitted with a controlled temperature-humidity chamber. The basic apparatus is described in Ref. 60. The moisture absorption measurement procedure consisted of the following steps. Prior to introduction of water vapor into the specimen chamber, the chamber was evacuated and back filled with dry air. Exposures were conducted over a saturated NaCl salt solution which provided a 76 percent relative humidity environment. The entire specimen chamber and microbalance assembly was enclosed in an oven maintained at 27 ± 1°C. Temperature and water vapor pressure were continuously monitored along with sample weight.

RESULTS AND DISCUSSION

Filament Characterization

A. Surface Texture

Observation of scoured 181 style Kevlar 49 fabric at low magnification revealed numerous randomly oriented fibrils and particle-like structures on and within the filament bundles. Closer examination of the particle shapes showed them to be tightly wound up fibrils or dust. The quantity of loose material on the surface is highly variable. Some areas are relatively clean while others are largely covered by fibrils.

Examination at higher magnifications shows the individual filaments to be quite variable in surface texture as well. Most Kevlar 49 filaments are quite smooth as seen in Fig. 2. Even magnifications of 20,000 times show very little surface texture. In other areas, however, numerous large defects are evident along the filaments. The defects appear to be regions where polymer has been pulled from the bulk fiber and solidified on the fiber surface during the spinning process.

A correlation was noted between the number of irregular filaments in an area of fabric and the amount of loose fibrils on the fabric surface. The fibrils are bundles of PPTA chains that have been torn from the filament surfaces during processing. The irregular filaments seem to be more prone to further damage during weaving.

After a 60 sec ammonia-plasma treatment, some of the loose fibrils on the fabric surface appeared to have been ablated off by the plasma. It was noted that many of the loose fibrils had been shortened by the plasma treatment. Although the filaments must

Fig. 2.   Surface texture or Kevlar 49 filaments from scoured 181
          style fabric.

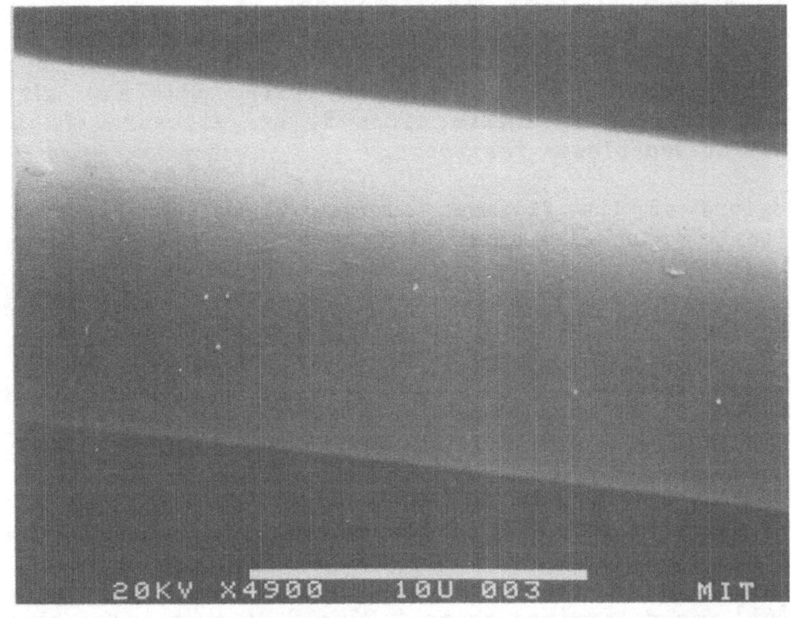

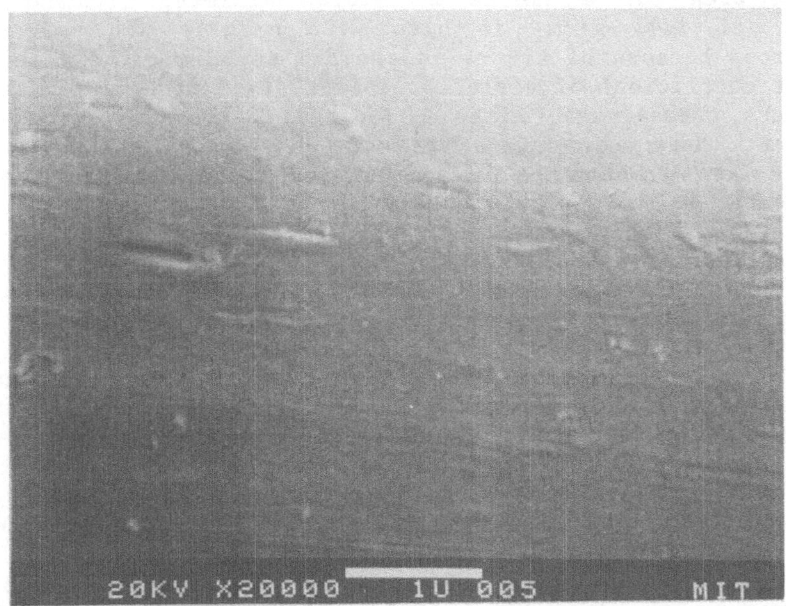

Fig. 3. Filament surface texture after 60 sec ammonia plasma (50 watts, 1 torr) exposure.

necessarily be ablated themselves, no evident change in surface texture is seen at high magnifications (Fig. 3). No change in surface texture was evident after exposures as long as 300 sec for initially smooth filament regions. These results are opposed to those of Wertheimer and Schreiber [32], who saw significant roughening of PPTA filaments after 30 sec exposures in air in a microwave driven plasma reactor.

Regions of the filament surface that had existing defects seemed to be smoothed somewhat by the 60 sec ammonia-plasma treatment. Elevated, rough regions are diminished and smoothed by the plasma. This result is not unexpected. Defect regions that lack the crystalline register of the bulk PPTA would have lower thermo--oxidative stability. In addition, a raised feature should experience higher molecular and photon fluxes that will increase the ablation rate.

B. Surface Area

The smooth appearance of the ammonia-plasma-treated filaments is verified by surface area results. The average surface area of Kevlar 49 181 style fabric for six measurements with the BET technique [61] was determined to be $0.23 \pm 0.04 \ m^2/g$. The theoretical surface area for a smooth 12 micron diameter filament with a density of $1.45 \ g/cm^3$ is also $0.23 \ m^2/g$. Such agreement is fortuitous because of errors in the BET technique (note the twenty percent coefficient of variation) and surface irregularities in the filaments themselves. After a lengthy (3600 sec) ammonia-plasma treatment, measured surface area was $0.24 \ m^2/g$. These data show that Kevlar 49 filaments are quite smooth and that ammonia-plasma treatments do not significantly change surface area. If the filaments were being extensively ablated, a large increase in measured surface area would be expected.

C. Surface Chemical Composition

X-ray Photoelectron Spectroscopy (XPS), also known as ESCA, has become a sensitive analytical method for the investigation of polymer surfaces [62-64]. Such surface sensitivity makes XPS an ideal method for the study of plasma-modified polymer surface [37,41,46,65,66]. XPS spectra were obtained from Kevlar fabrics and pressed 10 mm diameter pellets of model compounds before and after ammonia-plasma treatment. Atomic compositions were calculated from the peak areas adjusted for differences in individual element cross-sections (C_{1s} = 0.25; O_{1s} = 0.67; N_{1s} = 0.43 [67]) and normalized to 100 percent C, O and N.

The surface compositions of the model compounds examined are given in Table II. Compositions of the untreated model compounds

Table II. XPS Surface Composition of Model Compounds

Compound	%Carbon	%Oxygen	%Nitrogen
1,4-bis(n,N'-benzamido)benzene:			
Theoretical	83	8	8
Measured	80	10	10
After NH$_3$ Plasma[†]	71	12	17
p-terphenyl:			
Theoretical	100	0	0
Measured	100	0	0
After NH$_3$ Plasma[†]	76	10	14
4-amino benzanalide:			
Theoretical	81	6	12
Measured	83	8	10

[†] 50 watts, 1 torr, 60 sec

show good agreement with their theoretical bulk composition (Fig. 4). After ammonia-plasma exposure, substantial increases in nitrogen content are observed. Oxidation of the model surfaces after plasma exposure is also evident. Oxygen may have been in the plasma due to vacuum leaks or desorption from the reactor walls. Oxygen also readily combines with free radicals upon exposure to air, so some oxidation would be expected.

The increase in nitrogen content after ammonia-plasma treatment indicates that the model compounds are likely aminated to some degree. Examination of crystals of the 1,4-bis(N,N′-benzamido)benzene with the Ponceau 3R (Fig. 1) ion exchange procedure (Appendix A) showed 0.71 binding sites per 1 nm^2 of surface after ammonia-plasma exposure. Unexposed crystals did not bind the Ponceau 3R in measurable quantities.

Line shapes of the 1,4-bis(N,N′-benzamido)benzene and the 4-amino benzanalide were analyzed to determine the nitrogen binding states. High resolution C_{1s} and N_{1s} spectra are given in Fig. 5. The spectra have been corrected for charging by locating the main C_{1s} peak at 285 eV. 1,4-bis(N,N′-benzamido)benzene shows the following features of interest for the analysis of the PPTA

1,4-bis(N,N'-benzamido)benzene

4-amino benzanalide

p-terephenyl

Fig. 4. Chemical structure of aramid model compounds.

spectra: (1) the C_{1s} spectrum shows a separate carbonyl peak shifted up 3 eV from the main peak; (2) a well-defined C_{1s} shake-up structure shifted up 4.6-8.8 eV from the main peak. The intensity of the shake-up band is 3.9 percent of the main peak intensity; (3) the O_{1s} spectrum (not shown) shows one sharp peak with a shake-up band shifted up 4.0-8.0 eV; and, (4) the N_{1s} spectrum shows one sharp peak.

After a 60 sec, 50 watt ammonia-plasma exposure, the region between the carbonyl and the main C_{1s} peak has filled in, and the shake-up intensity has decreased to 0.4 percent of the main C_{1s} peak (Fig. 5). The main C_{1s} peak is slightly broadened (10%), while the O_{1s} (not shown) and N_{1s} peaks are broadened 30 percent. The N_{1s} peak also shows an asymmetry towards the lower binding energy tail.

4-amino benzanalide was prepared to determine the effect of an amine group on the C_{1s} and N_{1s} line shapes of a PPTA-type structure. Spectra of 4-amino benzanalide (Fig. 5) are similar to previous observations with the following exceptions: (1) the region between the main C_{1s} and carbonyl peaks is filling in due to a positive shift of the carbon attached to the amino group [68]; and, (2) a slight asymmetry has developed on the low binding energy tail of the N_{1s} peak similar to that observed in the ammonia-plasma treated 1,4-bis(N,N´-benzamido)benzene. All the 4-amino benzanalide peaks are slightly broader than the 1,4-bis(N,N´-benzamido)-benzene due to scattering, but the N_{1s} peak is broadened

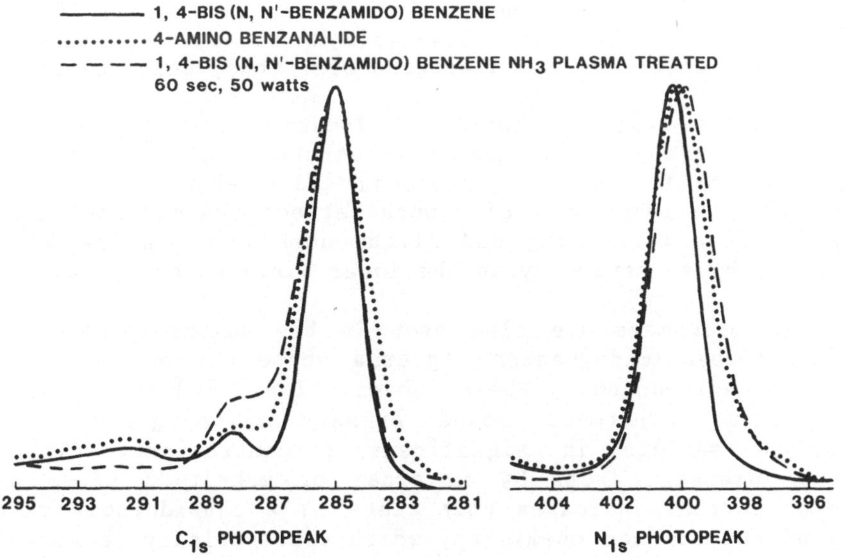

Fig. 5.   High-resolution ESCA C_{1s} and N_{1s} photopeaks of aramid model compounds from pressed bullets.

Table III.   XPS Surface Composition of Ammonia-Plasma-Treated Kevlar
             49 Filaments

Treatment	%Carbon	%Oxygen	%Nitrogen
PPTA Theoretical	78	11	11
Untreated	73	22	5
5 sec, 50 watts	75	19	6
15 sec, 50 watts	72	20	8
60 sec, 50 watts	70	20	10
150 sec, 100 watts	70	18	12
150 sec, 150 watts	73	14	13

significantly more. This indicates that the amino nitrogen has a
slightly different binding energy than the amide nitrogen and that
the N peak is composed of two sharp peaks. After correcting for
charge shift, the 4-amino benzanalide N peak maximum (400.0 eV)
is 0.3 eV less than the 1,4-bis(N,N´-benzamido)benzene. The
binding energies of the two nitrogens are, thus, quite close.

The model compound spectra indicate that the addition of
primary amine groups to an aramid structure results in only subtle
changes in the C and N photopeak line shapes. The primary
changes are the appearance of a peak between the carbonyl and main
C peak and a broadening and slight down-shift in the N peak
accompanied by an asymmetry in the lower binding energy tail.

These phenomena are also seen in the ammonia-plasma-treated
1,4-bis(N,N´-benzamido)benzene spectra where the new C peak is
especially pronounced. These observations indicate that the
ammonia-plasma treatment bound primary amine groups to the
aramid-type surface in significant concentrations. That the
shake-up intensity drops by an order of magnitude after plasma
treatment further indicates that there is a considerable restructuring of the surface chemistry, which substantially decreases the
aromaticity.

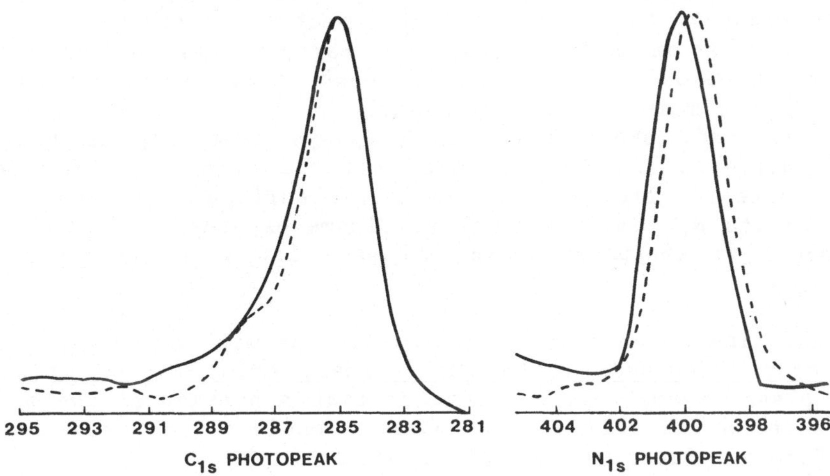

Fig. 6. High-resolution XPS C_{1s} and N_{1s} photopeaks of PPTA fabrics.

Surface chemical compositions of ammonia-plasma-treated PPTA fabrics are given in Table III. The fabric geometry caused low counting rates and long scan times (1-2 hrs) were required to obtain high-resolution spectra. The as-received filaments exhibit an oxidized hydrocarbon surface with less than half of the bulk PPTA nitrogen content in agreement with the findings of Penn and Larson [52]. There is a general trend of increasing nitrogen concentration with exposure time at 50 watts, while oxygen content decreases somewhat. Higher power exposures further increase nitrogen concentration and reduce oxygen content. That the oxygen content of the surface remains above that of the bulk polymer is an indication that considerable oxidation as well as amination occurs in an ammonia plasma under the conditions examined in this study, which was also seen with the model compounds (Table II).

Representative high resolution C_{1s} and N_{1s} photopeaks of PPTA fabrics are given in Fig. 6. The untreated PPTA carbon peak shows a high binding energy tail that is typical of carbonyl and ester linkages on oxidized surface [63,64]. There is also evidence of a weak shake-up structure around 293-295 eV with an intensity 0.8 percent of the main C_{1s} peak. The untreated PPTA O_{1s} peak (not shown) is quite broad (2.7 eV width at half max) which indicates that oxygen is bound in more than one binding state on the surface. Only one binding state is apparent in the nitrogen peak.

Ammonia-plasma treatments of PPTA filaments result in modifications of the high-resolution line shapes that develop as the plasma-exposure conditions increase in severity. After a 150 sec, 150 watt exposure (Fig. 6), the C_{1s} spectrum is sharpened into a step between the main hydrocarbon and carbonyl regions similar to the model compounds (Fig. 5). A weak shake-up structure 0.8 percent of the main C_{1s} peak intensity is also present. The N_{1s} peak is shifted down 0.2 eV and displays asymmetry in the low binding energy tail. Broadening of all the high-resolution peaks due to scattering from the uneven fabric surface precludes inferences from the N_{1s} peak width after plasma exposure. The O_{1s} peak (now shown) is sharpened by a 150 sec, 150 watt ammonia-plasma exposure.

These line shape changes are consistent with those taken from aramid model compounds containing primary amine groups or after ammonia-plasma exposures, and suggest that a portion of the nitrogen present on the PPTA surface after ammonia-plasma treatment is in the form of primary amine functionalities.

D. Surface Energetics

Plasma treatments of polymers generally cause large changes in their surface energy and chemistry [10,46,47,65,69-71]. It is of interest, therefore, to determine whether changes in surface energy of plasma-aminated PPTA filaments could contribute to improved adhesion.

Work-of-adhesion measurements for single-filament wetting were conducted by Dr. S. P. Wesson at the Owens-Corning Technical Center (Granville, OH). The technique consists of suspending the filament from the working arm of a microbalance while a precision elevator raises and lowers a liquid surface over 5 mm of fiber length. The apparatus and experimental technique are described in Ref. 72. Averaged work-of-adhesion results for wetting with diiodomethane and water showed that there is no discernable change in the surface energy of PPTA filaments after exposure to a 60 sec, 50 watt ammonia plasma. There is also no change in the dispersive and acid-base components of PPTA surface energy after ammonia-plasma treatment. Wetting of PPTA filaments with epoxy resin, polyester resin, and a pH 4 buffered solution also showed no apparent differences after surface treatment.

These results were substantiated by Dr. L. S. Penn at Ciba-Geigy Corp. (Ardsley, NY). Penn measured advancing and receding contact angles of untreated and 60 sec ammonia-plasma-treated PPTA filaments in a series of decreasing surface tension liquids and found no notable differences between the samples. Details of the filament wetting experiments will be given in a subsequent publication.

The surprising lack of change in surface energetics may be explained by examination of the surface chemical composition results given in Table III. Because of the oxidized surface layer on the untreated filaments, they have a substantial polar component. After a 60 sec ammonia-plasma treatment at 50 watts, nitrogen has been substituted for oxygen on the surface so that the total concentration of polar atoms on the surface is virtually unchanged. Since surface energetics are unchanged by plasma-amination, subsequent increases in adhesion may be attributed to covalent bond formation at the interface rather than improved wettability.

Filament Strength

Potential filament strength losses induced by plasma exposure are a major concern. Significant reductions in PPTA filament strengths have been reported after thermal [19] and microwave plasma [32] exposures. PPTA filament strength distributions have been shown to closely follow Weibull statistics [73]. It is of interest, therefore, to determine Weibull parameters of the plasma-treated filaments. Deviations from normal Weibull parameters could indicate changes in filament structure which could affect long-term aging behavior.

The Weibull distribution, which is based upon a "weakest link" chain rule, is often applied to the strengths of brittle materials and life testing applications. The two-parameter Weibull distribution takes the form [73]:

$$F(X) = 1 - \exp[-\ell (X/X_o)^\rho] \qquad (7)$$

where

X = filament strength, X_o = scale parameter at unit length, ℓ = filament length, and ρ = shape parameter.

Both X_o and ρ are constants in Eqn. (7). The shape parameter determines the shape of the Weibull distribution. As ρ increases, $F(X)$ is skewed towards the lower end of the distribution, which is the region of interest for reliability evaluation. The scale parameter approaches the mean strength as ρ increases. Kevlar 49 filaments have reported shape parameters between 7 and 9 [73].

Table IV. Weibull Parameters for Kevlar 49 Filaments

Treatment	Shape Parameter ρ	Scale g-f	Parameter X_0 MPa	Variation Coefficient Percent	Number of Tests
Untreated	11.3	36.0	3650	10.7	26
60 sec NH_3 Plasma (50 watts, 1 torr)	8.8	38.1	3860	13.2	21
600 sec NH_3 Plasma (50 watts, 1 torr)	8.1	35.6	3600	14.9	23
150 sec NH_3 Plasma (150 watts, 1 torr)	12.7	36.3	3680	9.5	21

Weibull parameters calculated from the measured strength distributions of 76.2 mm length specimens are summarized in Table IV. The shape parameter for the as-received filaments is higher than normally seen for Kevlar 49 filaments. The scale parameter agrees well with published strength results [73]. A high shape parameter may be caused by the relatively small sample size, or may imply a different flaw distribution for this lot of filaments.

The ammonia-plasma-treated filaments are characterized by Weibull parameters similar to the as-received filaments (Table IV). Scale parameters (characteristic strengths) are essentially unchanged by the ammonia-plasma treatments selected for analysis. The shape parameters of filaments treated at 50 watts are shifted down to the 8-9 range, a representative value for untreated Kevlar 49 [73]. At 150 watts, the shape parameter rises to 12.7. This increase may indicate that the plasma is healing small flaws in the filament surface structure, while larger flaws are unaffected [74]. The increased shape parameter could also be an artifact of the small sample size.

These data show no cause for concern about the effect of ammonia-plasma treatment on filament strength or aging characteristics; rather, the increased Weibull shape parameter seen for filaments treated at higher power levels raises the possibility that filament strength distributions might be altered by eliminating small surface flaws with plasma treatments. Clearly, more extensive filament strength testing is required to resolve these issues.

Plasma Amination Kinetics

In the analysis of surface amine content with the Ponceau 3R (Fig. 1) ion exchange technique (Appendix A), there are two key assumptions that affect the reported amine concentration. The first is that only one sodium sulfonate group per Ponceau 3R molecule binds with a surface amine group. If both sulfonates bind to surface amine sites, then the calculated surface site concentration is conservative. A doubly bound molecule may also require removal with a base of higher pH than 0.1 N NaOH. To check the extent of dye removal, some samples were extracted with 1.0 N NaOH after the standard procedure. No detectable amount of dye was removed with the stronger base. The second assumption is that the Ponceau 3R binds only to amine sites. This assumption is not strictly valid because sulfonates could also exchange with protonated carboxylic acid, carbonyl, hydroxyl, or amide groups as well. Only limited reaction with amide groups from the bulk PPTA would be expected. Amide groups are much less basic than amines, and would be even less so in the PPTA structure due to destabilization of the protonated nitrogen by the benzene ring as well as the carbonyl. The low reactivity of amide groups towards the Ponceau 3R exchange reaction is seen in the measurement of binding sites on 1,4-bis-(N,N′-benzamido)benzene. The extent of formation of sulfonic esters via hydroxyls should also be low [75]. These side reactions introduce some uncertainty into the measured surface-amine concentration; however, the majority of determined binding sites are amines.

The 0.005 correction to the measured optical density in Eqn. (A1) is the absorption of 40 ml of 0.1 N NaOH neutralized with 0.5 ml of 12 N HCl and concentrated to a 10 ml volume. A one square nanometer area was chosen to normalize amine concentrations in Eqn. (A3) because of ease of visualization. The average area of a PPTA repeat unit is 0.84 nm [76]. Amine groups per 1 nm^2 convert to moles NH_2/cm^2 by multiplying by 1.66×10^{-10}.

Analysis of as-received Kevlar 49 fabric with the Ponceau 3R procedure showed a small amount of dye consistently bound to the filaments. The washing procedure was varied extensively to prove that the dye actually bound to the filaments and was not merely trapped within the filament bundles. A mean concentration of 0.056 NH_2/nm^2 with a standard deviation of 0.026 was found for 15 samples taken randomly from the lot of fabric used in this study. The high coefficient of variation (46%) shows that the binding site concentrations are quite variable. There may also be a small contribution to the error from residual dye held within the fiber bundles.

XPS analysis of the as-received filaments shows that the filaments have a highly oxidized surface (Table III). Because of this oxide layer, it is unlikely that residual amine groups on chain ends, etc. are involved in the surface reaction. Rather, it appears that acid groups, etc. are present in the oxide layer that can bind with the Ponceau 3R sulfonate groups. Carboxylic acid surface groups have been shown to be the primary product of UV irradiation of aramids in air [49]. Those groups could also possibly form covalent bonds with an epoxide; however, they would yield a low interfacial bond density of approximately 1 bond per 18.00 nm^2. In addition, not all the species which might bind with Ponceau 3R are capable of reaction with epoxides (e.g., carbonyl), so the interfacial bond density would be even less.

Ammonia-plasma-treated filaments bind the Ponceau 3R in significant amounts, even for treatment times as short as 5 sec. Amine concentration versus exposure time for 50, 100, and 150 watt power inputs are given in Fig. 7. Measured amine concentration rises rapidly before reaching a steady-state value around 60 sec. The steady-state concentration increases slightly as input power increases, rising from 0.7 NH_2/nm^2 at 50 watts to 1.0 NH_2/nm^2 at 150 watts. Steady-state amine concentrations remain constant for ammonia-plasma exposures as long as 600 sec (not shown in Fig. 7). Amine concentration drops slowly with longer exposures, reaching a value of 0.55 NH_2/nm^2 after 1800 sec at 50 watts.

Reproducibility of the amination reaction at steady-state was examined by measuring the amine concentration of 18 separate 60 sec exposures at 50 watts and 1 torr. The mean and standard deviation (0.73 ± 0.04 NH_2/nm^2) of these results are plotted in Fig. 7. The six percent coefficient of variation is quite low for a batch

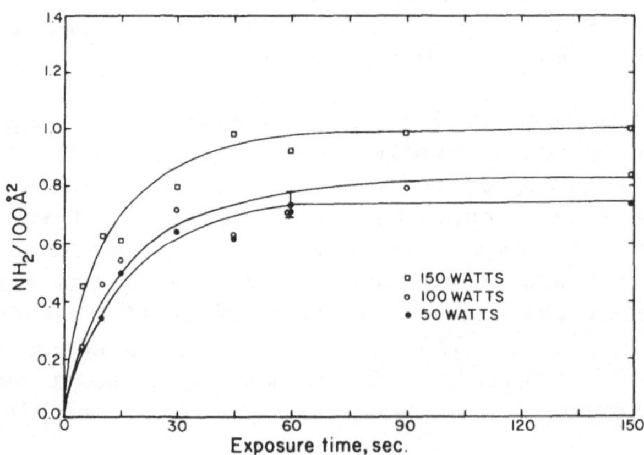

Fig. 7. Ammonia-plasma amination kinetics of PPTA filaments.

process reaction of this type. It is also expected that lot-to-lot variations in filament surface chemistry would cause significantly higher variations than observed here. More variability is seen before steady-state is reached (Fig. 7). This is not unexpected, since the plasma does not reach steady-state itself for 5-10 sec. In addition, the oxidized filament surface layer is simultaneously being removed, and may have regions of varying chemistry and thermo-oxidative stability.

The rapid increase in amine concentration with time (Fig. 7) is typical of many plasma reactions. For example, contact angle measurements show that plasma surface oxidation of poly(ethylene terephthalate) and polyethylene reach a steady-state after 10 sec [62,63]. The speed of the amination reaction is significant because it limits the exposure of the bulk filament structure. It also implies that a continuous feed system for yarn could be practical.

The dependence on plasma input power seen in Fig. 6 indicates a shift in the relative reaction rates of amination versus ablation. Higher power inputs appear to increase reactive surface sites or the impingement flux of reactive plasma species or both without substantially increasing the ablation rate.

To examine the contribution to measured binding site concentration from groups other than amines, Ar-plasma-etched (60 sec, 50 watts, 1 torr) filaments were also examined with the Ponceau 3R ion exchange procedure. The Ar-treated filaments bound the dye in concentrations equivalent to 0.1 site/nm^2. This value is approximately 15 percent of the measured amine concentration for an equivalent ammonia-plasma exposure. The binding sites are likely to be carboxylic acid groups. Acid groups can form covalent bonds with epoxides. Along with surface free radicals they are probably responsible for improvements in adhesion between PPTA filaments and epoxy after oxidizing-plasma treatments reported in Refs. 18, 32 and 33. Similar acid groups are undoubtedly formed to some degree with ammonia-plasma treatments, which will cause the amine concentration results reported in Fig. 7 to be overstated.

In analyzing these results, the question arises as to the uniformity of the amination reaction within the filament bundles. Gas absorption measurements on glass strands and fabrics show an apparent pore size of 100 to 1000 μm at filament contact areas [77]. The 181 style fabric used for the amination experiments is an open weave and should exhibit contact areas at least that large. These exclusion sizes are relatively insignificant compared to the atomic radii of plasma gases; however, a calculation shows the mean free path of an ammonia atom at 25°C and 1 torr pressure to be near 100 μm. Thus, while the interior filaments are not

excluded from the plasma gases, scattering losses may reduce the
impingement flux and average energy of molecular impinging on
interior filaments. The extent of this effect on amination
kinetics is unknown.

Amination kinetics were not affected by pressure over the
limited range from 0.5 to 3.0 torr; however, plasma oxygen concentration had a strong effect. As a simulation of the environment
that might exist in a continuous treatment operation, the standard
treatment procedure was modified to a pre-evacuation pressure of
0.4 torr before the 2 min ammonia bleed. The higher oxygen
concentration slowed the amination rate by 30 times as shown in
Fig. 8. Exposures of 1800 sec were required to reach amine
concentrations equivalent to those from a 60 sec exposure pre-
-evacuated below 0.1 torr. Oxygen may either poison the reactive
amine species, combine more readily with the active surface sites,
or both. This sensitivity may limit continuous processing with
ammonia plasmas and requires further investigation. A more controlled approach in which the ammonia and oxygen are simultaneously
introduced into the reactor in varying concentration ratios is
recommended.

In summary, surface amine concentration on PPTA filaments
rises rapidly in amine plasmas until a steady-state concentration
is reached at exposure times near 60 sec. Amination kinetics are
sensitive to plasma input power and oxygen concentration.

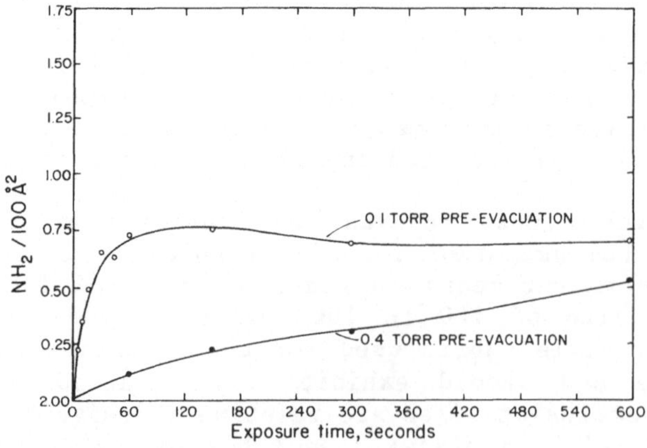

Fig. 8.   Effect of oxygen on the amination kinetics of PPTA filaments in ammonia plasma (50 watts).

Surface Amine Stability

Long-term stability of the plasma-bonded amine groups is of primary importance. Stable amine sites are required for creating permanent, hygrothermally stable bonds with epoxides, for subsequent conversion into alternative reactive species, or for storage prior to use. No change in measured amine concentration was observed after 18 months of ambient aging for ammonia-plasma-treated Kevlar 49 fabrics. Soaking in distilled water before the Ponceau 3R ion exchange reaction also had no effect on measured amine concentration.

The amine groups, thus, are stable and remain reactive for extended periods of time. Chain movements that could bury the amines in the bulk structure to lower the filament surface energy are likely inhibited by the interchain hydrogen bonding, or by cross-linking from the plasma exposure. Similar behavior has been observed in plasma surface oxidized polyamides [65]. These constraints could not prevent local rotation of chain segments, but would inhibit the long-range chain motion necessary to move a modified layer of finite thickness. Plasma-oxidized surface layers of polypropylene [65] and poly(ethylene terephthalate) [63] show increasing wetting contact angles with time, which indicates the importance of H-bonding in constraining chain movement. The rigidity of the PPTA backbone itself should also reduce long-range chain movement.

Composite Mechanical Properties

The effect of amination on the behavior of Kevlar 49 style 181 fabric/332-MPDA composites was examined with three test methods: (1) interlaminar T-peel, (2) interlaminar tension, and (3) longitudinal tension. Each test indicated that plasma amination substantially alters composite properties in directions indicative of increased filament-matrix adhesion. A summary of the results is given in Table V.

Two ply T-peel tests were run to measure adhesion between aminated PPTA fabrics and epoxy. Peel tests are difficult to interpret quantitatively because of the complex stress state at the crack tip and the contribution to measured peel force from deformation of the adherends [7,10]. A further complexity is added by the uneven fabric surface on the crack-front path. Care was taken to test specimens of constant thickness at a constant loading rate to reduce the effects of those variables; however, peel results should be viewed as qualitative indications of trends from modifying interphase adhesion rather than as engineering data. As shown in Table V, peel strength is increased significantly by plasma amina-

Table V. Properties of Aminated PPTA Fabric Composites

Test Method	Untreated	60 sec, 50 watt NH_3 Plasma
T-peel Force,		
g/mm width	57.0 (±10.4)	121.9 (±13.2)
Interlaminar Tension:		
strength, MPa	11.6 (0.7)	15.2 (1.2)
Longitudinal Tension:		
strength, MPa	580.6 (15.8)	544.9 (34.5)
stiffness, GPa	27.1 (1.0)	29.6 (1.8)
elongation, %	2.3 (0.3)	1.9 (0.1)

tion. The untreated peel specimens are characterized by bare filaments on the failure surface; however, examination of the filaments
at higher magnifications shows a thin resin layer adhering to the
filaments (compare Figs. 2 and 9). Failure, thus, occurs in an
interphase region rather than at the filament-matrix interface.
This result is not totally unexpected. Untreated PPTA wetting data
are similar to that of liquid epoxy resins [77], which should lead
to good wetting of PPTA by epoxy.

Peel fracture surfaces of aminated PPTA fabric/332-MPDA
specimens reveal a change in failure mode. They are characterized
by extensive filament splitting, fibrillation and skin-core
separation, all of which may be seen in Fig. 10. There is also
evidence of epoxy adhering to the filaments, and crack propagation
through the epoxy.

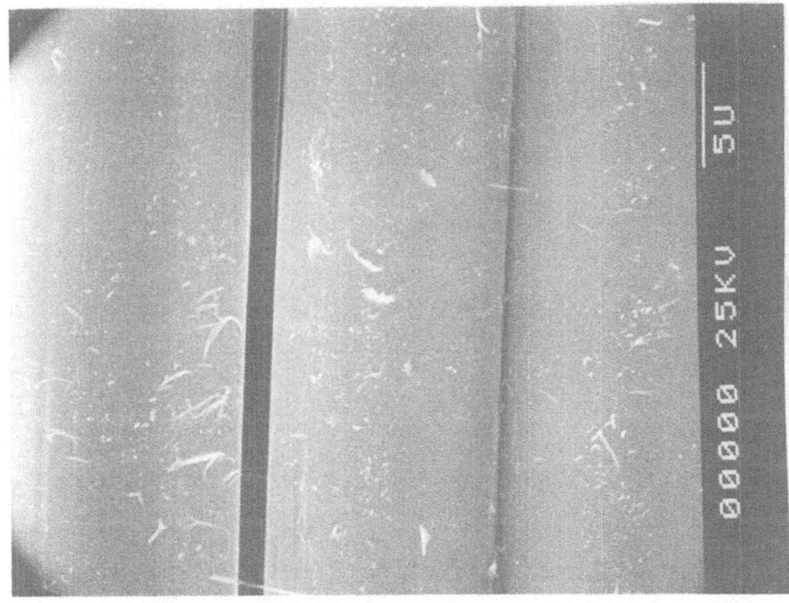

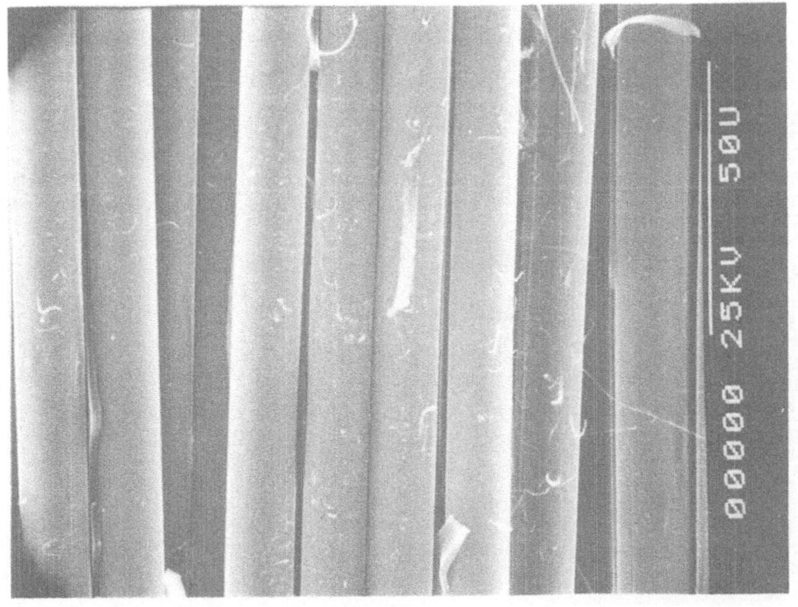

Fig. 9. T-Peel fracture surface of untreated Kevlar 49 fabric/332-MPDA laminates.

Fig. 10. T-Peel fracture surface of Kevlar 49 fabric/332–MPDA laminates with 60 sec NH_3 plasma treatment.

Amination, thus, appears to improve filament-matrix adhesion as evidenced by the failure mode change from interphase controlled to a mixture of filament splitting and matrix cracking. If the filaments split near the surface, it could be argued that a weak, plasma-modified surface layer was responsible for the failure mode change. Since the splitting is in the bulk filament, however, it can be concluded that covalent bonds formed between the surface amine groups and the epoxy network have strengthened the interphase region and shifted the failure mode to the next weakest link, i.e., the filament internal strength.

High magnification views of split filament internal morphology are shown in Fig. 11. The filament shown in Fig. 11 is split near the center of its diameter, so the morphology is not that of a plasma-modified surface layer, but rather that of the bulk filament structure. It can be seen that the filament splits into fine fibrils approximately 65-135 nm across. The split fibrils are much smaller than the 500-700 nm diameter fibrils characteristic of the filament microstructure [78]. The filament splitting morphology appears to be determined by defects between the crystallites in the natural fibrils rather than crack propagation between the fibril bundles.

Interlaminar tensile tests were performed on quasi-isotropic laminates to decrease stress concentrations from the laminate anisotropy. As given in Table V, surface amination of the PPTA filaments caused a modest 31 percent improvement in interlaminar tensile strength for laminates with a 332-MPDA matrix. Microscopic examination of the interlaminar tensile fracture surfaces revealed three pertinent points: (1) laminates fabricated from untreated PPTA filaments have a thin resin coating adhering to the filaments; (2) laminates with aminated filaments exhibit a general tearing of the filament surfaces; and, (3) the plasma-treated filaments did not consistently bond to the epoxy.

Interlaminar failure surfaces of laminates reinforced with aminated filaments show evidence of improved interfacial adhesion. Filament splitting and surface tearing are characteristic features of the fracture morphology. These features are expected for the condition where the interfacial bond strength exceeds the filament internal strength. In radial tension, the filaments tend to split into fine fibrils from the filament surface layers. It cannot be determined whether a plasma-modified surface layer is influencing the interlaminar tensile failure, or whether the fibrils are from a phenomena similar to that observed in the peel tests.

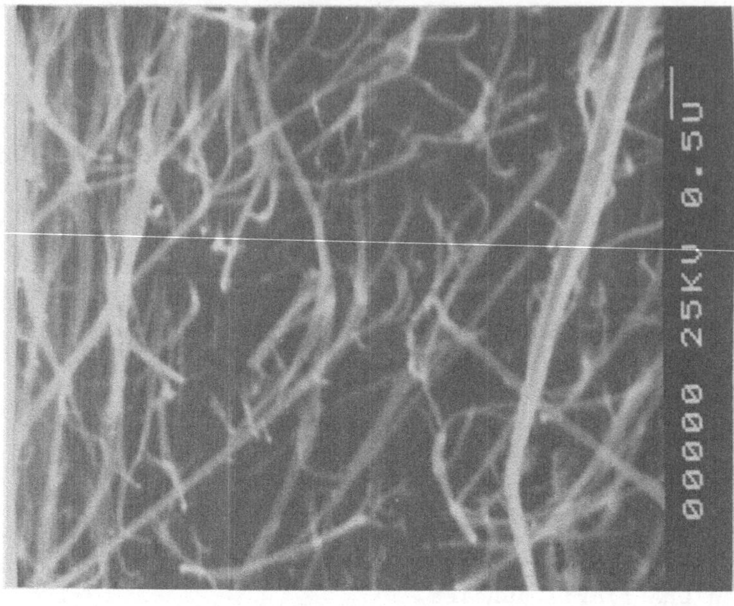

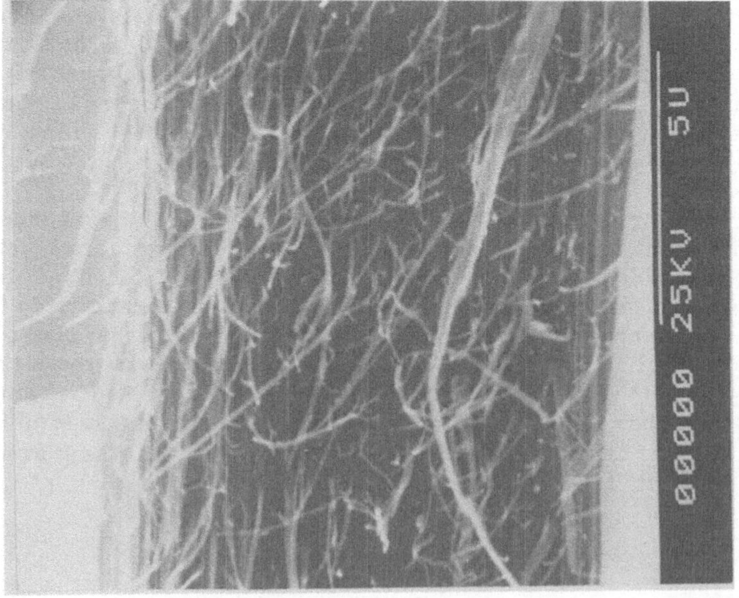

Fig. 11. High-magnification appearance of PPTA filament internal split fibrillation morphology.

Some filaments, however, show little or no evidence of improved bonding to the epoxy network. It appears that fracture still occurs in the interphase for a portion of the filaments after plasma amination. This effect could be caused by nonuniform amination, improper mixing of the resin or contamination of the fabric surface prior to lamination. Because the MPDA is crystalline at room temperature, it must be mixed with the epoxy at elevated temperature. Inhomogeneities in the local resin stoichiometry during mixing are the likely cause of what appears to be poorly bonded filaments. Amine rich areas in the epoxy would reduce the probability of reaction of the epoxide groups with filament surface amine groups.

The longitudinal tensile behavior of [0/90] fabric laminates with aminated filaments also shows several indications of increased filament-matrix adhesion when compared to laminates reinforced with untreated filaments (Table V). Increasing filament-matrix adhesion generally increases composite strength and stiffness, but decreases fracture toughness [4-6,13,79,80]. These trends are evident in the tensile behavior of the aminated laminates, except for the minor strength loss. Since the strengths of the individual filaments themselves have not changed, the composite strength loss may be due to a change in the composite internal stress state. Higher shear stress components on the longitudinal filaments, stress concentrations from split transverse filaments and a decreased ability of the longitudinal filaments to align with the applied load are possible degrading effects from increasing interphase adhesion in fabric laminates.

The drop in tensile strength is modest, however, as is the drop in composite fracture energy. Only a five percent drop in area under the stress-strain curve occurs after amination. In comparison, losses in fracture energy of up to 60 percent are seen in glass and graphite/reinforced composites as adhesion increases [79,80]. The retention of fracture toughness by PPTA-reinforced laminates as adhesion increases is due to the internal fibrillar structure of the filaments. Although energy absorption due to interfacial cracking is reduced as adhesion increases, the filaments absorb significant energy through internal splitting.

The mechanical properties and failure modes of laminates fabricated with aminated filaments indicate that the surface amine sites do improve interfacial adhesion with epoxy resins. Interlaminar tensile and peel strengths are increased, and only minor losses in longitudinal strength and fracture energy are observed. Property modifications are accompanied by failure mode changes from interphase controlled to one of filament splitting and matrix cracking as amine content increases. Because of the complications introduced by the fabric reinforcement geometry, these results should be regarded as trends in mechanical behavior resulting from improving filament-matrix adhesion, rather than as engineering data.

The observation of interphase-controlled failures in the untreated laminates implies that the benefits of amination may vary greatly between matrix resin systems. Resins which poorly wet the PPTA surface should benefit the most from aminating the filaments. Mechanical tests with specimens having better defined stress states, such as transverse tensile specimens from unidirectionally-reinforced composites, are required to draw further conclusions regarding the effects of amination on mechanical properties.

Composite Moisture Absorption

Moisture absorption rates provide an interfacial probe method to supplement interface-sensitive mechanical tests. In glass-reinforced composites with poorly bonded interfaces, moisture is absorbed preferentially at the interface at high rates due to wicking [8]. The interfacial wicking mechanism is prevented when interfacial bond strengths are increased. Moisture diffuses preferentially in the filament direction of Kevlar 49/epoxy by more than an order of magnitude [29]. The absorption rate is greater than that of the epoxy [29] or the PPTA filaments [81], which indicates that an interfacial mechanism is dominant. Moisture absorption rates should, thus, provide a sensitive means for monitoring changes in the interfacial bonding of PPTA-reinforced lamintes. Absorption data are also of interest because of the degrading influence of moisture on the mechanical properties of polyaramid/epoxy composites.

Interfacial absorption is a non-Fickian process more closely related to capillary phenomena, whereas matrix bulk diffusion does follow Fick's laws. The rapid absorption rates from wicking cause the interface to saturate with moisture at short times in small specimens. The microbalance apparatus used to measure moisture weight gains provides the sensitivity necessary to observe changes in absorption at relatively short times. Gains in sensitivity are due to the continuous measurement of weight gain at constant temperature and humidity conditions.

Initial moisture absorption curves of untreated and aminated Kevlar 49 fabric/332-MPDA laminates at 76 percent relative humidity are shown in Fig. 12. The data are plotted in g-moisture absorbed/g-composite versus square root hours. Diffusion mechanisms which obey Fick's second law are linear in square root time [82]. It can be seen that the absorption data for the untreated laminate are not linear, but rather define a curve with

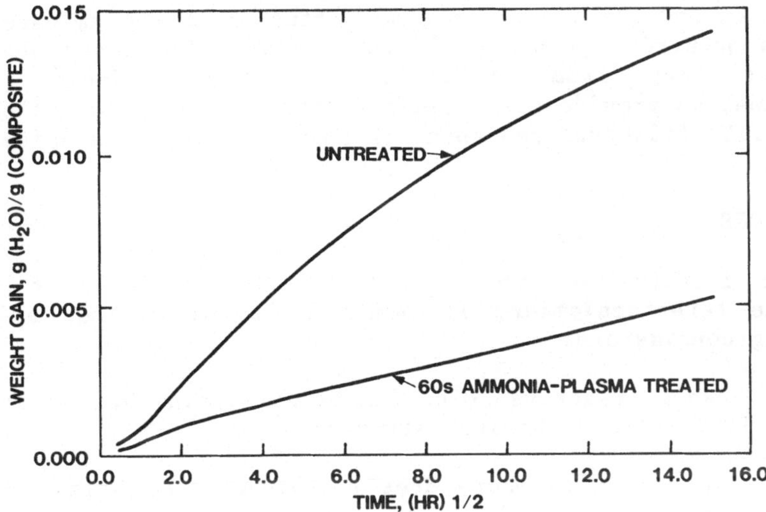

Fig. 12.   Initial moisture absorption curve for untreated and ami-
           nated Kevlar 49 fabric/332-MPDA laminates (76% relative
           humidity, 27°C).

constantly decreasing slope.   These observations are consistent
with a wicking mechanism which is saturating with time. The weight
gain due to wicking is superimposed upon the bulk diffusion into
the matrix and causes the combined absorption curve to be
nonlinear.

The laminates fabricated from aminated fabrics exhibit a
linear absorption curve versus square root time (Fig. 12). They
also absorb moisture at a much slower rate than the untreated laminates. After 196 h, the treated specimens have absorbed 4.5 mg of
moisture, whereas the untreated specimens have absorbed 13.5 mg.
After an equivalent amount of time, neat 332-MPDA resin specimens
with similar dimensions absorbed 7.0 mg of moisture. The composite
specimens used for the moisture absorption measurements (Fig. 12)
have virtually identical dimensions, densities, volume fractions
and process history, which allows their absorption curves to be
compared quantitatively. Likewise, the resin absorption specimens
are equivalent to the laminate specimens in size and process
history. They should, therefore, be representative of the bulk
matrix material.

The absorption slope of the aminated laminate specimens
(Fig. 12) is lower than that of neat resin specimens by an amount
approximately equal to the filament volume fraction. Such a volume

fraction dependence indicates that diffusion in the aminated laminates is occurring primarily in the matrix. The lower absorption
rates and change from non-Fickian to Fickian diffusion behavior
with amination provide clear evidence that the interfacial adhesion
between PPTA filaments and epoxy is improved by plasma amination of
the PPTA.

CONCLUSIONS

     The findings of this study on the exposure of poly(p-
phenylene terephthalamide) filaments to amine plasmas allow the
following conclusions:

     (1) Stable, reactive amine functional groups are incorporated
into the PPTA surface chemical structure.

     (2) The plasma exposure does not modify the bulk properties
of the filaments.

     (3) Significant amine concentrations result from plasma expo-
sure times between 5 and 60 sec.

     (4) Surface energetics of the PPTA filaments are unchanged by
plasma amination.

     (5) Epoxide matrix composite laminates reinforced with ami-
nated filaments show improved interlaminar strengths and resistance
to moisture penetration, which indicates that covalent bonds form
between the surface amine groups and the epoxy network at the
interface.

     (6) Composite failure modes are modified by improving inter-
facial adhesion, but the high-fracture energies typical of polyara-
mid/epoxy composites are retained.

APPENDIX A

<u>Procedure for Measuring Amine Concentration By Ion
Exchange</u>

     (1) Three, each approximately 2.5 cm square samples were cut
from the fabric swatches away from the edges. The three samples
from each specimen were grouped together and soaked overnight in
1.0 NHCl to protonate the amine groups.

     (2) The HCl was drained off and the samples covered with a 20
mg/ml solution of Ponceau 3R dye in water for an hour. The solutions were agitated occasionally to prevent a stagnation layer from
forming near the reaction zone.

(3) After draining off the Ponceau 3R dye, the fabric samples were thoroughly rinsed with 0.1 N CH_3COOH. The rinsing procedure consisted of washing with flowing acetic acid to remove the bulk of the unreacted dye, followed by agitation in a series of 0.1 N acetic acid solutions to remove dye entrapped between the fabric layers. Care was taken to prevent filaments being separated from the fabric swatch during the agitation steps.

(4) The fabric samples were then blotted dry and covered overnight with a solution of 0.1 N NaOH. After soaking, the fabric samples were individually rinsed with flowing 0.1 N NaOH followed by distilled water to remove dye entrapped in the fabric structure. A total combined effluent volume of about 40 ml resulted from the soaking and rinsing steps for each group of three fabric samples.

(5) The dye containing effluent was concentrated by heating the solution on a hotplate at 75°C to a volume near 20 ml. The effluent was then filtered under vacuum through a fritted glass funnel to remove loose fibrils which would influence the visible absorption measurement. After filtering, the frit was washed with 0.1 N NaOH and distilled water. The combined effluent (about 40 ml) was then concentrated at 75°C to 5-7 ml and cooled to room temperature. The concentrated solutions were then acidified with 0.5 ml of 12.0 N HCl and diluted to 10 ml with distilled water in a volumetric flask.

(6) Absorption of the solution was measured at 499 nm in a Carey Model 14C visible-ultraviolet spectrophotometer. Distilled, deionized water was run in the reference cell.

(7) The fabric samples were dried overnight under vacuum at 110°C, cooled, and weighed on an analytical balance.

(8) Amine concentrations were then calculated from the sample optical density, the Ponceau 3R calibration factor, the filament surface area and Eqns. (A1)-(A3):

$$\frac{\text{g-moles dye}}{\text{g-PPTA}} = \frac{(\text{O.D.} - 0.005)}{\text{weight PPTA}} \times \frac{(5.55 \times 10^{-8} \text{ g-moles dye}}{(\text{ml/unit O.D.}} \times \frac{10 \text{ ml}}{} \tag{A1}$$

$$\frac{\text{g-moles NH}_2}{\text{m}^2 \text{ PPTA}} = \frac{\text{g-moles dye}}{\text{g PPTA}} \times \frac{1 \text{ g PPTA}}{0.23 \text{ m}^2 \text{ PPTA}}$$

$$= \frac{(\text{O.D.} - 0.005)}{\text{weight PPTA}} \times (2.413 \times 10^{-6}) \tag{A2}$$

$$\frac{\text{NH}_2}{100 \text{ A}^2} = \frac{\text{g-moles NH}_2}{\text{m}^2 \text{ PPTA}} \times \frac{6.023 \times 10^{23} \text{ NH}_2}{\text{g-mole NH}_2} \times \frac{1 \text{ m}^2}{10^{20} \text{ A}^2}$$

$$= \frac{(\text{O.D.} - 0.005)}{\text{weight PPTA}} \times (1.453) \tag{A3}$$

ACKNOWLEDGEMENTS

Prof. Gary Wnek, MIT, and Dr. Larry Harrah, Sandia National Laboratories, contributed numerous technical suggestions throughout the course of this work. Lee Orear, Eli Perea, Gary Jones and Robert Martinez at Sandia Laboratories assisted in the preparation and testing of the composite samples and maintained the plasma equipment. John Martin at MIT was instrumental in obtaining the XPS spectra. Drs. Kay Hays, Richard Ericksen and Frank Gerstle at Sandia National Laboratories provided particularly helpful critical reviews of this manuscript. This work was supported by Sandia National Laboratories through the Department of Energy under contract number DE-AC04-76-DP00789.

REFERENCES

1. R. E. Allred, ScD Thesis, Massachusetts Institute of Technology, Cambridge, MA, February 1983.
2. L. J. Broutman, "Modern Composite Materials," L. J. Broutman and R. H. Krock, eds., Addison-Wesley, Reading, MA (1967) pp. 337-411.
3. L. B. Gresczuk, in "Interfaces in Composites," ASTM STP 452, Am. Soc. Testing and Materials, Phil, PA (1969) pp. 42-58.
4. C. C. Chamis, in "Interfaces in Polymer Matrix Composites," E. P. Pluddemann, ed., Academic Press, New York (1974), pp. 31-77.
5. G. B. McKenna, Poly.-Plast. Tech. and Engr., $\underline{5}$, 1 23 (1975).
6. S. W. Tsai and H. T. Hahn, in "Adhesion and Absorption of Polymers," L. H. Lee, ed., Plenum Press, New York (1979), pp. 463-472.
7. D. H. Kaelble, "Physical Chemistry of Adhesion," John Wiley & Sons, New York (1971).
8. D. A. Scola, in "Interfaces in Polymer Matrix Composites," E. P. Plueddemann, ed., Academic Press, New York (1974) pp. 217-284.
9. A. T. DiBenedetto and L. Nicolais, "Advances in Composite Materials," G. Piatti, ed., Applied Science, London (1978) pp. 153-181.
10. S. Wu, "Polymer Interface and Adhesion," Marcel Dekker, New York, (1982).
11. W. A. Zisman, in "Adhesion and Cohesion," P. Weiss, ed., Elsevier, Amsterdam (1962) pp. 176-208.
12. E. P. Pluddemann, "Interfaces in Polymer Matrix Composites," E. P. Pluddemann, ed., Academic Press, New York (1974) pp. 173-216.
13. D. M. Riggs, R. J. Shuford and R. W. Lewis, in "Handbook of Composites," G. Lubin, ed., Van Nostrand Reinhold, New York (1982) pp. 196-271.

14. N. B. Moore, P. S. Bruno and S. C. Browning, Proc. AIAA/SAE 10th Propulsion Conf., Am. Inst. Aero and Astro, New York, (1974), paper 74-1208.

15. L. L. Clements, R. L. Moore, E. T. Mones and T. T. Chaio, Lawrence Livermore Laboratory Report UCID-16747, Livermore, CA (April 1975).

16. L. L. Clements and T. T. Chaio, Composites, 87 (1977).

17. C. F. Griffin, L. D. Fogg, R. L. Stone and E. G. Dunning, Natl. Aero. and Space Admin. Report NASA CR-14370 (July 1978).

18. R. E. Allred, H. K. Street, and R. J. Martinez, Proc. 24th Natl. SAMPE Symp. and Exhib., Soc. Adv. Matl. and Process Engr., Azusa, CA, 31 (1979).

19. V. A. Chase, Whittaker Corp. Report, Naval Air Systems Command Contract N00019-74-C-0055 (Jan. 1975).

20. C. C. Chaio, R. L. Moore and T. T. Chaio, Composites 161 (1977).

21. L. L. Clements, Proc. 1977 Flywheel Tech. Symp., Dept. of Energy, Wash., DC (1977).

22. P. Ehrburger and J. B. Donnet, "New Fibres and Their Composites," The Royal Soc., London (1980) pp. 87-97.

23. R. J. Morgan, E. T. Mones, W. J. Steele, and S. B. Deutscher, "Proc. 12th Natl. SAMPE Tech. Conf.." Soc. Adv. Matl. and Process Engr., Azusa, CA (1980).

24. E. M. Wu, Wash. U., St. Louis, Report AMMRC-TR-80-19 (April, 1980).

25. R. E. Allred and D. K. Roylance, J. Mat. Sci., $\underline{18}$, 652 (1983).

26. L. T. Drzal, M. J. Rich and P. F. Lloyd, Polym. Preprints, Am. Chem. Soc., $\underline{22}$, 199 (1981).

27. J. H. Greenwood and P. G. Rose, J. Mat. Sci., $\underline{9}$, 1809 (1974).

28. S. V. Kulkarni, J. S. Rice and B. W. Rosen, Composites, 217 (1975).

29. R. E. Allred and A. M. Lindrose, "Composite Materials: Testing and Design," ASTM STP 674, Am. Soc. for Testing and Matls, Phil., PA (1979) pp. 313-323.

30. W. S. Smith, Proc. Conf. on Advanced Composites-Special Topics, Technology Conferences, El Segundo, CA (1979).

31. R. E. Allred, J. Composite Mat., $\underline{15}$, 100 (1981).

32. M. R. Wertheimer and H. P. Schreiber, J. Appl. Poly. Sci., $\underline{26}$, 2087 (1981).

33. E. M. Petrie and J. C. Chottiner, Proc. 40th Annual SPE Tech. Conf. and Exhib.: ANTEC '82, Soc. Plast. Engr., St. Louis, MO 777 (May 1982).

34. L. S. Penn, F. A. Bystry and H. J. Marchionni, Poly. Composites, in press (1983).

35. T. S. Keller, A. S. Hoffman, B. D. Ratner and B. J. McElroy, "Physicochemical Aspects of Polymer Surfaces," Plenum (1982).

36. J. R. Hollahan, B. B. Stafford, R. D. Falb, and S. T. Payne, J. Appl. Poly. Sci., $\underline{13}$, 807 (1969).

37. G. C. S. Collins, A. C. Lowe and D. Nicholas, European Poly. J., $\underline{9}$, 1173 (1973).

38. E. L. Lawton, J. Appl. Poly. Sci., $\underline{18}$, 1557 (1974).

39. J. C. Goan, U. S. Patent 3,776,829, Dec. 4, 1973.

40. V. P. Ivanova, G. D. Andreevskaja, J. Friedrich and J. Gahde, Acta Poly., $\underline{31}$, 752 (1980).

41. J. F. Evans and T. Kuwana, Anal. Chem., $\underline{51}$, 358 (1979).

42. A. T. Bell, "Techniques and Applications of Plasma Chemistry," J. R. Hollahan and A. T. Bell, eds., John Wiley & Sons, New York (1974) pp. 1-56.

43. B. Chapman, "Glow Discharge Processes," John Wiley & Sons, New York (1980).

44. R. F. Gould, ed., "Chemical Reactions in Electrical Discharges," Adv. in Chem. Series #80, Am. Chem. Soc., Wash., DC (1969).

45. M. Venugopalan, ed., "Reactions under Plasma Conditions," Vols. 1 and 2, John Wiley and Sons, New York (1971).

46. D. T. Clark, A. Dilks and D. Shuttleworth, "Polymer Surfaces," D. T. Clark and W. J. Feast, eds., John Wiley & Sons, New York (1978) pp. 185-211.

47. M. Hudis, "Techniques and Applications of Plasma Chemistry," J. R. Hollahan and A. T. Bell, eds., John Wiley & Sons, New York (1974) pp. 113-147.

48. H. Yasuda, in "Plasma Chemistry of Polymers," M. Shen, ed., Marcel Dekker, New York (1976) pp. 15-52.

49. D. J. Carsson, L. H. Gan and D. M. Wiles, J. Poly. Sci., Poly. Chem. Ed., $\underline{16}$, 2353 (1978).

50. R. d'Agostino, F. Cramarossa, S. DeBenedictis and G. Ferraro, Plasma Chem. and Plasma Processing, $\underline{1}$, 19 (1981).

51. M. Capitelli and E. Molinari, "Topics in Current Chemistry," $\underline{90}$, Springer-Verlag, New York (1980) pp. 59-109.

52. L. Penn and F. Larsen, J. Appl. Poly. Sci., $\underline{23}$, 59 (1979).

53. P. W. Morgan, Macromolecules, $\underline{10}$, 1381 (1977).

54. T. I. Bair, P. W. Morgan and F. L. Killian, Macromolecules, $\underline{10}$, 1396 (1977).

55. Standard Test Methods, Part 36, Am. Soc. for Testing and Matls., Phil, PA (1980).

56. W. S. Smith, E. I. duPont de Nemours, Wilmington, DE, private communication.

57. C. Hendrick, G. Grant and J. Howard, Millipore Corp., Bedford, MA, unpublished results.

58. R. E. Allred and N. H. Hall, Poly Engr. and Sci., $\underline{19}$, 907 (1979).

59. Standard Test Methods, Part 38, Am. Soc. for Testing and Matls., Phil., PA (1980).

60. K. D. Boultinghouse, Sandia National Laboratories Report SAND 80-0975 (June 1980), available NTIS.

61. A. W. Adamson, "Physical Chemistry of Surfaces," John Wiley & Sons, New York (1976).

62. D. T. Clark, in "Characterization of Metal and Polymer Surfaces: Vol. 2," L.-H. Lee, ed., Academic Press, New York (1977) pp. 5-51.

63. D. T. Clark, in "Polymer Surfaces," D. T. Clark and W. J. Feast, eds., John Wiley & Sons, New York (1978) pp. 309-351.

64. R. Holm and S. Storp, Surf. Interface Anal., $\underline{2}$, 96 (1980).

65. D. T. Clark and A. Dilks, in "Characterization of Metal and Polymer Surfaces: Vol. 2," L.-H. Lee, ed., Academic Press, New York (1977) pp. 101-132.

66. H. Yasuda, H. C. Marsh, S. Brandt, and C. N. Reilley, J. Poly. Sci., Poly. Chem., $\underline{15}$, 991 (1977).

67. J. A. Tyalor, Physical Electronics Laboratories, Eden Prairie, MN, communication to J. R. Martin, MIT, March 1981.

68. D. T. Clark and A. Harrison, J. Poly. Sci., Poly. Chem., $\underline{19}$, 1945 (1981).

69. C. Y. Kim, J. Evans and D. A. I. Goring, J. Appl. Poly. Sci., $\underline{15}$, 1365 (1971).

70. D. Briggs, D. G. Rance, C. R. Kendall and A. R. Blythe, Polymer $\underline{21}$, 895 (1980).

71. A. K. Sharma, F. Millich and E. W. Hellmuth, J. Appl. Poly. Sci., $\underline{26}$, 2205 (1981).

72. S. P. Wesson and A. Tarantino, J. Non-Cryst. Solids, $\underline{38/39}$, 619 (1980).

73. S. L. Phoenix and E. M. Wu, presented at the IUTAM Symp. on Mech. Compos. Matls, Blacksburg, VA, Aug. 16-19, 1982.

74. S. L. Phoenix, Cornell University, private communication.

75. C. R. Noller, Chemistry of Organic Compounds, W. B. Saunders Co., London (1966).

76. M. G. Northolt, European Poly. J., $\underline{10}$, 799 (1974).

77. S. P. Wesson, Owens-Corning Technical Center, Granville, OH, unpublished data.

78. P. Avakian, R. C. Blume, T. D. Gierke, H. H. Yang and M. Panar, Polym. Preprints, Am. Chem. Soc., $\underline{21}$, 9 (1980).

79. M. G. Bader, J. E. Bailey and I. Bell, J. Phys. D: Appl. Phys., $\underline{6}$, 572 (1973).

80. T. U. Marston, A. G. Atkins and D. K. Felbeck, J. Mat. Sci., $\underline{9}$, 447 (1974).

81. J. M. Augl, Naval Surface Weapons Center Report NKSWC/TR-79-51, Silver Spring, MD (March 1979), available NTIS.

82. J. Crank and G. S. Park, eds., "Diffusion in Polymers," Academic Press, New York (1968).

SPECTROCHEMICAL CHARACTERIZATION OF

CHEMICALLY MODIFIED SURFACES

D. E. Leyden and D. E. Williams

Department of Chemistry, Colorado State University
Fort Collins, CO 80523

Dow Corning Corporation, Midland, MI 48640

INTRODUCTION

The applications of surface immobilized chemical functional groups on various substrates have been steadily growing in industry, technology and pure sciences. Silylation is only one mechanism of immobilization, but one which is widely employed. The purpose of these investigations was to determine the effect, if any, of bonding organofunctional groups to siliceous surfaces upon the reaction properties of the functionality. In part, these properties were characterized by the use of a variety of chemical investigations. Any general conclusions necessarily oversimplify the results. However, it is frequently observed that the chemistry of the groups investigated does not deviate greatly from a comparable functional group or compound in solution. Therefore, the utilization of spectroscopic techniques commonly used in solution phase studies is suggested in order to elucidate subtle differences between bound and solution phase functionalities.

Because a variety of spectroscopic techniques are employed, these and their advantages in this study will be briefly described. Fourier transform infrared (FTIR) spectroscopy has become a standard laboratory tool for the investigation of organic compounds [1]. Silica substrates provide excellent samples for transmission FTIR. Although this transmission mode does not provide spectra restricted to the surface material, the ability to obtain high sensitivity and to subtract spectra of the substrate from those of modified materials provides a useful tool.

Many of the functional groups investigated yield ultraviolet or visible (UV-VIS) spectra. These spectra are low in information content; that is, they have little fine structure and the spectrum of one compound is often not readily distinguishable from another. However, when combined with variations in chemical environment, UV-VIS spectroscopy can be a powerful probe. The problem is to find a method to obtain such a spectrum. Photoacoustic spectroscopy (PAS) provides such a method [2]. The UV-VIS photoacoustic spectra may be obtained from particulate siliceous materials using rather simple optical components, and quantitative information about the degree of surface coverage or the composition of mixtures of treated and untreated substrates can be obtained [3].

A recently developed technique of magnetic resonance spectroscopy has proved to be an increasingly powerful tool for the investigation of organofunctional groups immobilized on silica and other substrates [4]. This technique of cross polarization (CP) nuclear magnetic resonance with magic angle spinning (MAS) has provided for the acquisition of high resolution nuclear magnetic resonance spectra from such samples. CP-MAS spectra of ^{13}C and ^{29}Si can be routinely obtained on samples of silylated substrates. Detailed studies of the silylation reaction using ^{29}Si NMR have been demonstrated [4]. The cross polarization technique involves an enhancement of the NMR signal of the desired nucleus by a transfer of the magnetization of the hydrogen nucleus to these more dilute nuclei. Because the bulk of silica contains less hydrogen than the surface, the technique becomes sensitive to the Si nuclei at the surface as a result of hydrogen atoms on the silane, or in silanol groups on the surface. One cannot say that ^{29}Si-NMR using CP-MAS is a surface technique, but the surface sensitization is a distinct advantage.

EXPERIMENTAL

Details of the experiments will not be given here as they are provided in the references given for each result discussed. The FTIR spectra were acquired with a Nicolet MX-1 FTIR spectrometer. Samples were either Cab-O-Sil 5-10 nm diameter fumed silica particles suspended in a halocarbon oil, or mixed with KBr and pressed into pellets, or silica gel platelets which were reacted from the vapor phase for 'surface modification. Photoacoustic spectra were obtained using a spectrometer constructed in our laboratory from components purchased from optical, electronic and scientific supply houses [3]. The instrument uses a 1kW xenon lamp for the ultraviolet source, a 400 W tungsten lamp for the visible source, and a double grating 1/4 meter monochrometer. Scans are acquired in the range 270-400 nm, and 400-800 nm with the separate gratings. Spectra are acquired by step scan under the control of an Apple II microcomputer with signal averaging at each step. A PAS cell constructed in our laboratory is used at room temperature. Both magnitude and phase data are acquired at modulation frequencies of approximately 30 Hz. Nuclear magnetic resonance spectra are acquired using the National Science Foundation Nuclear Magnetic Resonance Regional Center at Colorado State University.

RESULTS AND DISCUSSION

A significant interest in immobilized reagents in our laboratory has dealt with bonded metal chelating agents. These materials have been shown to be effective for the extraction of trace levels of metal ions from aqueous solution. If some of the problems of long term stability of these materials can be solved, they may be useful in the extraction of resources from mining and waste treatment operations. The fundamental behavior of these chelating reagents on surfaces is of importance to their further exploitation. Many studies of the stoichiometry of metal chelate reaction have been conducted in solution using ultraviolet and visible spectroscopy. Standard techniques, such as Job's plot and mole ratio method, have been used. Adaptation of these methods to the investigation of metal complex formation of surfaces is made possible using photoacoustic spectroscopic techniques [3,5].

N-2-aminoethyl-3-aminopropyltrimethoxysilane (Dow Corning Z-6020) is a silane which along with having many other important properties is an excellent chelating agent because of the 1,2-diamine functionality. This silane bonds well to silica materials such as silica gel and controlled pore glass to yield a surface coverage of approximately 2-4 micromoles/m^2 of surface area. The surface area is determined by N_2 BET measurement, and the functional loading by titration with acid or copper (II) ion, or by carbon and nitrogen microanalysis. This compound, when bonded to silica surfaces, forms metal chelates with stability ratios very analogous to ethylenediamine and other substituted 1,2-diamines in aqueous solution [6]. A plot of the log of the distribution coefficient of a series of metal ions between water and the silica immobilized N-2-aminoethyl-3-aminopropyltrimethoxysilane versus the log of the stability constant of the corresponding metal-ethylenediamine complexes in solution is linear, indicating a close correlation between the two complex formation processes. Most of these metal complexes have a ligand-to-metal ratio of 2:1. The restriction imposed when the ligand is bound to a surface could significantly affect this stoichiometry when immobilized reagents are considered. From the literature and from data acquired in our laboratory from studies of ethylenediamine and substituted 1,2-diamines in aqueous solution, it was determined that stepwise formation of 1:1 and 2:1 copper (II)-diamine complexes in solution was identified by absorbances in the visible region at 658 nm and 556 nm, respectively. The concentration of the two species could be followed at the above wavelengths. Samples were prepared by careful regulation of the copper (II) concentration applied to silica gel which contained immobilized N-2-aminoethyl-3-aminopropyltrimethoxysilane, and the visible PAS spectra of these samples were obtained. At low concentration, the photoacoustic absorption spectrum of the surface species formed is identical to the solution phase absorption spectrum of the bis copper (II) complex with N-2-aminoethyl-3-aminopropyltrimethoxysilane and is shown as spectrum A in Figure 1. The spectrum of the mono complex shown as C in Figure 1 is approximated by stripping the predominantly bis component from the spectrum obtained at high copper loading. A shift to longer wavelength is due in part to association with acetate ion. Careful analysis of the data by a variety of techniques showed that the immobilized diamine could form not only the expected 1:1 chelate with Cu(II), but the 2:1 chelate as illustrated in Figure 2. The evidence indicated that, in both cases, the diamine group was bidentate, bonding both nitrogen atoms to copper. Furthermore, it was concluded that there were two distinctive types of diamine functionalities on the surface. One type could form only the 1:1 chelate. These sites were very dilute on the surface. A second site, approximately 2.5 times more abundant, forms the 2:1 chelate easily, and could not be forced to the 1:1 complex even in contact with concentrated copper ion solution. No evidence was found for the conversion of the 2:1 to 1:1 sites with increasing Cu(II) concentration [5]. These results are in remarkable agreement with those of Pinnivaia who used electron spin resonance for a similar investigation [7].

Another type of interesting ligand is that which contain 1,2-diketones. These beta-diketones involve tautomeric equilibria between their keto and enol tautomers. Usually, the enol tautomer is the one which reacts with metal ions to form chelate compounds. Such compounds are important in the analytical chemistry and separation of metal ions. The tautomers show significantly different absorbances in the UV spectral region.

Tautomers of beta-diketones also exhibit significant differences in their spectral characteristics in the infrared region. An immobilized 3-benzyl-2,4-pentanedionetrimethoxysilane was investigated using UV-PAS and FTIR spectroscopy. For both of these studies, it was very helpful to investigate 3-benzyl-2,4-pentanedione as a model. In this case, the neat compound was found to be 46% enol by proton NMR. An even higher percentage of enol is present in dilute solutions in nonpolar solvents. The FTIR spectrum of the model compound adsorbed on silica gel is shown in Figure 3. The adsorbed compound shows absorption bands at 1721 and 1693 cm^{-1}, whereas the neat compound shows bands at 1725 and 1700

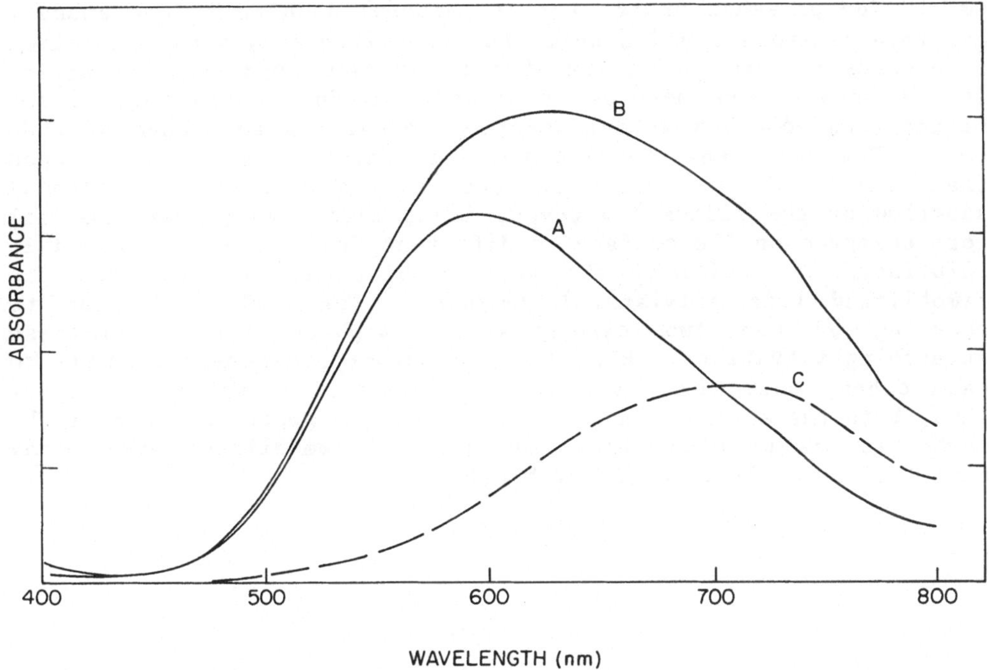

Fig. 1.   Scales photoacoustic absorption spectra for N-2-aminoethyl-
3-aminopropyltrimethoxysilane (Z-6020) modified silica gel
loaded with Cu(II) at an initial Cu(II) concentration of
(A) 0.001 mol/L and (B) 0.01 mol/L in 0.10 mol/L acetate
buffer.   (C) Spectrum of mono species found by difference.

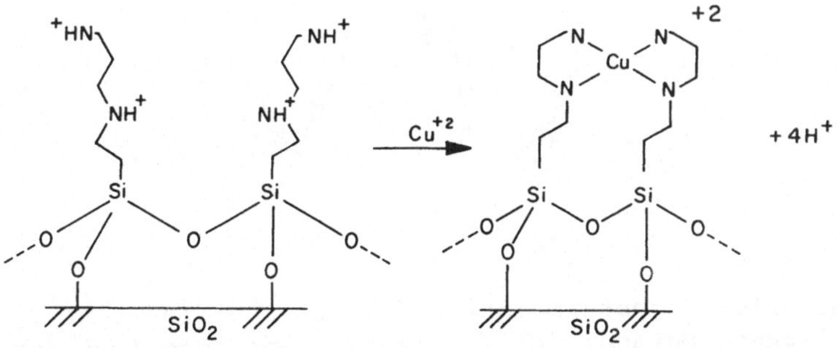

Fig. 2.   Schematic representation of the formation of the bis-
Cu(II) complex on a silica surface.

cm^{-1}. The presence of the ketone carbonyl doublet is not unusual for beta-diketones, and a shift to lower frequency with adsorption on a surface is an indication of hydrogen bond formation. However, in the case of the adsorbed compound, careful subtraction of the silica (and adsorbed water) spectrum reveals an enol band at 1630 cm^{-1}. The exact enol-to-keto ratio appears to be dependent upon the fraction of the surface covered in the case of the compound adsorbed on the silica. However, it is also likely that the enol form observed on the surface is different from the one observed in solution. In addition, the keto doublet is not seen for the immobilized (via silylation) compound. The doublet is due to coupling of the two carbonyls by in-phase and out-of-phase stretching vibrations. When the two carbonyl groups are trans to each other, only the out-of-phase vibration is infrared active. Thus, both the adsorbed and neat 3-benzyl-2,4-pentanedione have the carbonyls cis to each other, whereas the immobilized analog may have the carbonyls trans to each other.

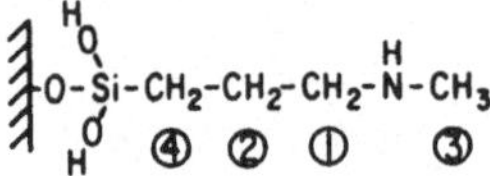

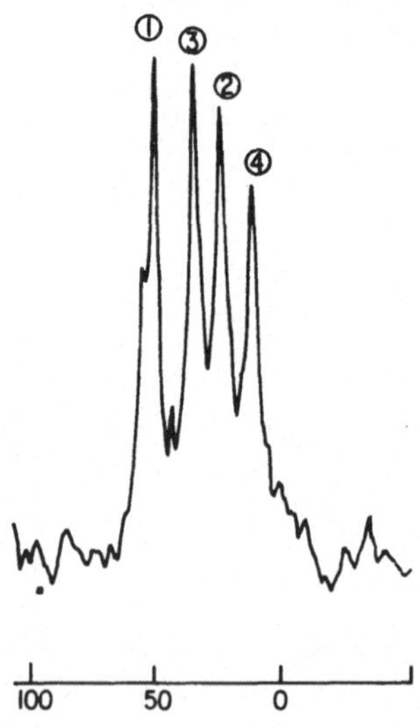

Fig. 3.   Fourier transform infrared spectra of (A) neat 3-benzyl-2,4-pentanedione, (B) 3-benzyl-2,4-pentanedione adsorbed on silica gel after subtraction of untreated silica gel spectrum, and (C) immobilized 3-benzyl-2,4-pentanedione silane analog after subtraction of untreated silica gel spectrum.

The infrared spectrum contains information only about ground electronic states, whereas the ultraviolet spectrum can provide information concerning excited electronic states as well. In cyclohexane, 3-benzyl-2,4-pentanedione has an absorption maximum at 287 nm, which is the enol absorption. The keto band, which is not observed, is expected to occur at a similar wavelength with a molar absorptivity an order of magnitude less. The photoacoustic spectra of the immobilized beta-diketone and the adsorbed model compound show maxima at 271 and 265 nm, respectively. The similarity of these two spectra, and the infrared results, support the conclusion that both the adsorbed, and immobilized beta-diketone are largely in the keto form, probably with hydrogen bonding to the residual silanol sites on the surface. Detailed analysis of the photoacoustic UV spectrum indicated that the remainder of the compound bound to the silica surface is in the form of an intermolecular hydrogen bonded enol. Studies of the immobilized beta-diketone in contact with basic solutions show that the enolate ion is formed on the surface, and that metal-enolate complexes with ions such as Fe(III), Cu(II), and Ni(II) can be formed. Infrared evidence indicates the presence of weaker metal binding with the keto form in more acidic media.

FTIR studies in progress have provided interesting results. Silica plates are prepared using a novel method of drying the plates at supercritical carbon dioxide pressure from a nonaqueous, but water miscible solvent such as acetonitrile or tetra-hydrofuran. Platelets prepared in this way may be considered as completely hydrated because the materials are never subjected to temperatures above 40°C. Silylation reactions were conducted in both aqueous and nonaqueous solvents. In the case of bonded aminosilanes, complete or partial protonation was accomplished by either: a) direct exposure of the treated plates to gaseous HCl; b) titration of HCl into a suspension of the plates in water, followed by extraction of the plates with acetonitrile and carbon dioxide drying. Both methods yield similar infrared spectra. These spectra are characterized by many interesting features, among which are bands at 1530, 1595 and 1570 cm^{-1} which are assigned to the protonated amine, nonprotonated amine and a hydrogen bonded or strained amine N-H, respectively. The band at 1570 has been previously assigned by Ishida and co-workers to the amines which are hydrogen bonded to untreated SiOH groups on the silane silicon atom [8]. A recent suggestion is that this band is a result of the formation of an amine-bicarbonate [9]. Either interpretation is consistent with the observation that immobilized triethoxy-aminopropylsilane reacted from aqueous solution, or samples reacted from nonaqueous solution, but exposed to water vapor, exhibit greater intensity at 1570 cm^{-1} than samples prepared from carefully controlled anhydrous conditions. An interesting observation is that these bands appear in silica gel or platelets that were prepared from anhydrous media, and then pressed into pellets at 2500 psi.

The use of ^{13}C CP-MAS NMR can provide information on the behavior of bonded functional groups, including confirmation of the structure of the bound organofunctional silane [10]. As a simple example, the ^{13}C NMR spectrum of N-methyl-3-aminopropylsilane bound to silica gel is shown in Figure 4. The chemical shift of the ^{13}C-NMR of carbon atoms adjacent or one carbon removed from an amine functional group in the ^{13}C-NMR spectrum is sensitive to the protonation of the amine. As a rule, the resonance of the carbon nucleus beta to the amine nitrogen atom shifts more upon protonation of the amine than does that of the alpha carbon. This result provides a probe for the investigation of the protonation of surface bound amine. Interesting preliminary results have been obtained. Bonded 3-aminopropyltriethoxysilane on silica gel was suspended in water and titrated with perchloric acid. The stoichiometry of the titration was confirmed using thermometric titration as potentiometric titrations gave poor endpoints. Aliquots of silica gel were removed during the titration and ^{13}C CP-MAS NMR spectra were acquired. If the silica samples were carefully dried, two NMR resonances were observed. These are assumed to correspond to the protonated and unprotonated amine groups, respectively, on the surface. During the titration, the former resonance grew in intensity at the expense of the latter. However, if the silica was not thoroughly dried, a single, broad resonance was observed which changed in shape as the titration progressed. Further detailed studies are underway. Such studies should provide valuable information for the design of useful bonded organofunctional groups.

CONCLUSIONS

This report has demonstrated the potential of several forms of spectroscopy for the characterization of silylated siliceous surfaces. Examples of the properties investigated are the behavior of metal chelating groups bonded to surfaces, keto-enol equilibria, quantitative determinations of bonded groups, and the details of protonation of bound aminosilanes. The potential of the methods has been adquately demonstrated. Further work will require detailed, systematic applications to answer fundamental questions.

ACKNOWLEDGEMENTS

This work was supported in part by Research Grant CHE-78-23123 from the National Science Foundation. The use of the Colorado State University Regional NMR Center funded by National Science Foundation Grant CHE-78-18581 is acknowledged. Contributions to this research by L. W. Burggraf, D. S. Kendall and F. J. Pern are also acknowledged.

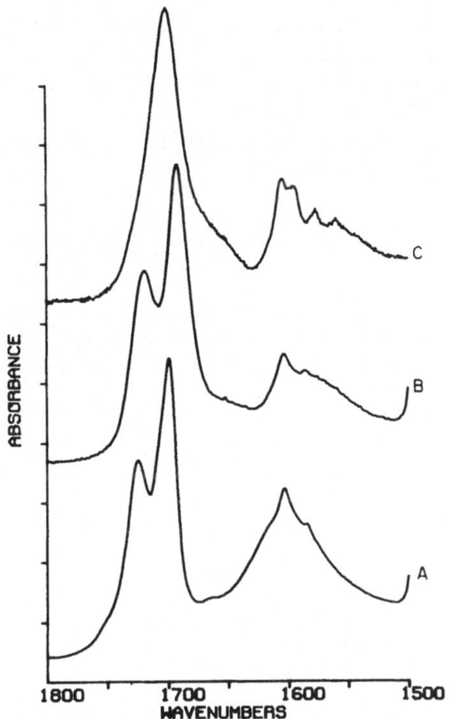

Fig. 4. Solid state ^{13}C NMR spectrum of silica gel treated with N-methyl-aminopropyltrimethoxysilane.

REFERENCES

1. P. R. Griffiths, "Chemical Infrared Fourier Transform Spectro-
     scopy," Wiley-Interscience Publishers, New York, NY, 1975.
2. Y.-H. Pao, "Optoacoustic Spectroscopy and Detection," Academic
     Press, New York, NY, 1977.
3. L. W. Burggraf and D. E. Leyden, Anal. Chem., 53, 759 (1981).
4. D. W. Sindorf and G. E. Maciel, J. Am. Chem. Soc., 103, 4623
     (1981).
5. L. W. Burggraf, D. W. Kendall, D. E. Leyden and F.-J. Pern,
     Anal. Chim. Acta, 129, 19 (1981).
6. D. E. Leyden and G. H. Luttrell, Anal. Chem., 47, 1612 (1975).
7. T. J. Pinnivaia, J. G. Lee and M. Abedeni, in "Silylated Sur-
     faces," D. E. Leyden and W. T. Collins, Eds., Gordon and
     Breach, New York, 333 (1980).

8. C.-H. Chaing, H. Ishida and J. L. Koenig, J. Colloid Interface
    Sci., $\underline{74}$, 396 (1980).
9. S. Naviroj, J. L. Koenig and H. Ishida, J. Macromol. Sci.-
    Phys., B22(2), 291 (1983).
10. D. E. Leyden, D. E. Kendall and T. G. Waddell, Anal. Chim.
    Acta, $\underline{126}$, 207 (1981).

DESORPTION OF WATER FROM GLASS FIBERS

Gary M. Nishioka and Janet A. Schramke

Research and Development Division
Owens-Corning Fiberglas Corporation
Technical Center
Granville, Ohio   43023

ABSTRACT

A sensitive electrolytic technique suitable for quantitative measurement of the different states of water bound to low surface area substrates (0.1 m^2/g) is described. The method involves the thermodesorption of water from a substrate with subsequent measurement of the desorbed water as a function of temperature using a P205 electrolytic cell.

This technique was used to study water on pure silica and silicas doped with metal cations. Strongly bound molecular water and water resulting from dehydroxylation reactions were detected and quantitatively measured. Results were consistent with previous studies of porous silicas. The presence of cation-aluminate sites on the silica surface greatly increased the amount of strongly bound water.

Studies of the desorption of water from glass fibers are also presented. It is concluded that water exists on glass fibers as either physically adsorbed water, chemisorbed molecular water, surface hydroxyl groups, near surface internal water, or bulk water.

INTRODUCTION

Numerous studies have indicated that adsorbed water controls significant properties of glass fibers. For example, a number of investigators found that the strength of E-glass fibers decreases with increasing humidity in the testing environment [1,2,3].

Static and dynamic fatigue of glass also increase with increasing testing humidity [4,5,6]. These stress corrosion effects are thought to be caused by the reaction of water with the silicate network at crack tips. This reaction causes bond rupture at the tip, exposing new bonds, which rupture after further water adsorption occurs. This process continues until the crack grows to the critical size required for spontaneous failure.

Adhesive failure at a polymer-glass interface is also induced by water attack. The strength of plastics reinforced with inorganic fillers is greatly reduced after immersion in water. Indeed, a standard test of glass reinforced composites involves measuring shear strength of a test composite before and after immersion in water [7]. The failure mode in compsites immersed in water has been shown to occur at the glass-resin interface [8,9]. Coupling agents are normally coated onto glass fibers to reduce this moisture induced degradation of composites; the efficacy of the coupling agent coating is inversely related to its water wettability [10].

Electrical conductivity is a third property affected by water. The conductivity of the glass surface increases with the amount of adsorbed water present [11,12]. Electrical conductivity presumably increases due to formation of an aqueous layer containing dissolved surface cations. A similar effect is responsible for the degradation in electrical resistance of glass fiber reinforced circuit boards when exposed to moisture [13,14].

Despite its importance, the water present on the glass fiber surface has not been studied extensively. This is probably due to the relatively low specific surface area of glass fibers. Whereas often studied substrates such as Cab-O-Sil M5 (a fumed silica) have surface areas around 200 m^2/g, glass fibers have specific areas around 0.2 m^2/g. All forms of infrared spectroscopy are insensitive for the study of adsorbed water on such a low area substrate. All but the most sophisticated gravimetric systems (sensitive to $\pm$ 1 ppm of weight) lack the requisite sensitivity for the measurement of adsorbed water. The inadequacy of standard techniques prompted our development of a thermodesorption method for the study of water on glass fibers. Reported herein is a study of water desorbed from glass powders and fibers, an interpretation of the data, and the variations encountered in the types and amounts of water on glass fibers.

MATERIALS

The glass fibers studied were formed under a variety of environmental conditions. All fibers were formed without the use of sizes (coatings).

The fumed silica studied was Cab-O-Sil M5, obtained from Cabot Corporation and used as received. The precipitated and doped silicas were supplied by R. K. Iler, prepared by the following procedures:

A. Precipitated Silica

An amorphous silica powder was prepared using a sample of Ludox AS-40 from DuPont. Ludox AS-40 is a silica sol containing particles of about 220 Angstroms in diameter stabilized with NH_4OH to pH 9. The Ludox was deionized with mixed ion exchange resin; warmed to 50 °C, and gelled. It was mixed in an equal volume of normal propyl alcohol, washed on filter with alcohol, air dried and dried in an oven at 150°C. BET area = 124 m^2/g.

B. Aluminate Modified Silica

Precipitated silica prepared in "A" was slurried in water. Keraco sodium aluminate solution (containing $NaAlO_2$ and NaOH) was slowly added with stirring to the silica suspension, stirred for an hour at 25°C and the pH reduced to 10 with acetic acid. The amount of sodium aluminate solution added was controlled so that two aluminate groups per square nanometer of surface should be produced.

C. Sodium Form of Silica Surface

Ten grams of precipitated silica "A" was slurried in 200 ml H_2O and titrated with 0.1N NaOH to pH 8.

D. Calcium Form of Silica Surface

Ten grams of precipitated silica "A" was slurried in 200 ml H_2O and 35 ml of 0.28 molar calcium formate was added. The pH increased from 3.7 to 7.5. The pH was raised from 7.5 to 8 by adding 8 ml of 0.1N NaOH and the silica was then filtered, washed, and dried at 125°C.

E. Sodium Form of Aluminate Modified Silica

Half of the sample prepared in "C" was washed to reduce the pH to 8.5 and filtered and dried at 125°C in air.

F. Calcium Form of Aluminate Modified Silica

Half of sample "C" was converted to the calcium form by ion exchange with an excess of calcium formate solution: 0.28M calcium formate solution was added to the aluminate modified silica slurry. The slurry was stirred one hour, filtered and washed until free from excess calcium ions. Just before filtering the solution,

0.1N NaOH was added to raise the pH from 7.2 to 8.1. The product was air dried at 125°C.

APPARATUS AND PROCEDURE

Figure 1 is a block diagram of the thermodesorption apparatus. An analysis involves placing a weighed sample into the oven. The oven temperature increases at a programmed rate, usually 4°C/min, causing water to desorb from the sample. Nitrogen carrier gas flows through a sieve dryer and into the oven, transporting the desorbed water to the electrolytic cell. The electrolytic cell is manufactured by DuPont Instruments and consists of a thin film of phosphorous pentoxide (P_2O_5), deposited between two helically wound electrodes. Water is absorbed by the P_2O_5, and electrolyzed by current flowing between the electrodes. Electrolysis changes the water to hydrogen and oxygen which discharge through a vent with the carrier gas, coulometrically regenerating P_2O_5. The charge required to regenerate P_2O_5 is related to the quantity of desorbed water, so the cumulative amount of desorbed water, oven temperature, and time are continuously recorded by a computer.

Weighed samples of powdered glass or pristine fiber were analyzed with this instrument. The specific surface area of the fibers was determined by measurement of fiber diameter and assuming a smooth cylindrical shape. The specific surface areas of the powders was determined by the one point BET method. All data were normalized on a surface area basis.

Areas of peaks occurring below 500°C are obtained by direct measurement of the cumulative water recorded between two temperatures. The water desorbed above 500°C is calculated by doubling the measured area up to the peak maximum (see Figure 2).

RESULTS AND DISCUSSION

Figure 2 is representative of the types and amount of water desorbed from glass fibers. The rate of water evolved with change in temperature, normalized for area, is plotted against temperature. The quantity of water evolved within any temperature region is the area of the curve between the temperature limits. Water does not desorb uniformly with increasing temperature, but rather in specific temperature zones, presumably corresponding to different types of adsorbed water. To better understand the meaning of these data, high area, well characterized substrates were first examined. It is assumed, for the samples studied in Section A, that all desorbed water originated from the surface. Since the samples in Section A were all high area powders, this is certainly a reasonable assumption. Some of the results in Section A will be used to interpret the data presented in Section B: studies of water associated with glass fibers.

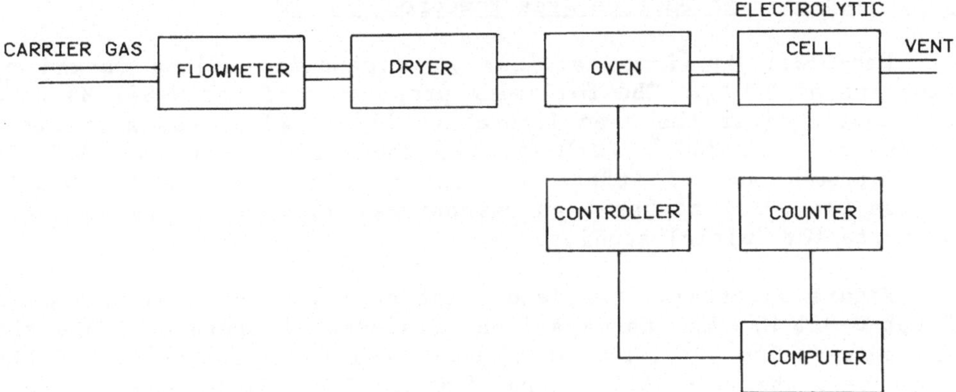

Fig. 1. A block diagram of the thermodesorption system. This apparatus measures desorbed water of gram samples with a precision to a few tenths of a microgram.

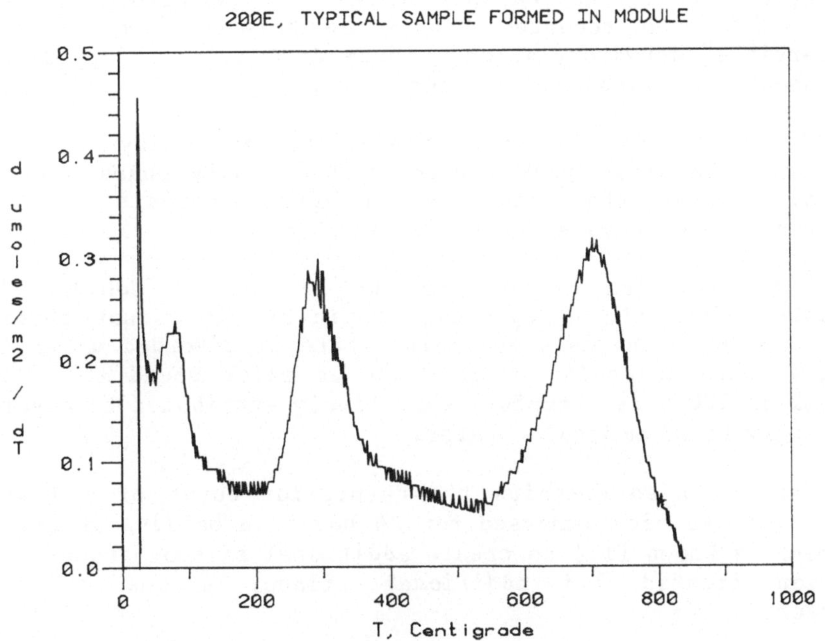

Fig. 2. Typical desorption trace of E-glass fibers.

A. Studies of High Surface Area Powdered Silicas

Cab-O-Sil M5 (fumed silica) is a powdered silica formed by oxidation of $SiCl_4$. The following properties of Cab-O-Sil M4 have been confirmed in the open literature [15]: (1) it has a specific surface area of 200 ± 20 M^2/g, (2) the surface contains 3.5–4.5 silanols/nm^2, and (3) there is little bulk water in Cab-O-Sil. Because Cab-O-Sil M5 has been extensively studied, it is an ideal substrate for initial studies.

Figure 3 contains two desorption curves of an ambient sample of Cab-O-Sil M5. The three salient features of these data are the peak between 25–120 °C, the broad peak from 150°C to 500° C, and the third peak between 500°C and 800°C. The first peak appears instantly upon insertion of the sample, and must be due to weakly bound water. The weakly bound, physically adsorbed water is not measured quantitatively by this technique.

Water desorbed above 150 °C can be wholly accounted for as the product of the dehydroxylation of surface silanols. The measured value of water desorbed above 150°C, equivalent to 3.3 silanols/nm^2, is in excellent agreement with reported values of clustered silanol concentration of 3/nm^2. An interesting effect, not reported in the literature, is the desorption of water in two temperature zones. The boundary between the two desorption regions is at 450°C, the reported transition between "reversible" and "irreversible" dehydroxylation. These data indicate approximately equal amounts of each type of clustered group.

If a sample of Cab-O-Sil is equilibrated in dry nitrogen for two hours, the large peak caused by the weakly bound water disappears. However, the solid curve in Figure 4 reveals that a small peak remains due to water desorbing at 100° C.

This water can be neither the product of dehydroxylation reactions, since these reportedly begin at 150°C, nor physically adsorbed water. The only remaining state of adsorbed water is the strongly bound molecular water, such as water bound to silanols. The peak at 100 °C is therefore tentatively attributed to desorption of strongly bound molecular water.

Figure 4 also contains the desorption curve of a Cab-O-Sil which had first been immersed for 24 hours in boiling water. This treatment is known [15] to create additional silanol groups. Water immersion created 1.3 additional silanol groups per square nanometer.

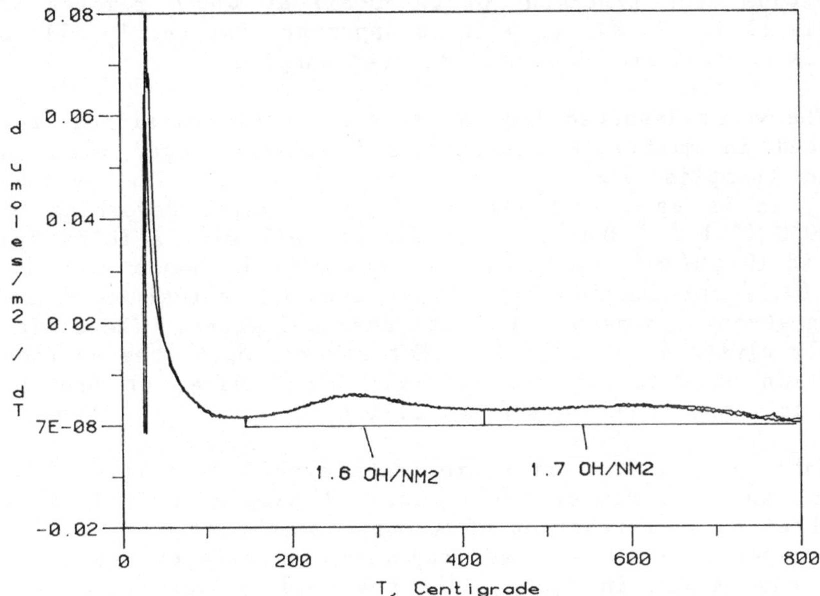

Fig. 3.   Two measurements of the water desorbed from CabOSil M5.

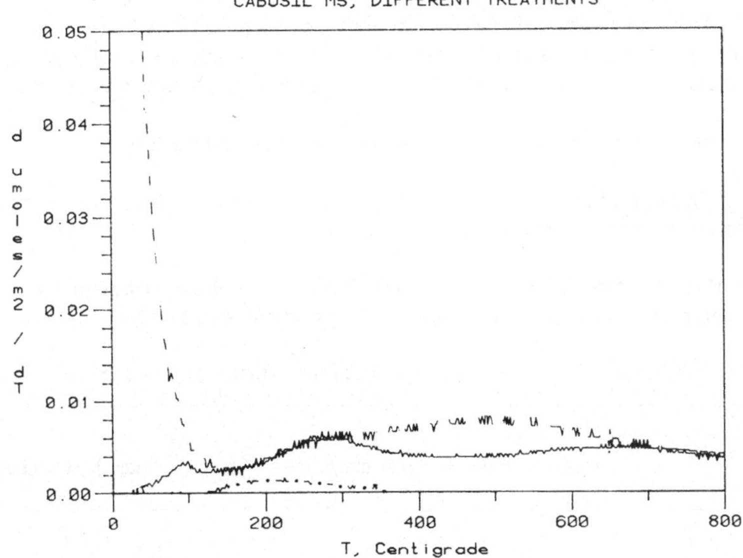

Fig. 4.   The effect of various treatments on the desorption of
water from CabOSil M5.

Vacuum heat treatment of Cab-O-Sil at 800°C removes surface silanols [15]. In Figure 4 it is apparent that nearly all surface water is removed from our heat treated samples.

The water desorbed from a fully hydroxylated silica (sample A described in materials section) and silicas doped with various cations (samples C-F) is shown in Figure 5. For hydroxylated silica, it is seen that the quantity of water desorbing between 200-800°C (5.1 ± 1 H_2O/nm^2) correlates well with a fully hydroxylated (8-10 OH/nm^2) surface. The quantity of water desorbing at 100°C ($0.5/nm^2$) implies that there are 0.5 sites/nm^2 capable of forming strong hydrogen bonds with adsorbed water. These sites are probably clusters of silanols. Not surprisingly, the surface concentration of silanols and strongly bound water on precipitated silica is higher than on a fumed silica.

Table I contains the approximate atomic densities of sodium, calcium, and aluminum on the surface of samples A-F. These values were obtained by extraction of a known weight of each silica in a known volume of 0.1N HCl, and solution analysis by atomic absorption. Also given in Table I is the surface water concentration obtained from the area of the desorption peak at 100°C.
It appears that only minor amounts of sodium or calcium adsorb onto silicate sites. Sodium adsorbs at roughly a 1:1 ratio on aluminate sites, while calcium adsorbs at roughly a 1:2 ratio on aluminate sites. Each sodium aluminate site attracts one water molecule, each calcium aluminate site attracts two or three water molecules.

The preceding studies allow us to conclude that:

(1) Physically adsorbed water cannot be measured since desorption occurs at room temperature;

(2) Water desorbing at 100°C is either strongly bound to cation-aluminate sites, or bound to paired silanols;

Table I. Estimated Surface Concentrations

Sample	Na (atoms/nm^2)	Ca (atoms/nm^2)	Al (atoms/nm^2)	H_2O (molecules/nm^2)
SiO_2 (A)	.28	.14	.006	0.5
SiO_2 + Na (C)	.12	.03	.004	0.6
SiO_2 + Ca (O)	.01	.37	.004	0.7
SiO_2 + AlO + Na (E)	2.00	.04	2.16	2.2
SiO_2 + AlO + Ca (F)	.31	1.02	1.73	2.7

(3) Cation sites adsorb more total water than silanol sites; and

(4) Water desorbing between 200–800°C is unaffected by surface cations and can be attributed to dehydroxylation reactions.

B. Studies of Water on Glass Fibers

Six replicate measurements of water desorbed from a single package of E-glass fibers are shown in Figure 6. This sample was formed under simulated plant conditions, however, no coatings (size) were applied during forming.

These fibers desorb more water per unit area than powdered silicas, too much water to be accounted for as solely the desorption of surface water.

The water desorbed between 500°C and 800°C must be in large measure due to diffusion of internal water out of the fibers. Several facts support this conclusion. First, this water cannot be entirely surface water since it is inconceivable that 245 molecules/nm^2 can be tightly bound to the surface. This would correspond to about 25 layers of chemisorbed water. Second, the quantity of water desorbed in this temperature range comprises 0.08 percent of the fiber weight, in reasonable agreement with literature values of bulk water in E-glass [16]. Third, at higher temperatures the diffusion distance of water in silica is greater than the fiber radius, implying that water diffuses out of the interior. The diffusion coefficient for water in silica is roughly:

$$D = D_O \exp -Q/RT$$

with $D_O = 2.5 \times 10^6$ cm^2/sec., and Q = 17 kcal/mole. At a heating rate of 4°C/min., the fiber temperature is 700°C or higher for over 30 minutes. Within this span of time, the diffusion coefficient for water in glass is at least:

$$D = D_O \exp - 17/R(973) = 3.8 \times 10^{10} \text{ cm}^2/\text{sec}$$

Actually, T $\geq$ 700°C, so D $\geq 3.8 \times 10^{10}$ cm^2/sec. Also, the diffusion coefficient for the lower density, alkali containing fibers is probably greater than that of pure silica.

The mean diffusion distance x is then:

$$x = 2Dt = 11\mu$$

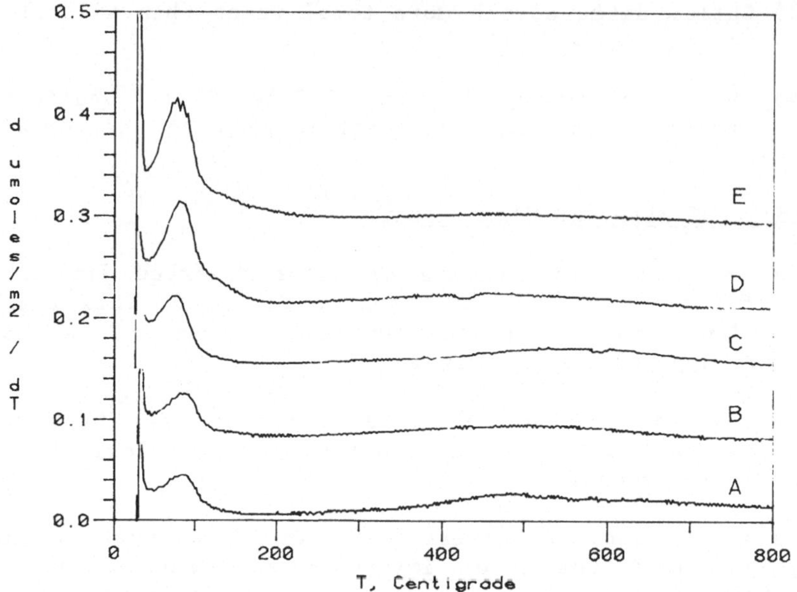

Fig. 5.   The desorption of water from silicas doped with sodium, calcium, or aluminum.

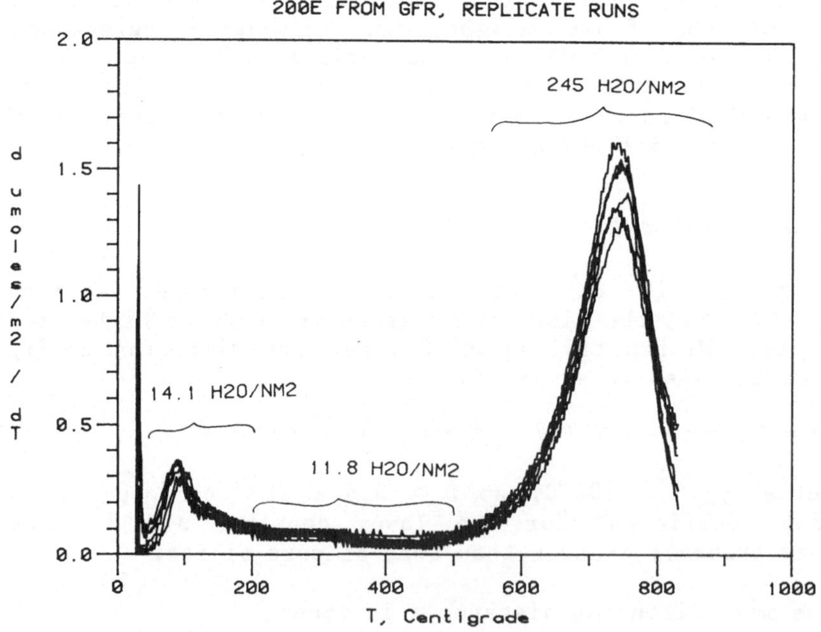

Fig. 6.   Six measurements of the water desorbed from a production E glass fiber.

with t = 30 min. The fiber radius is, however, 5 μm. This simple but conservative estimate indicates that there is sufficient time for internal water to diffuse out of fibers during the experiment.

The quantity of water desorbing between 200 and 500°C is twice that expected from dehydroxylation of surface silanols. As before, we believe internal water is responsible for part of this water in addition to surface desorption. Examination of Figure 6 reveals that this excess can be attributed to internal water (overlap of the 500-800°C peak). The water desorbing between 55-200°C is interpreted as strongly bound surface molecular water. Assuming a density of five high energy sites per square nanometer on this fiber then there are three chemisorbed water molecules per site.

For glass fibers, therefore, only surface water desorbing below 200°C can be detected by thermodesorption. The small amount of surface water desorbing above 200°C is swamped by water diffusing out of the interior. Since water produced by the condensation of surface hydroxyl groups is known to desorb above 200°C, the surface concentration of hydroxyl groups cannot be measured if significant quantities of internal water are present.

The various types of water associated with glass, and the temperatures at which they desorb are listed in Table II.

The effect of a humid environment during the forming process is shown in Figure 7. A stream of argon saturated with water was focused at various points on the fiber during forming. The interaction of the glass surface with water during forming is manifested by the growth of peaks at 100°C and 300°C. The amount of water desorbing at 300°C (23 H_2O/nm^2) is too great to be strongly held at the surface; we believe this water is near surface water, or water which had diffused into the fiber during forming.

The enormous difference in water content of fibers is seen in Figure 8. The dashed curve represents fibers formed under simulated

Table II. Summary of Water Associated with Glass Fibers

Water Type	Desorption Temperature
physically adsorbed water	25°C
strongly bound surface water	100°C
surface hydroxyl groups	200-800°C
internal water	300°C (near surface)
	700°

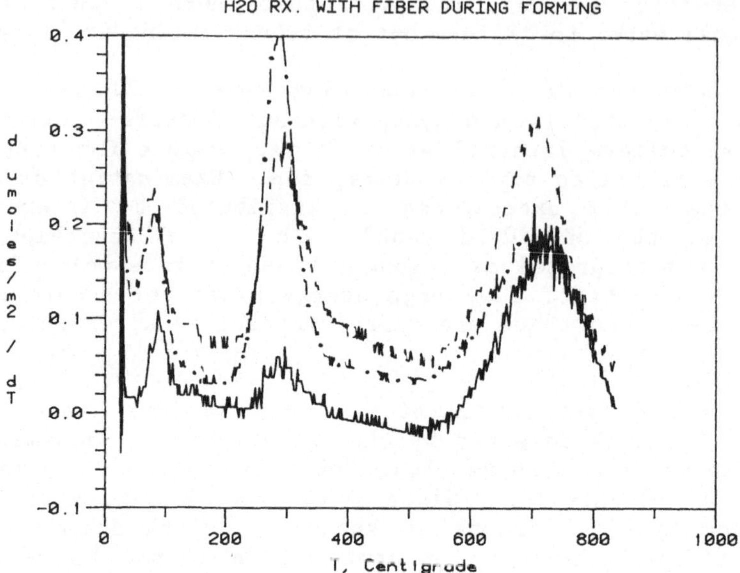

Fig. 7. The effect of humidity during the fiber forming process.
Note the growth of a peak at 300°C.

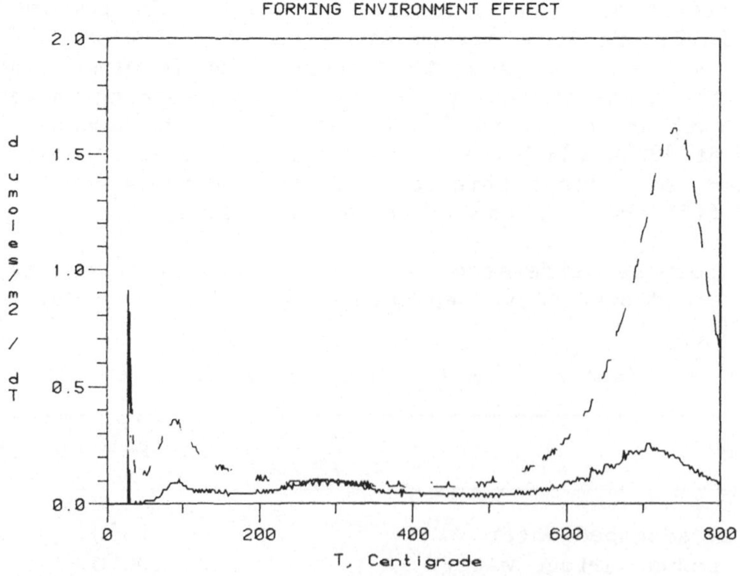

Fig. 8. Variations in the surface and bulk water for two fiber
samples of identical oxide composition.

plant conditions, the solid curve represents fibers formed in a small scale laboratory system. The oxide contents of both samples are identical. Figure 8 serves as a caveat to investigators studying glass fibers. Large variations in surface and bulk water on glass fibers probably occur. These normally unaccounted for variations might cause profound differences in the physical and chemical properties of composites containing glass fibers. The control and the effect of the various types of water described herein is not well understood and will be the subject of future investigations.

CONCLUSIONS

The thermodesorption device described herein identifies and quantitatively measures several of the different types of water associated with glass. Most physically adsorbed water cannot be quantitatively measured by thermodesorption. Thermodesorption does measure the strongly bound water desorbing at 100°C, which is bound to surface silanols, or cation-aluminate sites. The number of surface hydroxyls can be measured by following the dehydroxylation reaction occurring between 200°C-800°C if internal water is not present in the sample.

It appears that there are two types of internal water which can exist in glass fibers. The internal water desorbing at 700°C probably existed in the glass melt or raw materials before fiber formation. The internal water desorbing at 300°C, denoted as near surface water, appears to be due to diffusion of atmospheric water into the glass during the fiber forming process.

Different species of water associated with glass have been identified. The importance and effect of each species on important glass properties such as strength, stress corrosion, or adhesion need to be identified.

REFERENCES

1. J. A. Burgman, and E. M. Hunia, Glass Tech., 11, 147 (1970).
2. C. L. McKinnis, Fracture Mechanics, 4, edited by Bradt, Hasselman, and Lange (1978).
3. N. M. Cameron, Ph.D. thesis, University of Illinois, (1965).
4. D. L. Hollinger and H. T. Plant, Proceedings of the 19th Annual Conf., SPI, 11A, p. 1 (1964).
5. G. K. Schmitz and A. G. Metcalfe, I & EC Product R & D, 5, (1966).
6. A. G. Metcalfe and G. K. Schmitz, Glass Tech., 13, 5 (1972).
7. S. Brelant, in "Treatise on Adhesion and Adhesives," v. 2, Chap. 8, R. L. Patrick, ed., Marcel Dekker, N.Y. (1969).
8. H. Hojo, K. Tsuda, and M. Koyama, International Conference in Organic Coatings Science and Technology, 1, 221, (1979).

9. G. R. Kritchevsky and D. R. Uhlmann.
10. S. P. Wesson and J. S. Jen in Abstracts of the Fifth Annual Meeting of the Adhesion Society, 9, (1982).
11. M. Tomozawa and M. Takata, J. of Non-Crystalline Solids, 45, 141 (1981).
12. V. N. Pak and N. G. Ventou, Soviet J. of Glass Physics and Chem., 6, 223, (1980).
13. H. Yang and A. Tyler, Ind. Eng. Chem., Prod. Res. Dev., 16, 252 (1977).
14. M. Continaud, P. Bonniau, and A. R. Bunsell, J. Materials Sci., 17, 867 (1982).
15. R. K. Iler, "The Chemistry of Silica," Wiley, New York (1979).
16. L. Nemec and J. Gotz, J. Am. Cer. Soc.--Discussions and Notes, 53, 526 (1970).

PREDICTING ENTHALPIES OF INTERFACIAL BONDING OF POLYMERS TO
REINFORCING PIGMENTS

F.M.Fowkes, D.C.McCarthy and D.O.Tischler

Department of Chemistry
Lehigh University
Bethlehem, PA 18015

ABSTRACT

The adsorption and adhesion of polymers to inorganic pigments
depends very much on the degree of acid-base bonding between these
materials. The H of acid-base bonding at the surface of silica
particles has been measured calorimetrically during adsorption
from neutral solvents, and has been measured by infrared spectral
shifts of the carbonyl stretching frequency of esters and ketones
resulting from adsorption on silica. The results obtained by
these two methods are in excellent agreement, and allow accurate
determination of the Drago C and E constants for the acidic
silanol groups of silica (C =1.08 0.04 and E =4.36 0.2). With
these values, predictions of the energetics of polymer adhesion to
silica can be made quite reliably.

INTRODUCTION

The adsorption of organic compounds from solution in neutral
organic solvents (such as saturated hydrocarbons) onto inorganic
surfaces has been shown to be governed by acid-base interactions
between the compound and the surface[1-4]. In earlier times this
kind of adsorption was attributed to "polar" interactions which
were assumed to be related to dipoles. However, more detailed
study has shown that polar groups interact only when one is an
acid and the other a base; furthermore, the enthalpy of polar
bonding is found to be completely independent of dipole moment[4].

R.S.Drago and co-workers has related calorimetric enthalpy
changes of acid-base interactions in neutral solvents with

spectroscopically determined shifts of the OH stretching frequency
for phenol[5]:

$$\Delta H^{ab}=[3.08+0.0103\Delta\nu_{OH}(cm^{-1})]\ kcal/mole \qquad (1)$$

and have further correlated the ΔH values for a wide variety of
acids and bases with the "E and C" equation:

$$-\Delta H^{ab}= C_A C_B + E_A E_B \qquad (2)$$

in which the acid is characterized by two constants, C_A and E_A,
and the base by C_B and E_B. The C/E ratio is a measure of
"softness" of the acid or base; a high C/E ratio is a sign of a
soft acid or soft base[5].

In our recent infrared studies we have found that the shift
in the carbonyl stretching frequency of esters correlates very
well with the calorimetric ΔH^{ab} of these esters with unassociated
acids such as chloroform, iodine, antimony pentachloride, and
triethyl aluminum[6]:

$$\Delta H^{ab}= 1.0\Delta\nu\ (kJ/mole/cm^{-1})=0.236\Delta\nu\ (kcal/mole/cm^{-1}) \qquad (3)$$

In this paper we introduce the determination of heats of
adsorption of $\Delta\nu_{C=O}$ determination in mulls of silica in a
hydrocarbon containing low concentrations of esters, and compare
the results with ΔH^{ab} values determined by calorimetry.

The isotherms for adsorption of organic bases from dilute
solution in hydrocarbons appear to be of the Langmuir type which
can be represented by an equilibrium constant K:

$$K= \Gamma/(\Gamma_m- \Gamma)C \text{ or } C/\Gamma=(1/K\Gamma_m) + (C/\Gamma_m) \qquad (4)$$

in which C is the equilibrium concentration of solute in the bulk
phase when the surface concentration is Γ, and Γ_m is the surface
concentration for a fully saturated surface.

EXPERIMENTAL DETAILS

<u>Adsorption Isotherms</u>

The adsorption isotherms for pyridine and for triethylamine
adsorbing onto silica (HiSil 233) from decalin were determined at
$25\,^{\circ}$ and $50\,^{\circ}$C. The amines were dried with molecular sieves and the
silica was oven-dried at $150\,^{\circ}$C overnight. The decaline was used
as received. The suspensions were stirred overnight at constant

temperature and after 16 hours the solute concentrations were determined in the supernatant solution by UV absorption measurements.

Heats of Adsorption

A Tronac Microcalorimeter was used to determine heats of adsorption or pyridine and of triethylamine onto HiSil 233 from decalin. One gram of silica was stirred in 25.0 ml of decalin and after a good base line developed and a calibration for heat capacity was made, 1.5 ml of amine solution in decalin was added at a rate of about 0.1 ml per minute. The calorimeter was operated at constant temperature with constant cooling by a Peltier device plus pulsed additions of heat to keep at 25° C. The heat generated by the amine-silica interaction was determined by integration of the heat pulsed during the exotherm. The amount of amine adsorbed was calculated from the adsorption isotherms. Initial studies were made with silica oven-dried at 150° C overnight and with amines dried with activated molecular sieves(4A), but more consistent results were obtained when the decalin was freshly chromatographed on dry silica as well.

Infrared Spectra

The carbonyl stretching frequency of adsorbed esters and acetone were determined by FT-IR spectra, using Nicollet FT-IR spectrophtotometer on mulls of Aerosil 380 silica (380 m /g) in a white oil. The acetone spectrum was obtained with 200 microliters of acetone per grom of silica, and the polymethylmethacrylate (PMMA) spectrum was obtained with du Pont Lucite 4F adsorbed onto the silica (from methylene chloride) which was then dried and milled into the white oil.

EXPERIMENTAL RESULTS AND DISCUSSION

Adsorption Isotherms

The adsorption of pyridine and of triethylamine onto HiSil 233 from decalin give Langmuir-type isotherms which can be fitted easily to equation (4). The 25° C data are tabulated in Tables 1 and 2.

It can be seen that equation (4) fits the data very well, especially at the higher concentrations. These isotherms are used for the calculation of amounts adsorbed in the subsequent calorimetry measurements. The amounts adsorbed at 50° C were almost as great as at 25° C, suggesting that the heats of adsorption are quite small. However, as can be seen in the

Table 1. Adsorption of Pyridine from Decalin onto Hi Sil 233 at 25°C.

C(exp)	5	20	42	66	144	162	210
Γ(exp)	2.94	3.56	3.94	4.25	4.79	5.03	5.22
C/Γ(exp)		5.6	10.6	15.5	23.8	32.2	40.2
C/Γ(exp)	3.8	6.5	10.5	14.8	23.4	32.1	40.7

eq(4): $C/\Gamma = 2.93 + 1.83 \times 10^5 C$

$\Gamma_m = 5.46 \times 10^{-6}$ Moles/m^2 (30.4 Å^2/molecule)

(C in moles/m^3 x10^6, Γ in moles/m^{2}x10^6)

Table 2. Adsorption of Triethylamine from Decalin onto Hi Sil 233 at 25°C.

C(exp)	10	29	52	79	127	174
Γ(exp)	2.3	2.60	2.75	2.85	2.92	2.99
C/Γ(exp)	4.3	11.1	18.9	27.7	43.5	58.2
C/Γ(eq.4)	5.2	11.4	18.8	27.6	43.2	58.4

eq(4): $C/\Gamma = 1.96 + 3.26 \times 10^5 C$

$\Gamma_m = 3.07 \times 10^{-6}$ moles/m^2 (51.0 Å^2/molecule)

(C in moles/m^{3}x10^6, Γ in moles/m^{2}x10^6)

following section, the calorimetric heats of adsorption are quite appreciable, so these results are puzzling. We suspect that traces of adsorbed water may have caused this effect.

Calorimetry

The heats of adsorption measured at 25°C in the Tronac Microcalorimeter for pyridine and triethylamine adsorption from decalin onto HiSil 233 silica are summarized in Table 3.

Heats of Adsorption from Infrared Spectra

Enthalpy changes of acid-base complexation (ΔH^{ab}) have been determined for carbonyl-containing compounds with good accuracy from shifts of the carbonyl stretching frequency in solutions of constant surface tension[6] ($\gamma^d=27.0\pm1.5$ mJ/m^2), as shown in Figure 1 for ethyl acetate, acetone and polymethylmethacrylate. Since the adsorption of "polar" solutes from neutral solvents onto inorganic surfaces has been demonstrated to result entirely from acid-base interactions[1], it is reasonable to assume that $\Delta H_{ads}=\Delta H^{ab}$ and that spectral shifts are accurate measures of the heat of adsorption.

Table 3. Calorimetric Heats of Adsorption of Amines onto HiSil 233 from Decalin at 25°C.

g.HiSil	Amine	Moles added	Moles adsorbed	Calories	ΔH_{ads}(kcal/mole)
0	Pyridine	4.50×10^{-4}	0	+0.09	--------
0.95	"	"	2.40×10^{-4}	-2.72	-11.3
0.96	"	"	"	-2.71	-11.3
1.01	"	"	2.46×10^{-4}	-3.18	-12.9
1.03	"	"	2.49×10^{-4}	-3.17	-12.7
0	TEA	2.76×10^{-4}	0	-0.03	--------
1.00	"	"	1.69×10^{-4}	-2.85	-16.8
1.01	"	"	"	-2.77	-16.4

The carbonyl stretching frequency for acetone in dilute solutions (0.05%) in hydrocarbons is 1727-1728 cm^{-1}, but when adsorbed in mineral oil onto silica (Aerosil 380) the peak shifts to 1698 cm^{-1}, as shown in Figure 2. This shift of -29 cm^{-1} corresponds to ΔH_{ads}=-6.84 kcal/mole, according to equation (3).

Polymethylmethacrylate (PMMA) has a carbonyl stretching frequency of 1737 cm in toluene, a liquid with the same surface tension as mineral oil. However when adsorbed onto HiSil in mineral oil the frequency shifts to 1720 cm^{-1}, a -17 cm^{-1} shift which indicates ΔH_{ads}=4.0 kcal/mole.

These values of the heats of adsorption of oxygen bases on silica can be used together with the calorimetric heats of adsorption of nitrogen bases on silica to determine the C_A and E_A constants for the acid surface sites of silica.

Determination of C and E Constants

Figure 3 demonstrates a graphical technique for determining C_B and E_B constants for ethyl acetate and for PMMA from measured heats of acid-base interaction(ΔH^{ab}) with acids of known C_A and E_A. In these plots we have determined C and E for ethyl acetate (EtAc) and for PMMA from ΔH^{ab} with four test acids, two of them hard acids (chloroform and trimethyl aluminum) and two of them soft acids (iodine and antimony pentachloride). Equation (2) may be re-arranged to give:

$$E_B' = \frac{\Delta H^{ab}}{E_A} - C_B\left(\frac{C_A}{E_A}\right) \tag{5}$$

where E_B' and C_B' are trial constants related by the constants for the test acid and the ΔH^{ab} of its interaction with the base being investigated. The slope of each line is the C_A/E_A of the test acid and the intersections of the lines indicate the actual C_B and E_B for the base. In each of the two plots of Figure 3 there are six intersections, but only four significant intersections, where soft acid slopes cross hard acid slopes. In this case we find that for ethyl acetate, E_B=1.03±0.03 $(kcal/mole)^{1/2}$ and C_B=1.73±0.06 $(kcal/mole)^{1/2}$. Similarly for PMMA we obtain E_B=0.68±0.01 $(kcal/mole)^{1/2}$ and C_B=0.96±0.07 $(kcal/mole)^{1/2}$.

The data used for PMMA in Figure 3 were determined by FT-IR spectra[6], using the relation of equation (3) between enthalpies of interaction and the spectral shifts. The original spectral data for acid-base shifts of the carbonyl stretch for PMMA are shown in Figure 1 and used in Figure 3b to determine its C_B and E_B values.

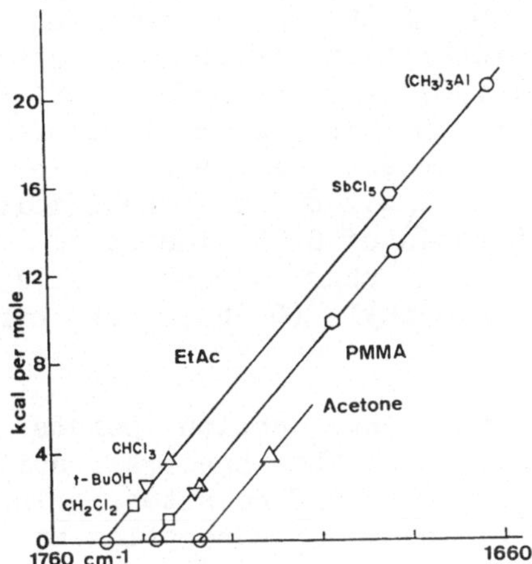

Fig. 1.   Calorimetric heats of acid-base interactions of test
acids with ethyl acetate, polymethylmethacrylate, and
acetone as a function of the carbonyl stretch frequency in
solutions of surface tension γ^d = 27.0 ± 1.5 mJ/m^2.

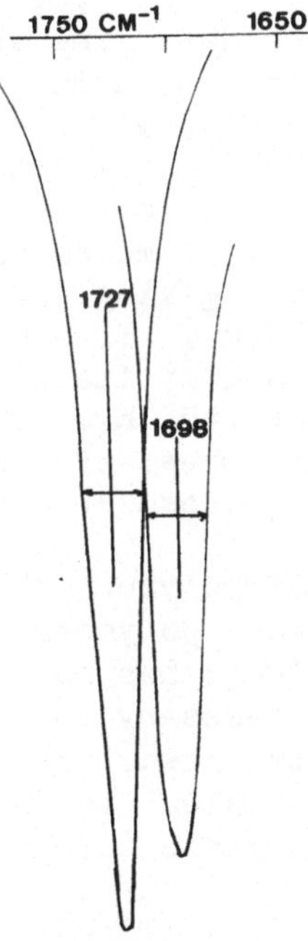

Fig. 2.   Carbonyl stretch frequency shift for acetone upon adsorbing from a hydrocarbon (of surface tension ~28 mJ/m^2)
onto Aerosil 380 silica.

Figure 4 is C and E plot for the acidic sites of silica, using the heats of adsorption of pyridine (C_B =6.40, E_B =1.17), and triethylamine (C_B =11.09, E_B =0.991) determined calorimetrically, and the heats of adsorption determined spectrometrically for acetone (C_B =2.33, E_B =0.987) and for PMMA (C_B =0.96, E_B =0.68). The four lines intersect in six places, five of which are the more significant (steeper intersections), and predict that the acidic sites of silica have C_A =1.08± 0.04 (kcal/mole)$^{\frac{1}{2}}$ and E_A =4.36± 0.2 (kcal/mole)$^{\frac{1}{2}}$. The C/E ratio of 0.25 indicates that the silanol groups are softer acids than alcohols (0.15), about the same softness as boron trimethyl (0.26), but harder than boron trifluoride (0.39).

The closeness of the intersection points in Figure 4 is evidence in support of the spectral shift technique for determining ΔH_{ads} and ΔH^{ab} of surface sites. This technique may become widely used for heats of adsorption and for determining C and E value of surface sites.

Prediction of Heats of Adsorption of Polymers onto Inorganic Fillers

Technique described in this paper allow determination of the C and E constants for polymers and for surface sites on fillers. These data can now be used to determine the heat of adsorption per mole of monomer groups in polymers adsorbed on fillers. For example, PMMA is a base with C_B =0.96 and E_B =0.68, and when it adsorbs on silica (C_A =1.08, E_A =4.36) the ΔH_{ads} equals ΔH^{ab} which calculates from equation (2) to be -4.0 kcal/mole. In other studies we find that up to 6×10^{-6} moles of methacrylate groups can adsorb on each m^2 of silica, giving a heat of adsorption of -2.4×10^{-5} kcal/m^2, or -116 mJ/m^2. If a more basic polymer such as polyvinylpyridine were used and if its C_B and E_B constants were the same as pyridine (C_B =6.40, E_B =1.17), it should adsorb with about 5×10^{-6} moles/m , giving a heat of adsorption on silica of -12.0 kcal/mole of pyridine groups, or -289 mJ/m^2.

In practical problems of polymer adhesion the intervention of water at the interface between polymers and inorganic surfaces is a major problem. Although Drago has published the C_A and E_B for water, the C_B and E_B vlues necessary to calculate water adsorption on silica have yet to be determined. When these are in hand it will become practical to calculate competitive adsorption of polymers versus water at inorganic interfces.

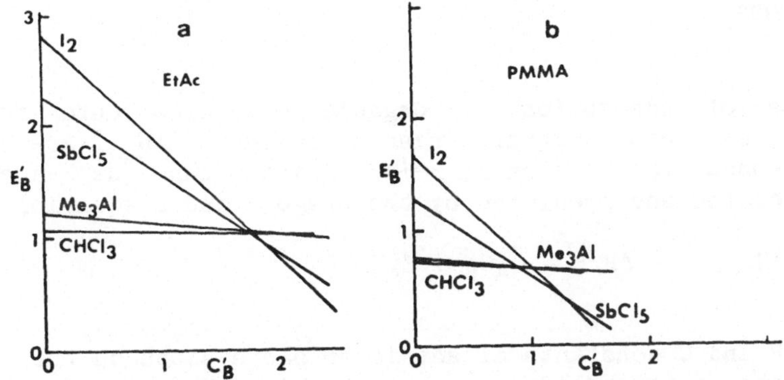

Fig. 3. Graphical technique for determining C_B and E_B of oxygen bases from calorimetric ΔH^{ab} value for ethyl acetate, and for spectrometric ΔH^{ab} values for polymethylmethacrylate.

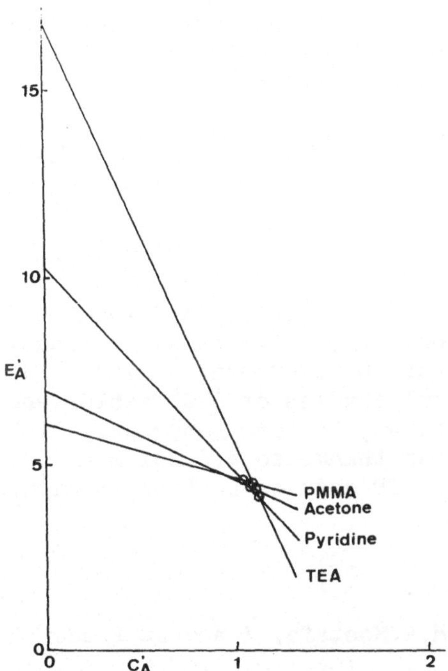

Fig. 4. C and E plot for determining C_A and E_A values for silica from heats of adsorption, determined calorimetrically for pyridine and triethylamine, and spectrometrically for acetone and polymethylmethacrylate.

CONCLUSIONS

1. Heats of adsorption of organic molecules onto inorganic surfaces from neutral organic solvents (ΔH_{ads}) are heats of acid-base interaction (ΔH^{ab}) which can be accurately correlated and predicted by the Drago E and C equation:

$$-\Delta H_{ads} = -\Delta H^{ab} = C_A C_B + E_A E_B$$

2. The E and C constants of acidic or basic sites on the surface of inorganic solids can be determined by measuring heats of adsorption of test acids or bases of known E and C constants.

3. The C and E constants for the acidic silanol sites on silica are 1.08 0.04 and 4.36 0.2, respectively.

4. Heats of adsorption of carbonyl compounds onto inorganic acidic fillers can be determined accurately by infrared shifts of the C=O stretching frequency:

$$\Delta H_{ads} = 0.236 \ k \, cal/mole \ cm^{-1} \ \Delta \nu_{ads}$$

ACKNOWLEDGEMENTS

The RCA Research Laboratories in Princeton, NJ and the LNP Corporation in Malvern, PA have generously supported the calorimetry studies of D.C.McCarthy. IBM (East Fishkill) has supported the spectral studies of D.O.Teshler and Air Products and Chemicals in Trexlertown, PA have generously let him use their FT-IR equipment. Our thanks to J.A.Wolfe at Wright Patterson Air Force Base for the FT-IR data on acetone in dilute solutions.

REFERENCES

1. F.M.Fowkes and M.A.Mostafa, I and EC Prod. R. & D., _17_, 3 (1978).

2. F.M.Fowkes, C.Y.Sun and S.T.Joslin, Enhancing Polymer Adhesion to Iron Surfaces by Acid-Base Interaction, Corros. Control Org. Coat., (Pap. Conf.), 1 (1981).

3. F.M.Fowkes, Acid-Base Interactions in Polymer Adhesion, Microscopic Aspects of Adhesion and Lubrication, $\underline{7}$, 119 (1982).

4. F.M.Fowkes, in Physicochemical Aspects of Polymer Surfaces, K.L.Mittal, ed., Plenum, New York (1983) Vol.2,p.583.

5. R.S.Drago, G.C.Vogel and T.E.Needham, J. Am. Chem. Soc., $\underline{93}$, 6014 (1971).

6. F.M.Fowkes, D.O.Tishler, J.A.Wolfe, L.A.Lannigan, C.M.AdemuJohn and M.J.Halliwell, J. Poly. Sci., in press.

WETTABILITY OF REINFORCING FIBERS

H. W. Chang, R. P. Smith, S. K. Li, and A. W. Neumann

Bendix Advanced Technology Center, Columbia, Md and
Department of Mechanical Engineering, University of
Toronto, Toronto, Canada

INTRODUCTION

In the production of fiber reinforced composite materials a
solid-liquid interface is generated between the fiber and the matrix which is in a liquid or molten state. The strength of the
final composite material will, in part, depend on how well the matrix material wets the fibers in this production phase. Poor wetting of the fibers may result in gaps between the fiber and the
final solidified matrix. These gaps will be flaws in the final
composite which will reduce its ultimate strength.

A measure of how well the liquid matrix wets a fiber can be
obtained by considering the contact angle between the fiber and the
liquid. For complete wetting of the fiber it is required that this
contact angle be zero. A contact angle greater than zero is indicative of a fiber-matrix system in which a state of less than complete wetting exists.

The contact angle between the fiber and the polymer melt can,
in principal, be measured using the Wilhelmy technique [1]. However, in practice some difficulties arise due to the 200-400°C
elevated temperature required for some polymers. At these temperatures, convection currents are generated at the liquid surface, and
they can affect the stability of the measurements. Oxidation of
the matrix material also becomes a problem in some cases.

Therefore the objective of this work was to determine whether
the Wilhelmy technique can be successfully used to measure the
contact angle between fibers and the thermoplastic polymer melt.

EXPERIMENTAL PROCEDURE

The Wilhelmy technique has been described in detail by Neumann and Good in Ref. 1; an account is included in the Appendix for completeness. The technique may be used to measure the surface tension of a liquid if the contact angle is known (normally the contact angle will have to be zero) or to measure the contact angle if the surface tension of the liquid is known. While the technique has been used for these purposes for some time, typically with measuring probes in the form of a plate, it can, in principle be used whenever the probe has vertical sides. Early application of the technique in this mode was made by Neumann et al. [2] to determine contact angles on treated and untreated glass fibers. More recently, the technique was used by Miller [3,4] and Hedvat [5].

The fibers in this study are graphite fibers (Hercules HMS type 100-2-bats), Kevlar 49 and glass fibers (Owens Corning S-2, 449 AA-1250). Since the fibers were flexible, it was not possible to suspend a long section of these fibers from the electrobalance. To overcome this problem, a 1 to 2 cm segment of the fiber in each case was attached to the end of a thicker glass rod, approximately 200 μm diameter. The glass rod was then suspended from the electrobalance.

Attachment of the fiber to the end of the glass rod was achieved by using a small drop of epoxy adhesive. Working under a microscope, a drop of adhesive was placed on the end of the glass rod. By placing the fiber between two glass slides and holding the glass rod fixed, it was possible to move the fiber around on the microscope stage until it aligned and made contact with the glass rod. At this point the fiber and glass rod were held in place until the adhesive had cured sufficiently to allow further handling without weakening the adhesive joint. The short segment of fiber attached to the glass rod was sufficiently stiff to resist any bending away from the liquid surface which could have been caused by minor vibrations or air currents.

Since the polymer matrix materials studied become liquid only at elevated temperatures, it was necessary to design a high temperature chamber meeting several requirements. First, it had to be small enough to fit on the stage of our current Wilhelmy apparatus (see Figure 1) and be capable of heating the polymer materials to above 300°C. Also, since some of the polymer materials may oxidize at 300°C temperatures, an argon atmosphere had to be provided in the test chamber.

Figure 2 shows a schematic diagram of the high temperature cell. The core of cell was made of stainless steel with a conical chamber machined in it. Use of stainless steel allowed for easy

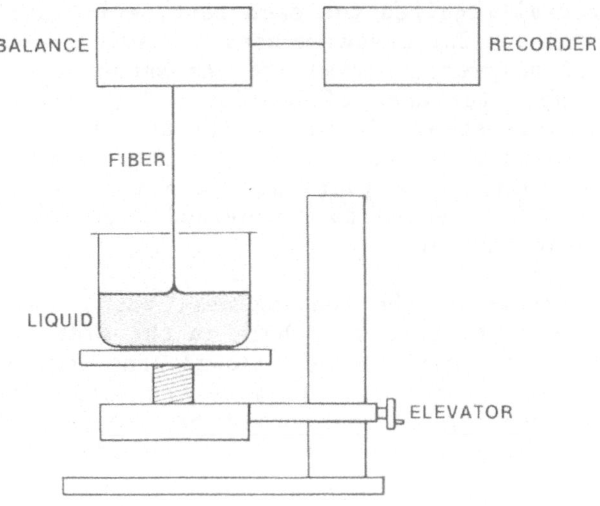

Fig. 1.   Apparatus for Wilhemy technique.

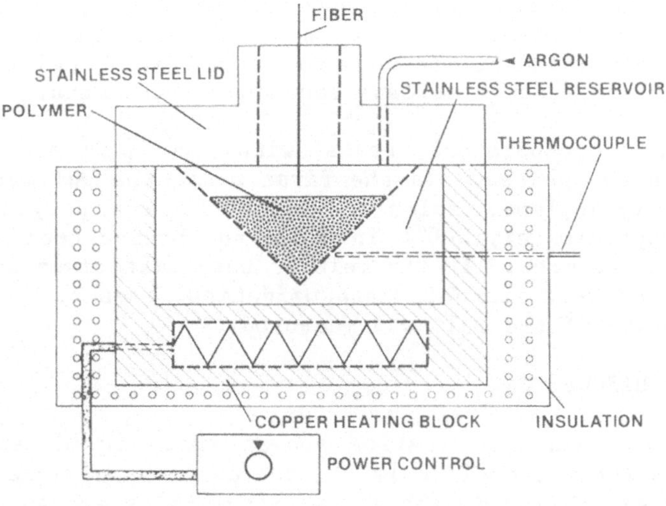

Fig. 2.   High temperature cell for surface tension and contact
angle measurements with polymer melts.

cleaning. The conical shape of the chamber helped to minimize the amount of material required for each test, while still providing a large liquid surface for measurements. Because of the poor thermal conductivity of polymers, minimizing the amount of material needed was also desirable for ease of heating. To seal off the melting chamber a stainless steel lid was constructed to fit on top of the chamber. Two holes were machined in the lid. A hole in the center allowed a test fiber or a platinum plate to pass through to the liquid surface. The second hole provided a passage for the argon gas into the heat chamber.

The temperature of the heating cell was controlled manually. A thermocouple was inserted in a hole in the side of the stainless steel heating reservoir which allowed the temperature to be recorded by a chart recorder. To vary the temperature, the power control to the heating element could be changed manually. With this simple arrangement it was possible to produce the desired test conditions.

The central stainless steel test chamber was surrounded by a copper heating block, which was in turn surrounded by insulation. This section of the apparatus was placed on the test stage of the Wilhelmy balance. The power control for the heater was located on a separate bench beside the balance to allow for easy access and to prevent the operator from disturbing the balance when varying the control.

During an experiment the thermocouple was connected to a chart recorder, the argon feed to a gas bottle and the fiber hung from the electrobalance. In this form the heating cell is ready to perform high temperature surface tension measurements.

The high temperature tests with the polymer melts were performed in two phases. In the first phase the polymer materials such as polypropylene, Nylon 6, Nylon 12 and polysulfone were heated to approximately 300°C in order to observe their properties at these temperatures. If the melt of these materials was suitable with regard to melt viscosity and oxidation, then the liquid–vapor surface tension of the melt was measured.

RESULTS AND DISCUSSION

The first phase of testing showed that liquid–vapor surface tension measurements could only be made on polypropylene. Polypropylene is a clear liquid with a low viscosity at around 300°C. Both nylons and polysulfone, however, were still extremely viscous at temperatures up to 400°C. Problems also arose with oxidation of these materials at such high temperatures. Even in the presence of an argon atmosphere, a black crust would form on these melts after prolonged exposure to high temperatures. When surface tension measurements were attempted on these materials, the fiber would deflect as it touched the surface. After sitting on the surface of the melt for several seconds, it would slowly sink through the surface. At this time any readings obtained by the electrobalance fluctuated randomly and were not reproducible. Thus, no surface tension information could be obtained for nylons or polysulfone at high temperatures using this approach.

Figure 3 shows the results of surface tension measurements as a function of temperature for polypropylene using a platinum plate. The slope of fibers in this plot is approximately $d\gamma_{LV}/dT = -0.043$ mJ/m^2°C. The random fluctuations observed in the curve are likely due to convection currents and other background noise which affects the measurements made by the electrobalance.

As a means of independently characterizing the surface energy of the fibers, contact angle measurements were performed with several liquids at room temperature. Tables 1 and 2 show the surface tension for carbon and Kevlar fibers calculated from these contact angles using the equation of state [6]. The fact that there is good agreement in the surface tension values obtained for each fiber using different liquids indicates that the surface tension values obtained are indeed characteristic of the fiber. Table 3 gives results for three independent contact angle measurements made with the same liquid, ethylene glycol. These results demonstrate the reproducibility of this type of measurement.

Because polypropylene has a lower surface tension than any of the fibers used, the contact angle in all cases was found to be zero degrees as shown in Table 4.

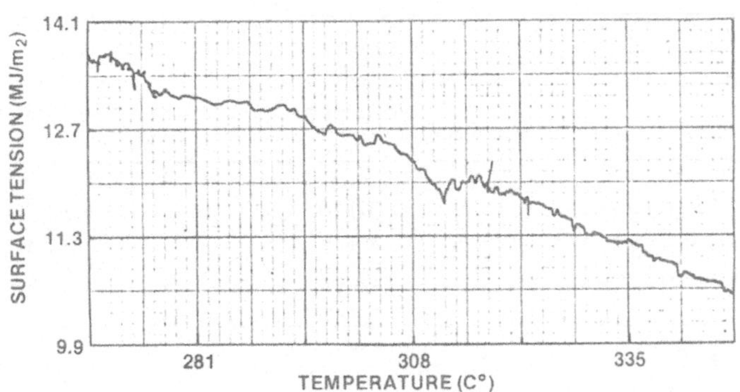

Fig. 3. Surface tension as a Function of Temperature (polypropylene, Shell DP 5072).

Table 1.   Contact Angle Measurements on Carbon Fibers at 20°C.

Liquid	Contact Angle [degrees]	γ_{SV} [mJ/m^2]
Water	66.0	44.0
Glycerol	58.9	40.4
Ethylene Glycol	31.7	41.6

Average Fibre Diameter 7.4 $\pm$ 0.4 µm

Table 2.   Contact Angle Measurements on Kevlar Fibers at 24°C.

Liquid	Contact Angle [degrees]	γ_{SV} [mJ/m^2]
Water	73.5	39.1
Glycerol	52.8	43.8
Ethylene Glycol	26.7	42.9

Average Fibre Diameter 12.4 $\pm$ 0.7 µm

Table 3.   Contact Angle Measurements on Glass Fibers at 20°C.

Liquid	Contact Angle [degrees]	γ_{SV} [mJ/m^2]
Ethylene Glycol	30.0	41.3
Ethylene Glycol	28.7	42.5
Ethylene Glycol	34.7	40.2

Average Fibre Diameter 11.4 $\pm$ 0.1 µm

Table 4. Contact Angle Mesurements on Fibers with Liquid
          Polypropylene

Fibre Material	Fibre Diameter [μm]	Temperature [°C]	Liquid Surface Tension [mJ/m²]	Contact Angle [degree]
Carbon	8	200	16.9	0
Kevlar	12.4	200	16.9	0
Glass	11.4	200	16.9	0
Glass	22.5	237	15.3	0

In conclusion, this work shows that the Wilhemy technique can be used successfully to measure contact angles between reinforcing fibers and polypropylene melt. Further work is needed to perfect the technique for polymers with high viscosity.

APPENDIX

<u>The Wilhelmy Technique</u>

A smooth solid surface, either a plate or a fiber, as in this work, is suspended above a liquid surface from an electrobalance. In these particular tests a CAHN 2000 electrobalance was used. From this point the liquid surface is slowly raised until the plate or fiber makes contact with the liquid surface.

If the contact angle is less than 90 degrees, the liquid will rise onto the solid as contact is made. For contact angles greater than 90 degrees the liquid surface will be depressed as the solid is lowered further into the liquid. If the contact is exactly 90 degrees, the level of the liquid surface will be unchanged due to contact with the solid. However, regardless of how the liquid surface reacts to the presence of the solid, the electrobalance will measure the mass of liquid which either rises onto the solid or is depressed by it. If the liquid surface is depressed a negative mass reading will be recorded. A schematic of the apparatus to perform these measurements is given in Figure 1.

Once the change in mass measurements has been obtained using the electrobalance, either a contact angle or the surface tension of the liquid can be calculated from the relation:

$$\cos \theta = \frac{mg}{P\gamma_{LV}} \qquad\qquad [1]$$

where θ is the contact angle between the liquid and the solid, g is the gravitational constant, m is the mass reading from the electrobalance, P is the perimeter of the solid at the point of contact with the liquid and γ_{LV} is the surface tension of the liquid. To determine the surface tension of a liquid, it is necessary to have a high energy material for the measuring plate such that a zero degree contact angle between the liquid and the solid is assured. Then the surface tension of the liquid can be calculated from the following equation :

$$\gamma_{LV} = \frac{mg}{P} \qquad\qquad [2]$$

In order to obtain the contact angle between a solid and a liquid, Eq. [1] is used in the following form:

$$\theta = \cos^{-1} \frac{mg}{P\gamma_{LV}} \qquad\qquad [3]$$

and it is necessary to know the surface tension of the liquid used in the measurement.

From Equation [1] it is seen that one limiting factor in the accuracy of the results is the perimeter of the solid. If it is not sufficiently large, the expected mass reading may be less than the sensitivity of the electrobalance. Even if this is not the case, the effect of mass measurement errors caused by such factors as vibrations and air turbulence will be more severe when dealing with smaller perimeter solids. Thus, measurements made with a large plate are more accurate than similar measurements made with small diameter fibers.

Temperature dependent measurements are performed in the same manner as those measurements mentioned above. However, after solid-liquid contact has been made, the temperature of the liquid is slowly and continuously changed. The temperature of the liquid is measured using a thermocouple and the surface tension of the liquid is directly proportional to the mass reading obtained on the electrobalance. Thus, by connecting the output from these two devices to the separate channels of an X-Y plotter, it is possible to obtain a continuous plot of the variation of the liquid surface tension with temperature. The slope of this plot is then the temperature coefficient for the liquid-vapor surface tension $d\gamma_{LV}/dT$.

REFERENCES

1. A. W. Neumann, and R. J. Good, "Surface and Colloid Science,"
    Vol. 11, Plenum, New York (1979) pp. 31-91.
2. A. W. Neumann, and W. Tanner, Vth Intern. Congr. Surface Acti-
    vity, Barcelona 1968, Vol. 2, p. 727, Ediciones Unidas,
    Barcelona (1969): A. W. Newumann, Chemie Ing. Tech. $\underline{42}$,
    969 (1970).
3. B. Miller and R. A. Young, Textile Research Journal, $\underline{45}$, 359
    (1975).
4. B. Miller, "Surface Characteristics of Fibres and Textiles,"
    M. J. Schick, ed., Marcel Dekker, New York (1977), p. 417.
5. S. Hedvat, Ph.D. thesis, Department of Chemical Engineering,
    Princeton University, Princeton, New Jersey (1981).
6. A. W. Neumann, R. J. Good, C. J. Hope, and M. Sejpal, J. Col-
    loid Interface Science, $\underline{49}$, 291 (1974).

ISS/SIMS ANALYSIS OF GRAPHITE FIBER SURFACES AND THE THERMO-

OXIDATIVE STABILITY OF GRAPHITE FIBER/PMR-15 POLYIMIDE COMPOSITES

Daniel A. Scola and Bruce L. Laube

United Technologies Research Center

East Hartford, Connecticut 06108

INTRODUCTION

The long term durability of graphite fiber reinforced PMR-15 polyimide composites at 316°C and above, is dependent on two major aspects of these composite systems. These are (1) the thermal and thermo-oxidative stability of the individual components and (2) the influence of the interface region between fibers and matrix on matrix stability. In studies carried out at 316°C (600°F) on several fiber/PMR-11 and PMR-15 polyimide systems, in flowing air, it has been demonstrated that the thermo-oxidative stability of these systems is effected by fiber surface impurities, such as the alkali metal ions, sodium and potassium [1]. Gibbs et al. [2] have shown that HTS graphite fibers contain a large concentration of sodium ions on the surface, while Dryzal and Hammer [3] have demonstrated that AS graphite fibers also contain large concentrations of sodium ions relative to other graphite fibers. The overall objectives of this investigation are (1) to determine the mechanism of the thermo-oxidative process in flowing air, of several composite systems consisting of the addition polyimide PMR-15 in combination with the graphite fiber reinforcements, and (2) to determine the upper temperature limit of these composite systems.

In the present paper the surface composition of several graphite fibers as determined by ISS/SIMS analysis and the relationship of the surface composition to component properties and composite thermo-oxidative stability will be discussed. Thus, the surfaces of the neat fibers and fibers from composites exposed as a result of the thermo-oxidative process were analyzed. Weight loss studies of the component materials and each graphite/PMR-15 composite system were conducted. Optical microscopy and mechanical property measurements of the "as fabricated" and isothermally exposed composites were also carried out.

EXPERIMENTAL

Fiber Materials

Fiber materials were obtained commercially; AS from Hercules Inc., Thornel 300 from Union Carbide Corp., Celion 6000 from Celanese Corp., Fortafil 5 from Great Lakes Carbon Corp., and Panex 30 from Stackpole Carbon Corp. The following fibers were investigated: Panex 30 (unsized) (Roll #18-00506), Fortafil 5 (unsized) (Lot #SeRS-1528-B), Fortafil 5 (P.I. compatible size, not PI) (Lot #2237-29-15), Celion 6000 (unsized) (Lot #HTA-7-8422), Celion 6000 (NR150B2 size) (Lot #HTA-7-9821), Thornel 300 (water sized) WYP 30 1/0, AS4W6 (water size) (Lot #205-1), and Celion 6000 (Celanese PI size) (Lot #HTA-7-2832).

PMR-15 Polyimide Resin

The polyimide resin system used in this study was prepared and processed from a methanol solution of monomers as described previously [4,5].

ISS/SIMS Analysis of Fiber Surfaces

The equipment and procedures used in the surface analysis by Ion Scattering Spectroscopy (ISS) and Secondary Ion Mass Spectroscopy (SIMS) have been described previously [6]. A schematic of the ion generation and detection technique is illustrated in Fig. 1. Table 1 lists the conditions for the analysis. For the ISS analysis, samples were bombarded by a primary ion beam of $^3He^+$ at an energy of 2.5 keV and a beam current density of 4.6 μamps/cm^2. For the SIMS analysis, samples were bombarded by a primary ion beam of $^{20}Ne^+$ for a near surface analysis (5-10Å) and again at the approximate sputtered depth of 100Å, at a beam energy of 2.5 keV and a beam current density of 4.6 μamps/cm^2. Surfaces were sputtered using a Perkin Elmer PHI sputter ion gun at a beam energy of 5 keV and a current density of 51.1 μamps/cm^2. The sputter rate was approximately 5Å/min. The SIMS data for each fiber are reported in relative atomic ratios. These were calculated as follows: The values of counts per second (c/s) of each element on the fiber surface were adjusted to relative atomic values by dividing each c/s value by the sensitivity factor for that element. These relative atomic values were then normalized by dividing the relative atomic value of each element on a given fiber surface by the relative atomic value of the element present in the smallest

Table 1. ISS/SIMS Analysis

<u>Conditions:</u>

Energy of primary ion beam	2.5 keV
Beam current	40 nAmps
Current density	4.6 μamps/cm^2
Rastered area	(1x3) mm^2
Gated area	70% of rastered area
Sputter rate	~2Å/min

<u>Analysis:</u>

ISS – First scan with ^{3}He (2 scans, 3 minutes each) to detect scattered ^{3}He$^\oplus$ ions only; sensitive to low atomic no. elements.

SIMS– Second scan with ^{20}Ne (2 scans) to detect $\oplus$ and $\ominus$ secondary ions; sensitive to higher atomic no. elements and molecular fragments.

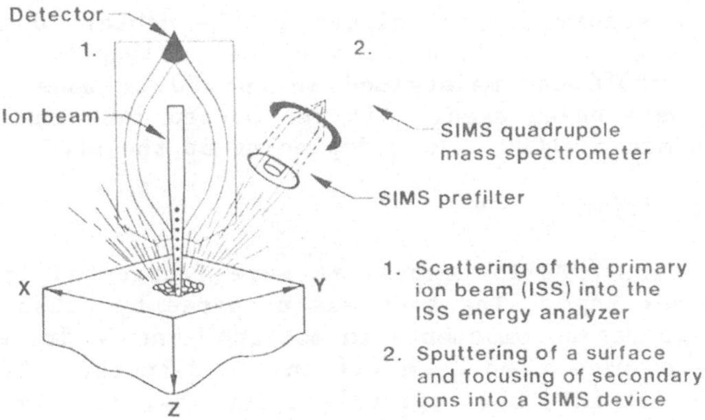

Fig. 1. ISS/SIMS Ion Generation and Detection.

amount. For the data at the 5-10Å level, hydrogen atomic ratios were calculated but not reported since the hydrogen ratios are high for all fibers, as expected. In addition, since high levels of C^+, CH^+, CH_2^+, CH_3^+ and C_2H^+ species were observed in most fibers, these were not considered in the atomic ratio calculation. Moreover for species other than ^{12}C, correction factors do not exist. The data can be considered as relative atomic ratios of elements on a carbonaceous substrate.

To compare the elemental surface compositions between fibers, the relative atomic values for each element determined as described above were divided by the smallest relative atomic value observed for that element in the series of fibers analyzed. Such comparisons assume that substrate effects on the sputtering of easily ionized elements are negligible.

PMR-15 composites of each fiber were fabricated and subjected to isothermal aging at 350°C in flowing air, 100 cc/min or 1000 cc/min for time periods up to 700 h. The surfaces of the fibers, exposed as a result of this thermo-oxidative process, were also examined by the ISS/SIMS technique.

Isothermal Weight Loss Measurements

Fibers, neat PMR-15 resin and composites were weighed at various time periods to determine weight changes due to the 350°C, air flow (100 cc/min or 1000 cc/min) exposure.

Isothermal Aging Procedure

Component materials and precut flexural and interlaminar shear composite specimens were placed in the center 20 cm portion of a ceramic tube 5.0 cm diameter x 61 cm in length. A constant temperature of 350°C was maintained in the 20 cm zone area where the specimens were being aged. Air was passed into the tube at a rate of 100 cc/min or 1000 cc/min depending on the study.

Composite Fabrication

Graphite fiber/PMR-15 composites were fabricated from PMR-15 impregnated fiber tape. The tape was prepared by brush coating a 50 wt% of PMR-15 resin components in methanol onto a dry wound tape 11.3 cm (4.5 in) wide x 43.2 cm (17 in) in diameter. The impregnated tape was cut into the appropriate lengths, 25.4 cm (10 in), then 19 layers were stacked one over the other in a mold. The mold was subjected to a heat treatment at 200°C for 1 h to form an imide prepolymer. The female portion was positioned into the mold and the assembly was placed in a preheated press (232°C). Pressure (6.9 MPa, 1000 psi) was applied immediately to consolidate the layers, released at 232°C, then reapplied. The temperature was raised to 316°C, and the system was maintained at these conditions (316°C, 6.9 MPa, 1000 psi) for 1 h. The composite was cooled to RT under pressure and postcured in air in the free state by a graduated increase in temperature to 316°C over an 8 h period, followed by a hold at this temperature for 12 h.

The nominal dimensions of each composite were 11.4 cm x 25.4 cm x 0.254 cm (4.5 in x 10 in x 0.10 in). Interlaminar shear specimens 0.64 cm wide x (4 x thickness + 0.25 cm) and flexure specimens (0.64 cm wide x (20 x thickness + 0.25 cm)) were cut from each composite for mechanical tests and isothermal aging studies.

Mechanical Property Measurements

Flexural strength measurements were made by the 3 point bend method at a span-to-depth ratio of 20/1 (ASTM-D-790) and crosshead speed of 0.127 cm/min (0.05 in/min). The interlaminar shear strength measurements were made by the 3 point bend method at a span-to-depth ratio of 4/1 and crosshead speed of 0.127 cm/min (0.05 in/min) (ASTM-D-2344).

RESULTS AND DISCUSSION

ISS/SIMS Analysis of Fibers

Typical ISS spectra are shown in Figs. 2b, 3b and 4b. ISS spectra for the other samples are not shown. All samples reveal the presence of carbon, oxygen, sodium and potassium within 5-10A of the surface. In addition to these elements, unsized Fortafil 5 shows a high concentration of chromium (Fig.2b) whereas the sized Fortafil 5 fibers shows the complete absence of the chromium ion (Fig.3b). Celion 6000 NR-150B2 sized shows a high concentration of fluorine (Fig.4b) as expected since the NR-150B size is a fluorinated material. The high chromium and fluorine levels for these fibers are also revealed in the SIMS spectra (Figs.2a and 3a). Data from the ISS analysis were not used semi-quantitatively in this study, but were used to corroborate the SIMS results.

The relative atomic ratios of the elements present on the fiber surfaces at near surface (5-10Å) and after sputtering to a depth of 100Å are listed in Tables 2 and 3. Typical SIMS spectra are shown in Figs.2a, 3a and 4a. It should be mentioned that all fiber surfaces at 5-10Å show the presence of organic species, based on masses corresponding to C^+, CH^+, CH_2^+, CH_3^+, C_2^+ and C_2H^+ which sputtered from the surface. In addition, all show a high concentration of hydrogen which is a reflection of the presence of hydrocarbons, water, and hydrogen-containing graphite fibers. The carbon and hydrogen concentrations were not considered in the

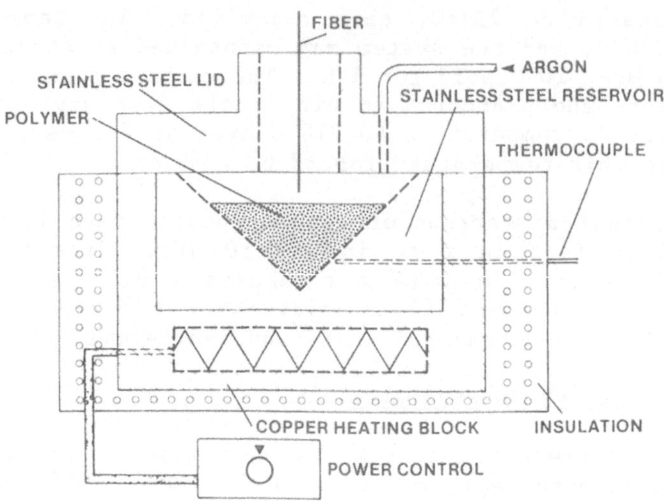

Fig. 2.   ISS/SIMS Spectra of Fortafil 5 Unsized Fiber ~10Å.

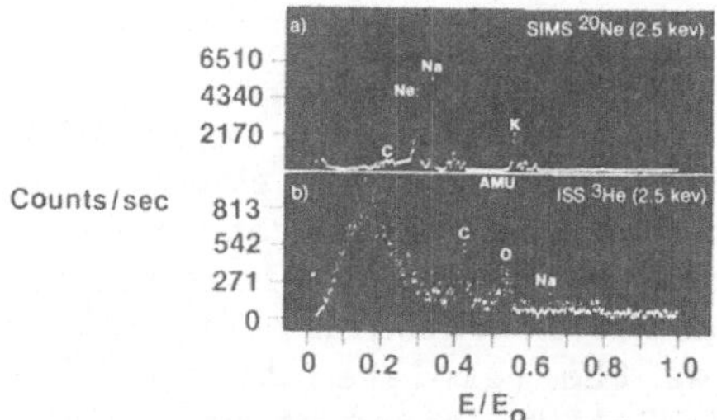

Fig. 3.   ISS/SIMS Spectra of Fortafil 5 Sized Fiber ~10Å.

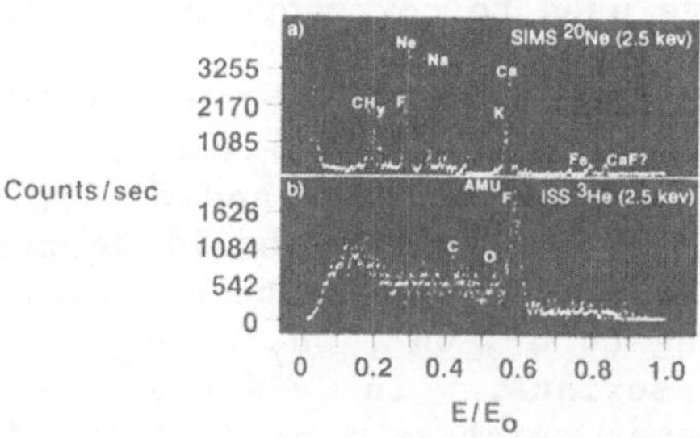

Fig. 4.   ISS/SIMS Spectra of Celion 6000 NR150B2 Sized Fiber ~10Å.

Table 2. SIMS Analysis of Graphite Fiber Surfaces

	Na	K	Ca	Cr	Fe	F
Relative Atomic Ratio ~10Å ^{20}Ne						
Panex 30 (u)	110	1.0	---	7.9	23	---
Fortafil 5 (u)	2.0	1.0	---	9.4	21	---
Celion 6000 (u)	8.4	1.0	33	4.4	15	---
Thornel 300 (u)	3.1	1.0	1.7	---	18	---
AS4 (u)	2.8	1.0	3.2	---	---	17
Celion 6000 (PI)	3.5	1.0	20	---	15	---

Table 3. SIMS Analysis of Graphite Fiber Surfaces

	Na	K	Ca	Cr	Fe	F
Relative Atomic Ratio ~100Å ^{20}Ne						
Panex 30 (u)	11	1.0	---	3.2	5.2	---
Fortafil 5 (u)	1.6	1.0	---	5.1	3.7	---
Celion 6000 (u)	2.2	1.0	5.3	4.8	6	---
Thornel 300 (u)	1.7	1.0	1.7	5.2	7.5	---
AS4 (u)	6.0	2.6	3.2	2.7	5.3	1.0
Celion 6000 (New PI)	2.0	1.0	1.4	2.4	5.1	---

calculation of atomic ratios for the 5-10 and 100Å levels. The ions of special interest at both the 5-10 and 100Å levels are sodium, potassium, calcium, chromium and iron. At the 5-10Å level (Table 2), sodium is present at a very high concentration on the Panex 30 (unsized) fibers, (relative atomic ratio 110), and on the unsized Celion 6000, Thornel 300, and AS graphite fibers at relative atomic ratios of 8.4, 3.1 and 2.8 respectively. The potassium concentrations on all fibers are lower than the sodium concentrations and much lower than the iron concentrations, except for AS fibers, which do not contain iron. A high chromium concentration is noted for unsized Fortafil 5, Panex 30 (unsized) and Celion 6000 unsized and PI sized. The ISS spectrum (Fig.2b) for Fortafil (unsized) also shows the presence of chromium. Except for Magnamite AS4(u) fibers, all contained considerable quantities of iron. The presence of fluorine on AS4(u) is noted. This element is also present at the 100Å level. The high concentration of calcium on unsized and PI sized Celion 6000 is also noted. This element persists at the 100Å level, but at much lower concentrations.

Table 4.   SIMS Analysis of Graphite Fibers

	Relative Atomic Ratio (Counts/sec[1]) $\sim$10Å ^{20}Ne					
	Na	K	Ca	Cr	Fe	F
Panex 30 (u)	43	1.0	-----	2.0	1.0	---
Fortafil 5 (u)	1.0	1.3	-----	3.2	1.3	---
Celion 6000 (u)	5.2	1.6	18.4	1.8	1.1	---
T300 (u)	1.9	1.6	1.0	---	1.4	---
AS4 (u)	1.5	1.4	1.6	---	1.6	1.0
Celion 6000 (PI)	1.5	3.8	1.3	1.0	---	---

[1] actual counts/sec for element in lowest count/sec

Fortafil 5 (u)	Na	1348
Panel 30 (u)	K	475
T300 (u)	Ca	186
Celion 6000 (PI)	Cr	37
Panex 30 (u)	Fe	127
AS4 (u)	F	492

At the 100Å level (Table 3), all but Panex 30(u) and Fortafil 5(u) contain calcium. AS4(u) is the only fiber which shows the presence of fluorine. It should be noted that there is a significant reduction of the sodium content at the 100Å level (Table 3) on all fiber surfaces, except for AS4(u) where an approximate two-fold increase is found. Iron persists in all fibers at approximately one-third to one-quarter the levels exhibited at the ~10Å level. Chromium is now detected on Thornel 300(u), AS4(u) and Celion 6000 PI size, whereas this element was not detected at the 10Å level in these fibers.

In comparing elements present on each fiber surface from fiber to fiber, the relative atomic ratios based on counts/sec of a given element were calculated as described in the experimental section. The results are listed in Tables 4 and 5 for near surface ~10A and ~100Å depths, respectively. At the ~10Å level, a high relative atomic ratio of sodium is noted on Panex 30(u) and Celion 6000(u); a high atomic ratio of calcium is also noted for Celion 6000(u). The potassium atomic ratio for Celion 6000 PI size is high relative to the other fibers. The high atomic ratio of chromium on Fortafil 5(u) and Panex 30(u) should be recognized. With these exceptions, the relative atomic ratio of elements listed are essentially equivalent. At the 100Å depth (Table 5), two fibers, Panex 30(u) and AS4(u), show very high relative atomic ratios of sodium compared to the other fibers. Moreover, AS4 fiber also contains a very high level of potassium, calcium, chromium, and iron relative to the other fibers. The high relative atomic ratio

Table 5. SIMS Analysis of Graphite Fibers

	Relative Atomic Ratio (Counts/Sec[1]) ~100Å ^{20}Ne					
	Na	K	Ca	Cr	Fe	F
Panex 30 (u)	27.5	6.12	----	1.7	2.0	---
Fortafil 5 (u)	2.52	3.94	----	1.7	1.1	---
Celion 6000 (u)	2.22	2.40	2.39	1.0	1.1	---
T300 (u)	2.13	3.17	1.00	1.3	1.3	---
AS-4 (u)	71.3	75.0	17.6	6.5	9.2	---
Celion 6000 (PI)	1.00	1.00	4.58	---	1.0	1.0

[1] actual counts/sec for element in lowest count/sec

Celion 6000 (PI)		Na	1814 counts/sec
Celion 6000 (PI)		K	670
T300 (u)		Ca	500
Celion 6000 (u)		Cr	161
Celion 6000 (PI)		Fe	127
Celion 6000 (PI)		F	866

of calcium on Celion 6000 unsized and sized is also apparent. With these exceptions the levels of elements present on the surface in comparing fiber to fiber are essentially equivalent.

ISS/SIMS Analysis of Fibers From Isothermally Aged Graphite Fiber/PMR-15 Composites

Exposure of the graphite fiber/PMR-15 polyimide composite to flowing air (100 cc/min) or (1000 cc/min) at 350°C for time periods of 250 h or longer caused surface oxidation of the polyimide resin, thereby exposing graphite fibers on the periphery of the composite in the process. The exposed fibers were then analyzed by the ISS/SIMS technique in the same manner as described for the neat fibers. Composite samples from each series were handled with care in order to prevent contamination of the fiber on the top surface of each sample.

Examples of ISS spectra of fibers from two isothermally aged composites, Celion 6000(u)/PMR-15 and Panex 30(u)/PMR-15 are shown in Fig.5. Of significance in these spectra is the large peak (counts/sec) due to sodium on the Panex 30(u) fiber relative to the Celion 6000(u) fiber. A high concentration of sodium was also found in Panex 30 fibers before composite fabrication. Examples of SIMS spectra of fibers from Celion 6000(u)/T300(u)/AS4(u) composites at the 10Å level are shown in Fig. 6. The high counts/sec of sodium on T300(u) and AS4(u) relative to Celion 6000(u) and Fortafil 5(u) should be noted. The ISS/SIMS analyses were conducted after 250 h at 350°C in an air flow of 1000 cc/min. The relative atomic ratios of elements present on each fiber surface are compared at two depths of penetration, near surface ≈10Å and at ≈100Å and are listed in Tables 6 and 7 respectively. For an appreciation of the actual atomic counts/sec which these numbers represent, the actual counts/sec for each specific element, for example sodium, present in the lowest atomic count/sec from fiber-to-fiber is given as a footnote to the tables. Therefore, it can be seen that a high value for a relative atomic ratio for a specific element does not necessarily mean that the fiber or composite contains a high concentration of that element on the surface. However, it is clear that at near surface ~10Å and at ~100Å, sodium is by far the element present in the greatest concentration; approximately 5 times potassium, 30-60 times calcium, 200-800 times chromium, 200-400 times iron. Fluorine was found to be present in only AS4(u) fiber.

The relative atomic ratios of sodium for each fiber and composite material are compared in Table 8. Of particular interest is the significant change in sodium concentration from the near surface≈10Å, to the 100Å level in the AS4(u) neat fiber and in the Panex 30(u) fiber from the composite. Also, the higher concentration of sodium is noted on the T-300(u), and Fortafil 5(u)

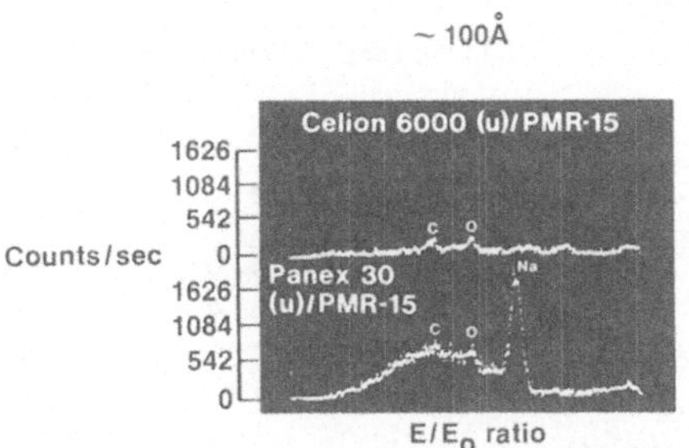

Fig. 5.   ISS Spectra of Isothermally Aged Composites (350°C, 220 hrs, 1000 cc/min airflow).

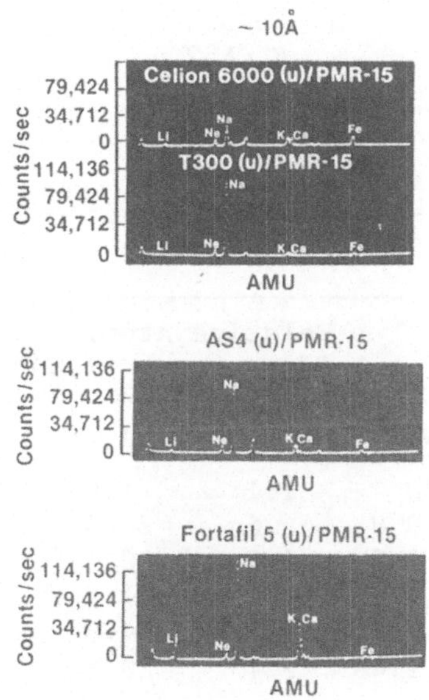

Fig. 6.   SIMS Spectra of Isothermally Aged Composites (350°C, 220 hrs, 1000 cc/min airflow).

Table 6. SIMS Analysis of Graphite Fiber/PMR-15 Composites After
          250 hrs at 350°C, 1000 cc/min (airflow)

	Relative Atomic Ratio (counts/sec[1]) ~10Å ^{20}Ne					
	Na	K	Ca	Cr	Fe	F
Panex 30 (u)	9.58	5.4	1.0	4.4	1.0	----
Fortafil 5 (u)	5.41	10.1	11.4	1.0	29	1.78
Celion 6000 (u)	1.02	1.95	17.6	8.0	17.2	1.0
Thornel 300 (u)	4.20	1.0	2.4	3.4	33	----
AS4 (u)	3.52	2.0	6.3	3.1	59	1.13
Celion 6000 (Celanese PI size)	1.00	4.0	16.7	16	127	----

[1] actual counts/sec for element in lowest count/sec from fiber to
fiber.

Celion 6000 (PI)	Na 21,958
T-300 (u)	K 4076
Panex 30 (u)	Ca 37.3
Fortafil 5 (u)	Cr 25
Panex 30 (u)	Fe 51
Celion 6000 (u)	F 652

Table 7. SIMS Analysis of Graphite Fiber/PMR-15 Composites After
          250 hrs at 350°C, 1000 cc/min (airflow)

	Relative Atomic Ratio (counts/sec[1]) ~100Å ^{20}Ne					
	Na	K	CA	Cr	Fe	F
Panex 30 (u)	30	1.1	1.0	1.0	1.0	----
Fortafil 5 (u)	8.8	6.6	25.6	3.1	12.5	3.36
Celion 6000 (u)	1.0	1.0	30.1	5.4	18.5	1.00
Thornel 300 (u)	10.3	1.1	11.5	5.0	14.8	----
AS4 (u)	5.45	1.6	14.6	4.7	11.8	7.66
Celion 6000 (Celanese PI size)	1.48	2.7	29.5	9.7	23.9	----

[1] actual counts/sec for element in lowest counts/sec from fiber to
fiber.

Celion 6000 (u)	Na 4839
Celion 6000 (u)	K 1686
Panex 30 (u)	Ca 152
Panex 30 (u)	Cr 25
Panex 30 (u)	Fe 25
Celion 6000 (u)	F 136

Table 8. SIMS Analysis of Fibers and Aged Composites for Sodium

| | Relative Atomic Ratio (counts/sec[1]) | | | |
| | Fiber | | Composites 250 hrs at 350°C 1000 cc/min (airflow) | |
	10Å	100Å	10Å	100Å
Celion 6000 (u)	5.2	2.2	1.02	1.0
Celion 6000 (PI)	1.5	1.0	1.0	1.5
Thornel 300 (u)	1.9	2.1	4.2	10
AS4 (u)	1.5	71	3.5	5.5
Fortafil 5 (u)	1.0	2.5	5.4	8.8
Panex 30 (u)	43	28	9.5	30

[1]See Tables 4, 5, 6, and 7 for actual counts/sec.

fibers from the composite than was observed on the neat fiber at the 10Å and 100Å levels.

Component and Composite Weight Loss

The fiber, resin and composite weight losses after exposure at 350°C in an air flow (1000 cc/min) for several hours are listed in Table 9. Composite weight losses at 350°C in an air flow of 100 cc/min were almost identical to the 1000 cc/min air flow weight losses. The component and composite weight losses over approximately 700 h are shown graphically in Figs.7 and 8. The high weight loss (78%) of the PMR-15 neat resin after 385 h is particularly revealing since several of the composites show very low weight loss after 520 h. This suggests that after the surface polyimide resin is oxidized, the fiber forms a protective blanket over the composite, acting as a scavenger for the oxygen present. The formation of a passive layer on the resin from components of the fiber thereby inhibiting oxidation of the matrix is also a possibility. The poor thermo-oxidative stability of the Thornel 300(u) and Panex 30(u) relative to the other fiber is also noted. Moreover, the poor thermo-oxidative stability of the Fortafil 5/PMR-15 composite relative to the excellent thermo-oxidative stability of the Fortafil 5(u) fiber itself suggests that the interface region in this system is the weak link and the point at which the composite system is thermo-oxidatively degraded. The higher weight loss of the AS4(u)/PMR-15 composite system relative to the neat AS4(u) fiber itself also suggests that interaction

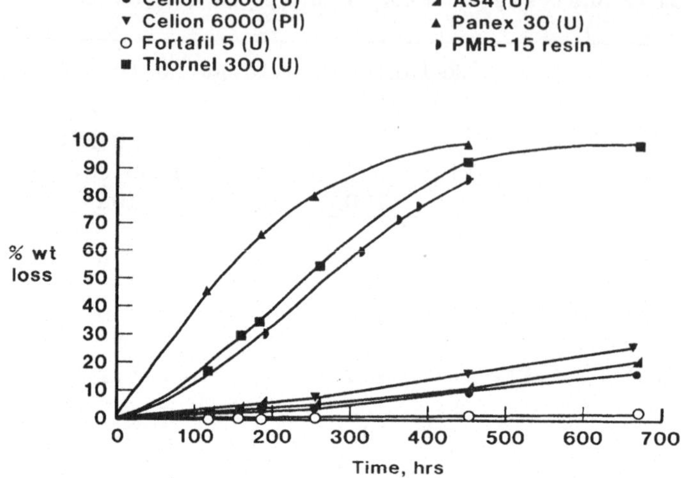

Fig. 7.   Weight Loss of Component Materials (350°C, 1000 cc/min airflow).

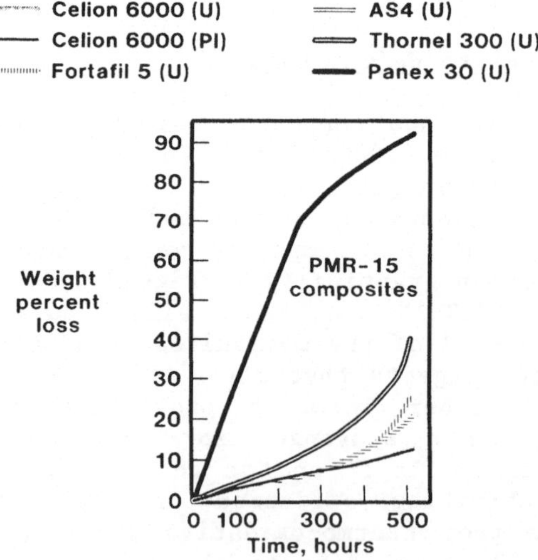

Fig. 8.   Composite Weight Loss (350°C, 1000 cc/min airflow).

Table 9. Component and Composite Weight Losses at 350°C (1000 cc/
         min airflow).

Fiber/PMR-15 Composite	Fiber[1] Wt% Loss after 450 hrs	Composite Wt% Loss after 520 hrs
Celion 6000 (u)	11.2	12.5
Celion 6000 (PI)	16.0	12.0
Thornel 300 (u)	94.2	40.0
AS4 (u)	10.4	21.1
Fortafil 5 (u)	0.43	24.3
Panex 30 (u)	99.5	95
PMR-15 resin[1]	77.8[1]	----

[1] PMR-15 resin is listed in this column (after 385 hrs).

effects in the interface region of the composite may be responsible
for this behavior. The relatively good agreement in weight losses
of the Celion 6000 unsized and PI sized composites with weight
losses of the neat fibers indicates that this interface region is
not being degraded and, as mentioned above, that the exposed fiber
is protecting the underlying composite system.

Mechanical Properties and Optical Microscopy of Isothermally Exposed Composites

     Elevated temperature flexural and shear strengths of these
composite systems after aging at 350°C for several hours in flowing
air are listed in Tables 10 and 11 respectively. The "as
fabricated" properties are listed for comparison.

     The higher flexural properties after aging at 350°C for 250 h
is due to a postcure effect [7], which causes an improvement in the
properties over the initial unaged composite. Inspection of Table
10 shows that only two composites retain a reasonable fraction of
the unaged or 250 h-aged shear properties. These are the Celion
6000 unsized and PI sized fiber/composite systems. The behavior of
these composites in shear (Table 11) is similar to that observed
with the flexural properties. There is an improvement in the shear
strength for most composite systems after 250 h exposure at 350°C,
due to the postcure effect. However, only two composite systems
retain a reasonable fraction of the unaged and 250 h-aged
properties. These are the Celion 6000 unsized and PI sized

Table 10. Flexural Properties of Isothermally Aged PMR-15 Composites (350°C, 1000 cc/min airflow)

Fiber/PMR-15 Composite System	350° Flexural Properties											
	Unaged				After 250 hrs				After 520 hrs			
	Strength		Modulus		Strength		Modulus		Strength		Modulus	
	MPa	(ksi)	GPa	$(10^6$ psi)	MPa	(ksi)	GPa	$(10^6$ psi)	MPa	(ksi)	GPa	$(10^6$ psi)
Celion 6000 (u)	269	(39)	49.7	(7.2)	373	(54)	58.7	(8.5)	421	(61)	87.6	(12.7)
Thornel T-300 (u)	366	(53)	42.8	(6.2)	366	(53)	96.6	(14)	82.8	(12)	11.0	(1.6)
Celion 6000 (PI)	276	(40)	39.3	(5.7)	656	(95)	159	(23)	297	(43)	60.7	(8.8)
AS4 (u)	297	(43)	46.9	(6.8)	469	(68)	145	(21)	124	(18)	65.5	(3.7)
Fortafil 5 (u)	379	(55)	48.9	(7.1)	814	(118)	138	(20)	221	(32)	82.8	(12.0)
Panex 30 (u)	393	(57)	69.0	(10)	173	(25)	19.3	(2.8)	completely degraded		------	

Table 11.   Interlaminar Short Beam Shear Strength of Isothermally
            Aged PMR-15 Composites (350°C, 1000 cc/min airflow)

Fiber/PMR-15 Composite System	350°C Shear Strength, MPa					
	Unaged		After 250 hrs		After 520 hrs	
Celion 6000 (u)	19.3	(2.8)	34.3	(4.97)	33.1	(4.8)
Thornel 300 (u)	28.9	(4.2)	28.3	(4.10)	6.76	(0.98)
Celion 6000 (PI)	26.9	(3.9)	29.7	(4.3)	13.4	(1.94)
AS4 (u)	12.4	(1.8)	23.8	(3.45)′	2.85	(0.413)
Fortafil 5 (u)	13.1	(1.9)	23.5	(3.40)	4.41	(0.64)
Panex 30 (u)	20.0	(2.9)	mostly fiber remaining		mostly fiber remaining	

Table 12.   Comparison of Component and Composite Properties (after
            350°C, 1000 cc/min airflow) - 520 hrs

Fiber/Composite Systems	Fiber Wt Loss Rate 10^{-4}g/hr	Composite Wt Loss Rate 10^{-4}g/hr	Fiber Composite Sodium Content Rel. Atomic Ratio[1] ~100Å	350°C Flexural Strength ksi
Celion 6000 (u)	11.0	3.1	1.0	61
Celion 6000 (PI)	15.5	3.7	1.5	43
Fortafil 5 (u)	0.40	3.9	8.8	32
AS4 (u)	11.0	5.4	5.5	18
Thornel 300 (u)	86	10.9	10	12
Panex 30 (u)	140	25	30	completely degraded
PMR-15 resin	5.7	----	----	

[1] where 1.0 = 4839 counts/sec of sodium

composites, with the unsized fiber system displaying superior
thermo-oxidative resistance relative to the PI sized Celion 6000
composite, and vastly superior oxidation resistance relative to the
Thornel 300(u), AS4(u), Fortafil 5(u) and Panex 30(u) composite
systems. A comparison of the component and composite weight loss
rates, the sodium content on fiber/composite systems and 350°C
flexural strength after 520 h is listed in Table 12. The stability
of the Celion 6000(u)/PMR-15 composite, and by inference, the
apparent stability of the fiber-matrix interface region is clearly
shown when the rate of oxidation of the fiber, resin and composite
and the flexural strength of the aged composite are considered.
The protective effect of the fiber on the surface of the composite
is illustrated by the strength of the aged composite, and is
further illustrated by the photomicrograph of the cross-section
(transverse to the fiber direction) of the Celion 6000(u)/PMR-15
composite (Fig.9). The photomicrograph of a shear specimen 1.52 cm
(0.6 in) in length reveals that no degradation occurs at the
interface, and that there is no apparent pyrolysis or
thermo-oxidative degradation of the matrix in the bulk of the
composite. Surface cracks which develop at the ends of the
composite during aging do not penetrate the composite. This was
illustrated by the observation that after grinding and polishing

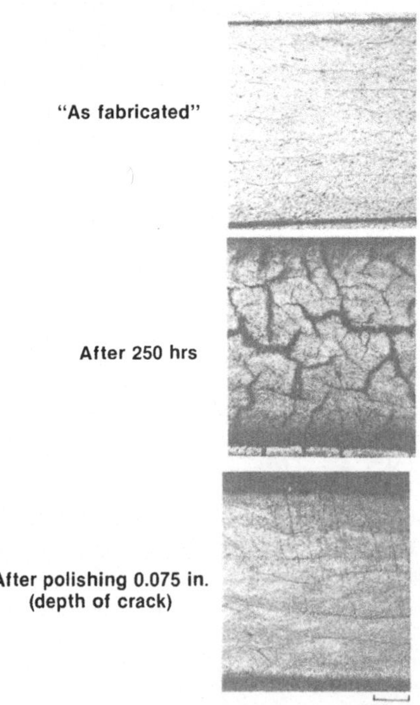

Fig. 9.   Cross Section of Celion 6000 (u)/PMR-15 Composite (350°C,
          1000 cc/min airflow).

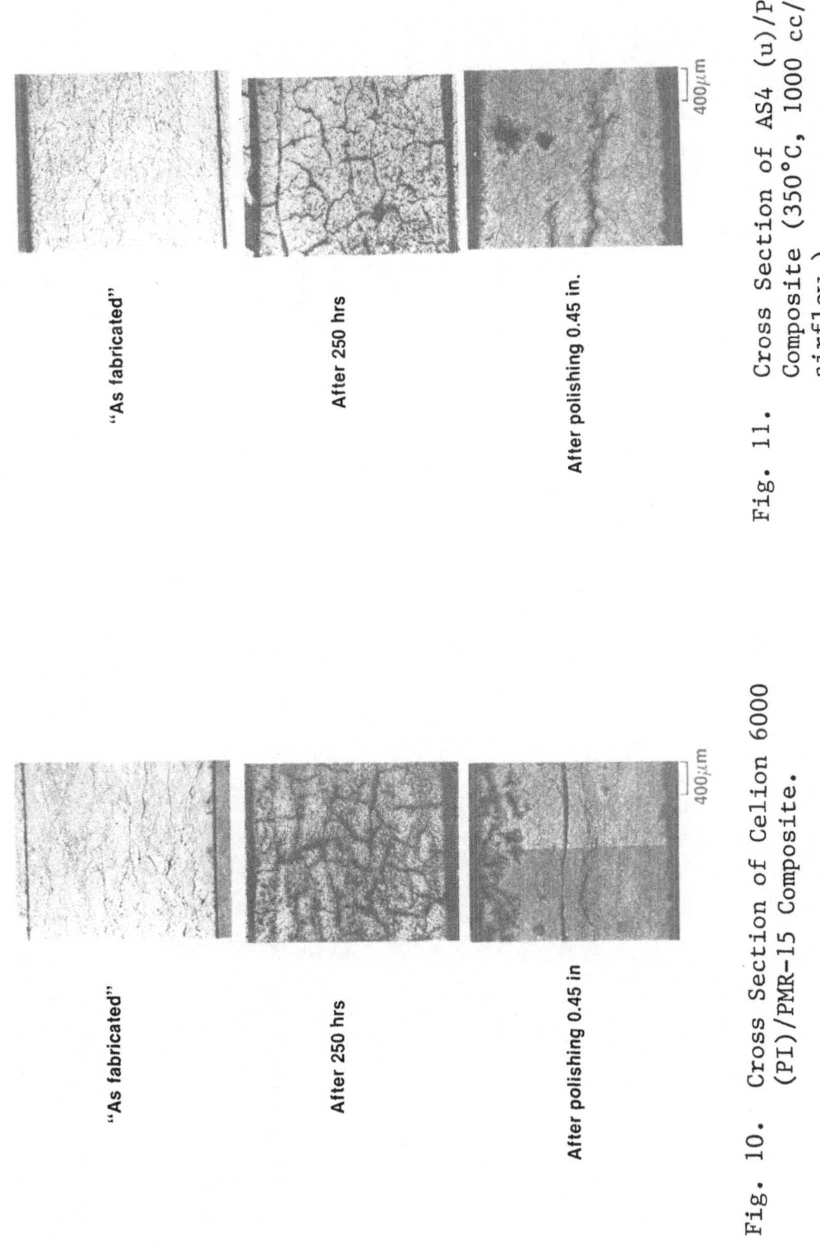

Fig. 11.   Cross Section of AS4 (u)/PMR-15 Composite (350°C, 1000 cc/min airflow.)

Fig. 10.   Cross Section of Celion 6000 (PI)/PMR-15 Composite.

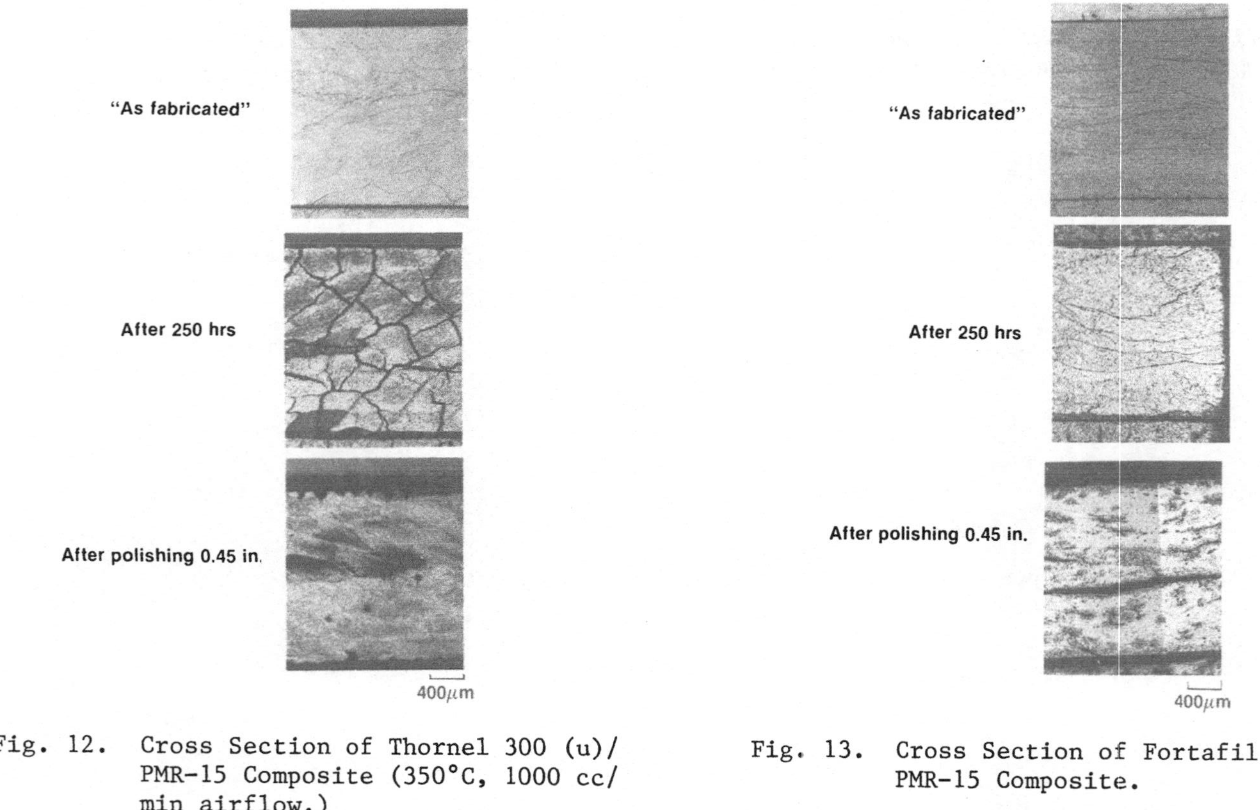

Fig. 12.　Cross Section of Thornel 300 (u)/ PMR-15 Composite (350°C, 1000 cc/ min airflow.)

Fig. 13.　Cross Section of Fortafil 5(u)/ PMR-15 Composite.

the aged composite to a depth of 0.19 cm (0.075 in), all evidence of surface cracks was removed (Fig.9). Micrographs of shear specimens of the other composites (Figs.10-13) reveal the presence of at least one major flaw or crack through the entire composite and the presence of voids in the interface regions of the composites.

CONCLUSIONS

The surfaces of commercial graphite fibers contain sodium, potassium, calcium, chromium and iron in varying relative atomic ratios. For 350°C applications in flowing air (100 cc or 1000 cc/min), the fiber rating in PMR-15 composites is as follows: Celion 6000(u) > Celion 6000 (PI Celanese size) > Fortafil 5(u) > AS4(u) > Thornel 300(u) > Panex 30(u). The thermo-oxidative stability of the above fibers in the neat form and the PMR-15 resin under the same conditions is as follows: Fortafil 5(u) >> Celion 6000(u) > AS4(u) > Celion 6000 (PI Celanese size) > PMR-15 > Thornel 300(u) > Panex 30(u). The greater thermo-oxidative stability of composites relative to the fiber and polyimide matrix can be attributed to an overall protective blanket provided by the fiber after the resin is oxidized. The formation of a passive layer on the matrix due to deposition of components from the fiber can also be the method by which thermo-oxidative process is retarded. The formation of these protective barriers appear to be the mechanism of degradation for composites which exhibit internal thermo-oxidative stability and no internal pyrolysis of the matrix, namely the Celion 6000(u) and Celion 6000 (PI)/PMR-15 systems. For composites which show good fiber thermo-oxidative stability, such as Fortafil 5(u) and AS4(u), two factors may be responsible for the poor thermo-oxidative stability of the composite and subsequent poor mechanical properties. The first is (1) the aging process may generate poor bonding at the fiber/matrix interface and (2) impurity components on the fiber, such as sodium or potassium ions, may accelerate thermo-oxidation of the resin and fiber in the interface region. Those composite systems which exhibited the poorest thermo-oxidative stability contain substantial quantities of sodium and potassium on the fiber, relative to the more thermo-oxidatively stable composite systems.

The fiber degradation rate appears to have a significant influence on the thermo-oxidative stability of the PMR-15 composite. The major degradation process occurs via thermo-oxidation of the surface of the composite. There is no evidence of internal oxidation or pyrolysis of the polyimide matrix under these conditions.

REFERENCES

1. D. A. Scola, SAMPE, 27th National SAMPE Symp. and Exhibit., San Diego, CA 27, 923 (1982).
2. H. H. Gibbs, R. C. Wendt, and F. C. Wilson, Proc. 33rd Ann. Tech. Conf., Reinforced Plastics/Composites Inst., SPI Section 24-F (1978).
3. L. T. Drzal, and G. E. Hammer, ALWAL-TR-80-4143, April, 1981.
4. T. T. Serafini, P. Delvigs, and G. R. Lightsey, U.S. Patent 3,745,149, July, 1973.
5. T. T. Serafini, P. Delvigs, and G. R. Lightsey, J. Appl. Poly. Sci., 16, 905 (1978).
6. A. T. DiBenedetto, and D. A. Scola, J. Colloid Interface Sci., 64, 480 (1978).
7. R. H. Pater, 13th National SAMPE Tech. Conf., 13, 38 (1981).